CURRENT TOPICS IN

DEVELOPMENTAL BIOLOGY

VOLUME 21

NEURAL DEVELOPMENT
PART IV
Cellular and Molecular Differentiation

CURRENT TOPICS IN
DEVELOPMENTAL BIOLOGY

EDITED BY

A. A. MOSCONA

CUMMINGS LIFE SCIENCE CENTER
THE UNIVERSITY OF CHICAGO
CHICAGO, ILLINOIS

ALBERTO MONROY

VOLUME 21

NEURAL DEVELOPMENT
PART IV

Cellular and Molecular Differentiation

VOLUME EDITOR

R. KEVIN HUNT

THE SALK INSTITUTE
SAN DIEGO, CALIFORNIA

1987

ACADEMIC PRESS, INC.
Harcourt Brace Jovanovich, Publishers

Orlando   San Diego   New York   Austin
Boston   London   Sydney   Tokyo   Toronto

ACADEMIC PRESS, INC.
Orlando, Florida 32887

United Kingdom Edition published by
ACADEMIC PRESS INC. (LONDON) LTD.
24–28 Oval Road, London NW1 7DX

LIBRARY OF CONGRESS CATALOG CARD NUMBER: 66-28604

ISBN 0–12–153121–X (alk. paper)

PRINTED IN THE UNITED STATES OF AMERICA

87 88 89 90   9 8 7 6 5 4 3 2 1

CONTENTS

CHAPTER 4.   The Insect Nervous System as a Model System for
              the Study of Neuronal Death
              JAMES W. TRUMAN

CHAPTER 5.   Brain-Specific Genes: Strategies and Issues
              ROBERT J. MILNER, FLOYD E. BLOOM,
              AND J. GREGOR SUTCLIFFE

CHAPTER 6.   Changes in Intermediate Filament Composition
              during Neurogenesis
              GUDRUN S. BENNETT

PREFACE

Volume 21 of *Current Topics in Developmental Biology* is the fourth in a series of special volumes devoted to Neural Development, the first three of which appeared as Volumes 15 and 16 (1980) and Volume 17 (1982) of the series. The present group of chapters focuses on the problems of cellular and molecular differentiation in the developing nervous system. The aim of the editors was to consider the cell biology of developing neurons with respect to a broad range of issues, from the molecular and supramolecular specializations that characterize particular classes of developing neurons to the more global problems of growth, tissue assembly, and cell patterning. Accordingly, the 15 chapters in the present volume offer an eclectic treatment of neurogenesis, from the chemistry of growth cones and gene expression in the fetal brain to clonal growth mechanisms and the morphogenesis of neurochemically defined circuits in the cortex.

Three quarters of a century has passed since Harrison's development of a tissue culture method for embryonic neurons and his classic observations proving that the neural circuitry was, after all, made up of cells and cell processes. Those watershed experiments redefined the study of brain development as a problem in cell biology. Over the 10 years since we first conceived of a set of neural development volumes within the *Current Topics in Developmental Biology* series, neuroembryology has become an increasingly cellular discipline. The powerful techniques of immunocytochemistry and molecular biology have been increasingly focused on major unresolved problems in neural development, and it seems likely that, within a very few years, we will be able to describe at least some important features of neurogenesis at the level of genes and gene products. Equally encouraging are the recent approaches to multicellular problems in the developing nervous system—growth, cell–cell interactions, selective cell association, and tissue assembly—that have begun to recast these critical processes in the language and grammar of ordinary cell biology. We think Harrison would be pleased at the large and vigorous body of current research on cellular and molecular differentiation in neural development.

We are grateful for the flexibility of production schedules afforded us by Academic Press, for without such flexibility we would not have been able to accommodate the varied schedules of our many contributors. Most importantly, we would like to thank the contributors for taking time from their busy scientific lives to share their thoughts on research topics important to them. The usefulness of the volume is the sum of their individual contributions.

R. Kevin Hunt

CHAPTER 1

CELL PATTERNING IN VERTEBRATE DEVELOPMENT: MODELS AND MODEL SYSTEMS

Lawrence Bodenstein

PROGRAM IN CELL AND DEVELOPMENTAL BIOLOGY, HARVARD MEDICAL SCHOOL, AND
DEPARTMENT OF NEUROSCIENCE, CHILDREN'S HOSPITAL
BOSTON, MASSACHUSETTS 02115

and

Richard L. Sidman

DEPARTMENT OF NEUROPATHOLOGY, HARVARD MEDICAL SCHOOL, AND
DEPARTMENT OF NEUROSCIENCE, CHILDREN'S HOSPITAL
BOSTON, MASSACHUSETTS 02115

I. Introduction

The basic formative event in vertebrate embryonic development is the progressive organization of very large numbers of cells into integrated, multicellular tissues and organs. The mechanisms and rules which govern this assembly are poorly understood. Ultimately, we would like to explain the generation of multicellular form as a consequence of the pooled behaviors of individual cells. In order to dissect this complex process, it is necessary to determine the behavior of individual cells and elucidate the principles which guide this behavior. Methods have been sought to mark small numbers of cells (and their clonal descendents) and monitor their behavior in the context of developing anlagen. The requirements of a valid marking system have been discussed (McLaren, 1976; West, 1984) and more and more powerful systems are under development (Weisblatt *et al.*, 1978; Hirose and Jacobson, 1979; Oster-Granite and Gearhart, 1981; Rossant *et al.*, 1983; Gimlich and Cooke, 1983; Ponder *et al.*, 1983, 1985; Gardner, 1984; Schmidt *et al.*, 1985b; Kimmel and Warga, 1986). Of particular importance, markers must be readily scorable by the experimenter but transparent to the developing system.

A powerful strategy for marking cells during development involves

1

the production of experimental chimeras and mosaics[1] (review Stern, 1968; Le Douarin and McLaren, 1984). Constructed by genetic or surgical means, these are composite animals which contain an admixture of phenotypically distinguishable (i.e., marked and unmarked) cells. We attempt, by examining the spatial and temporal distribution of these cells, to deduce how tissues are constructed from smaller, generally clonal, units (review Jacobson, 1978, 1985; in insects, Bryant and Schneiderman, 1969; Bryant, 1970; in amphibians, Conway et al., 1980; in birds, Le Douarin, 1980; in mammals, review McLaren, 1976). Studies of invertebrates (notably *Drosophila*) and cold-blooded vertebrates (notably frogs) have revealed a mode of development in which tissues are constructed as spatial mosaics, precisely erected from a clonal blueprint. In *Drosophila* this form of development is strikingly evident, and cells rarely violate highly stereotyped spatial boundaries which define *compartments* (Garcia-Bellido et al., 1973; Garcia-Bellido, 1975; review Martinez-Arias and Lawrence, 1985). The frog has yielded evidence of a similar clonal construction, although somewhat more variability is present, and cells can be experimentally manipulated into dramatic departures from the normal plan of development (review Conway et al., 1980; Jacobson, 1985). In higher vertebrates (notably mice), such stereotyped boundaries are conspicuously absent (Sanyal and Zeilmaker, 1976, 1977; Mintz and Sanyal, 1970; West, 1976a,b; Mullen, 1977, 1978; Oster-Granite and Gearhart, 1981; Wetts and Herrup, 1982) and an understanding of the complex spatial patterns seen in mouse chimeras and mosaics has not been forthcoming (review West, 1978; Schmidt et al., 1986). Indeed, the high degree of variability in these patterns has made analysis difficult and largely unrewarding. This apparent difference between warm-blooded vertebrates on the one hand and cold-blooded vertebrates and certain invertebrates on the other hand may reflect fundamental differences in the rules which govern embryonic development. Alternatively, the fundamental mechanisms may be quite similar (Wolpert, 1971), and our inability to decode these mechanisms, given the complexity and variability of mammalian cell patterns, may mask this similarity. In ei-

---

[1]The distinction between chimeras and mosaics is discussed by Ford (1969) and Benirschke (1970). Chimeras are composite animals derived from more than one zygote; mosaics derive from a single zygote but are nonetheless phenotypically heterogeneous. We will refer to cell patterns found in chimeras and mosaics as "chimeric patterns" (i.e., the type of pattern found in chimeras). We believe this usage is reasonable since the evolution and gross form of these patterns in both types of animals are fundamentally similar. In addition, we use the term "mosaic" sparingly since it is quite charged with meaning from its historical use in experimental embryology (Harrison, 1933).

ther case, it is of paramount importance for us to develop insight into the meaning of mammalian cell patterns.

A. DECODING MAMMALIAN CELL PATTERNS

Analysis of chimeric patterns in mammals has centered about the concept of the *coherent clone* (Nesbitt, 1974). In a composite tissue (one composed of marked and unmarked cells), cells which are clonally related will tend to lie side by side. A cohort of contiguous, clonally related (and therefore like-marked) cells forms a coherent clone. In many instances, two or more like-marked coherent clones will abut by random assortment to produce a larger polyclonal *patch* or *clump*.[2] The size and distribution of these larger groupings (patches composed of coherent clones) can be ascertained experimentally; a theoretical approach is then used to determine coherent clone size.

Coherent clone size is increased by cell division (since daughter cells tend to lie side by side) and decreased by local cell movements, long-range migrations, and cell death (Lewis, 1973). Patch size is a function of the size of individual coherent clones and the number of like-marked coherent clones which abut. If the tissue has a high proportion of marked cells relative to unmarked cells, then the average size of the marked patches will be large since many coherent clones will be marked and these clones will frequently abut (West, 1975; Whitten, 1978). In contrast, if the proportion of marked cells is very small, coherent clones will rarely abut and individual patches consist of single clones (Whitten, 1978; Schmidt *et al.*, 1986).

Analysis of chimeric patterns in experimental material has involved (1) measuring average patch size in a tissue, and (2) calculating (a theoretical) coherent clone size from this patch size (review West, 1978). Although the details of measuring average patch size are specific for each individual tissue, several generalizations are valid. When a higher dimensional tissue, e.g., two-dimensional sheet, is evaluated in

---

[2]The term "patch" has historical precedent in the study of mammalian chimeras and mosaics (Nesbitt, 1974); the term "clump" derives from the often-cited mathematical treatment of random clumping phenomena (Roach, 1968). We use the terms interchangeably, although patch seems most appropriate in the two-dimensional case and clump best in the three-dimensional case.

The term "polyclone" derives from compartment analysis in insect development (Crick and Lawrence, 1975) and has been used to describe surgical grafts in amphibian chimeras (Conway *et al.*, 1980).

Patches are aptly described as polyclones. This is a particularly useful nomenclature since it serves to emphasize that while the experimentally observed patches may be single clones, they usually are not.

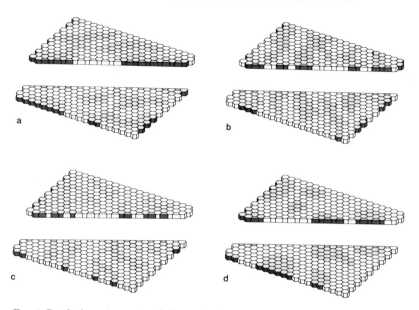

FIG. 1. Patch shape in sectioned tissue. A schematized two-dimensional cell sheet has been sliced in half to demonstrate the relationship between the cell patterns seen in tissue sections (cut edge, dark shading) and the cell patterns which may exist in whole preparations (surface, light shading). If the surface pattern is anisotropic, the edge pattern is dependent on the orientation of sectioning. Sections cut parallel to the long axes of the surface patches yield long edge patches (a), whereas sections cut perpendicular to the long axes of the surface patches yield shorter edge patches (b). If the surface patches are "ragged" (c), then the edge pattern will contain smaller patches than if the surface patches are more cohesive (d). Thus, caution must be used when trying to infer whole tissue cell patterns from sectioned material.

a lower dimension, e.g., tissue section taken perpendicular to the sheet (Deol and Whitten, 1972; Oster-Granite and Gearhart, 1981; West, 1976a,b), inaccuracies arise if the higher dimensional patches are "ragged" or anisotropic (Fig. 1) (West, 1976a); therefore, this expediency should be avoided and either whole mounts of the specimen prepared (Schmidt *et al.*, 1985a, 1986) or careful reconstruction undertaken. Such reconstruction from tissue sections is laborious and fraught with difficulty (Sanyal and Zeilmaker. 1977); it may require new methodology, since cellular reconstruction must be done within the larger tissue reconstruction (e.g., Pearlstein *et al.*, 1986; Pearlstein and Sidman, 1986). Patch size calculations are most accurate when the population under examination (e.g., marked) is dilute compared to the surrounding population (e.g., unmarked) (West, 1975; Whitten, 1978).

In particular, edge effects can significantly bias the results and, if possible, natural tissue boundaries are preferable to arbitrary cut edges (see Section VI).

Analytical and computational methods exist for calculating the number and size of coherent clones from patch data (West, 1975; Whitten, 1978; see Roach, 1968, for a mathematical treatment). The prototype analysis is based on a formulation in which a period of random cell assortment (organ formation) is followed by uniform clonal growth with minimal cell movement or death. The tissue should be homogeneous and isotropic since the basic analysis scheme (West, 1975; Whitten, 1978) does not handle conditions of regional variation and is based on an isotropic model of random cell assortment. These constraints severely limit the general usefulness of the method; indeed, these requirements and the formulation of tissue growth inherent in the analysis have not been verified in any mammalian system examined to date (e.g., cerebellum, Mullen, 1977; liver, West, 1976b) and have been proved invalid in most systems (skin, West and McLaren, 1976; Green *et al.*, 1977; axial skeleton, Moore and Mintz, 1972; retina, Sanyal and Zeilmaker, 1977; Schmidt *et al.*, 1986; intestine, Schmidt *et al.*, 1985a). In addition to the biological restrictions, a number of conditions inherent in the mathematical analysis (including well-defined coherent clone shape and an absence of edge effects) have not been formally addressed. Indeed, we believe the significance of the requisite a priori assumptions generally exceeds the expected yield from the analysis (see Section VI).

Finally, the usefulness of this form of patch size analysis in understanding how tissues are constructed and cell patterns generated, especially in mammals, is questionable. In all mammalian systems examined to date, patch size and calculated coherent clone size are remarkably small (e.g., retina, West, 1976a; Schmidt *et al.*, 1986; skin, Mintz, 1967; cerebellum, Mullen, 1977; Oster-Granite and Gearhart, 1981; peripheral nervous system, Peterson and Bray, 1984; liver, West, 1976b). The inference has been that much cell mixing occurs during mammalian development. Unfortunately, this important and now well accepted finding seems to have signaled an impasse in the analysis of mammalian cell patterns. The methodology of patch size analysis is severely limited, perhaps because reduction of these complex patterns to a single number (coherent clone size) has necessitated a wholesale elimination of information (Whitten, 1978). Salvaging this information to glean more from these remarkable cell patterns is the current challenge.

B. The Measurement of Mixing

Chimeric patterns are the result of multiple events of several classes (most notably cell division, cell movement, and cell death) occurring over long time periods. To infer these events from final patterns is a difficult task. Patch size analysis is not sufficient, in part because it requires a particular and generally unrealistic set of assumptions (e.g., little cell movement, no cell death, homogeneous clone size). The most important legacy of this approach is the understanding that cell mixing (and hence cell movement) is a dominant phenomenon in mammalian development. In such a setting, a search for "coherency" in clonal growth is apt to be unrewarding. Other methods of characterizing chimeric patterns (e.g., Schmidt *et al.,* 1985c) and insights into the cellular mechanisms which generate pattern archetypes (e.g., elongated clonal patches) are needed. The notion of *average patch size* is useful since it imparts a sense of the degree of variegation in a cell pattern. *Weighted average patch size* (see Section VI) is somewhat less sensitive to unusual pattern anomalies not representative of the tissue in general. Schmidt and co-workers (1986) have emphasized that the distribution of patch sizes may be revealing (also see Whitten, 1978) and have noted that measures which reflect this distribution are preferable to simple averages.

Cell mixing is the result of a composite of forces which act to cause like cells to be associated (e.g., cell division) and forces which act to cause like cells to be less associated (e.g., cell movement). Recently we have introduced the notion of a *mixing index* (Bodenstein, 1986) as a quantitative measure of cell mixing. We define the mixing index as the proportion of cells which abut at least one unlike neighbor, i.e., border cells. This index is useful in quantitating cell mixing over time. It includes a measure of degeneracy in that widely divergent patterns may have the same index, especially when the tissue is not spatially homogeneous. All attempts to describe cell patterns in some mathematical "shorthand" suffer from a degree of degeneracy, and caution must be used when attempting to infer biological mechanisms from these artificial parameters.

We suggest that the complexity of mammalian chimeric patterns will prevent a straightforward analytical or computational analysis. We have attempted instead a different approach which makes use of special techniques in computer modeling. In these systems, characterizations such as average patch size and size distribution and the mixing index, do not provide a developmental script; rather, they serve to compare theory and experiment.

II. A Synthetic Model of Tissue Growth

We have constructed a dynamic simulation model designed to aid in the study of tissue growth and cell patterning during embryogenesis (Bodenstein, 1986). This model is synthetic in that it provides the framework for a theoretical study of morphogenesis without postulating any new mechanism of interaction (see below).

Within the model, schematized cells carry out stereotyped behaviors under the guidance of simple rules. Cells possess size and shape, and a behavioral repetoire which includes the ability to divide (symmetrically or asymmetrically), move, or die. In addition, cells maintain a series of internal states which include age (time since last division), generation (number of prior divisions), and state of differentiation (e.g., mitotically active or postmitotic). Behaviors are selected on a cell-by-cell basis under guidance of a series of rules which take into account the internal state (characteristics) of the cell and its local environment (e.g., the position and state of neighboring cells). The complete array of cells forms a model tissue which grows and matures over time; the nature of this tissue maturation is a function of the rules governing the behavior of individual cells (Fig. 2).

The model has been implemented in FORTRAN on a Digital Equipment Corporation VAX 11/780 with dynamic graphics output provided by a Megatek 7255 color raster display. This setup allows observation of tissue development in real time; the user may observe intermediate states and interactions which are often of considerable importance in understanding the generation of complex patterns.

The model has been designed for maximum flexibility. The behavioral repertoire, the list of internal states, and the measures of local environment can all be augmented or modified. The rules which govern the selection of behaviors are also easily modified. The model can therefore be adjusted to the specifications of a wide variety of developmental systems.

As an initial test of the model, we examined clonal growth in two-dimensional, flat cell sheets (Bodenstein, 1986). The tissue was modeled with simple characteristics, including cells all of equal size, no cell death, and the probability of cell division fashioned as a simple function of cell age. Orientation of division was not constrained (except that both daughters had to remain on the two-dimensional sheet) but the spatial distribution of mitotic activity within the tissue was varied. These simulations demonstrate that the topography of mitotic activity, independent of all other parameters, can exert a significant effect on the degree of cell mixing.

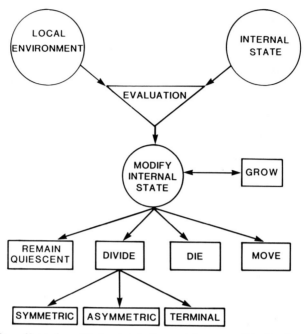

Fig. 2. Flow of logic for cell behavior. In the model, every cell in the tissue is accessed at random and evaluated based on a list of internal states and local environmental conditions. This process is repeated until some specified criterion is met, e.g., total cell number. The evaluation biases the probability of occurrence of various behaviors (move, die, divide, or remain quiescent). The "move" behavior is active and includes migration and local movement. Cells other than the one which is being evaluated may move passively in response to a neighbor's movement. Cell size can increase through a growth algorithm if appropriate.

CELL DIVISION AND CELL MIXING

When a model tissue, begun as a two-polyclone composite, is allowed to grow under different conditions, the relationship between cell division and cell mixing becomes evident (Fig. 6). When cell division is restricted to the edge of the tissue, little cell mixing occurs and the tissue matures in a systematic fashion. However, when cell division occurs throughout the tissue, cell mixing becomes significant and marked disruption of clonal boundaries is obvious. Thus, the movements associated with cell division may be sufficient to drive cell mixing. There is experimental evidence for an association between cell mixing and cell division in development (Jacobson and Klein, 1985), although a causal relationship has not been demonstrated empirically.

Further, regions of low mitotic activity may act as barriers to cell mixing (O'Brochta and Bryant, 1985).

The balance between cell division (acting to enlarge patches) and cell movement (acting to disrupt patches) has been treated in strict mathematical fashion by Lewis (1973). We have suggested that these actions may not be wholly independent, and that some cell movement proceeds directly from cell division. In a tissue undergoing interstitial cell division, the sum of these two actions (cell division and division-dependent cell movement) is to produce larger and larger patches (the former effect), with more and more disrupted (ragged) boundaries (the latter effect).

This link between mitotic activity and cell mixing follows from the nature of the division event. In the absence of specific cell–cell interactions[3] (e.g., differential adhesion), all cell movements act to increase cell mixing. Cell division (especially if followed by growth of the daughter cells) results in net movements of the daughter pair and surrounding cells. In our model, a range of cell displacement occurs consequent to cell division. At the site of cell division, daughter cells are generated and grow to mature sizes; this results in small but significant displacements of themselves and immediate neighbors. Cells at greater distances are slightly displaced as a ripple proceeds outward from the site of division; the more distant from the division event, the less the displacement. In the model, most cell mixing results from the former (local interactions) rather than from the latter (distant effects); this is reassuring, since we doubt that the motions of a single cell *in vivo* ramify to distant cells. When cell division occurs at patch borders, its effect on mixing will be maximal. Interstitial growth causes substantially more cell mixing than edge growth because more division events occur in boundary regions (Fig. 3). For the same reason, mixing secondary to cell division will be more prominent if patch size is small (i.e., the ratio of total patch border length to total patch area is greater if patch size is smaller). In accord with this notion, Lewis (1973) concluded that "the tendency to mingle . . . will be dominant when clones are small, and the tendency to apartheid will be dominant when they are large." In the analysis by Lewis, movement to produce mingling was secondary to cellular "diffusion." Cell division necessarily results

---

[3]If more than one cell type is present, differential adhesion may act to distribute the two types spatially in very nonrandom fashion (Steinberg, 1963; Moscona, 1968). However, in a homogeneous population in which one subpopulation bears an experimental marker, differential adhesion cannot be involved since (by definition) a good marker is transparent to the system.

a

b

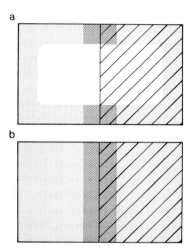

FIG. 3. Topography of cell division and cell mixing. A schematized tissue is divided into two polyclones (cross hatched and blank). Cell division activity (light shading) occurs either at the tissue edge (a) or throughout the tissue (b). Cell mixing occurs when the boundary between polyclones coincides with the region of mitotic activity. The greater the region of overlap (dark shading), the more cell mixing is possible. Therefore, less cell mixing occurs in the edge growth case (a) than in the interstitial growth case (b). This difference is enhanced because in the edge growth case each section of the polyclonal boundary overlaps the zone of division for only a short period of time. In contrast, in the interstitial growth case the entire polyclonal boundary coincides with the mitotic zone throughout tissue development.

in some cell movement as the centers of the daughter cells cannot coincide with that of the original (mother) cell. Our model demonstrates that this movement is sufficient to cause cell mixing; of course, other causes of cell movement and hence cell mixing are not excluded. The absolute contribution to mixing made by cell division will depend on the exact nature of the division event, i.e., the amount of displacement associated with each division. By varying our model description of the division event, we can vary the amount of associated cell mixing (Fig. 6).

III. An Experimental Cell Sheet

It is of particular interest that our model is capable of generating a spectrum of clonal boundary types, from smooth to very ragged, since this range is seen in amphibian and mammalian chimeras and mosaics. The retinal pigment epithelium (PE) is a model system for the study of cell patterning and has been extensively investigated (in frogs, review Conway *et al.*, 1980; in mice, West, 1976a; Sanyal and

Zeilmaker, 1977; Schmidt *et al.*, 1986). We have focused on the disparity between frog and mouse patterns, and in particular the role of cell division in the formation of these patterns. We have attempted to define the normal morphogenesis of these cell sheets, and mimic tissue development by the use of experimentally determined parameters in our synthetic simulation model.

Autoradiographic and colcemid studies have shown that the bulk of frog retinal growth (including PE) occurs by accretion of cells in a germinal zone at the retinal periphery (ora serrata); as new cells are generated, postmitotic cells are displaced centrally and the germinal zone comes to lie further and further from the optic nerve head (Straznicky and Gaze, 1971; Hollyfield, 1971; Beach and Jacobson, 1979). These studies have been directly confirmed in the frog PE by serial photography *in vivo* (Hunt and Cohen, 1984), a method which uniquely allows one to follow the evolution of cell patterns over time in an individual animal.

In contrast, autoradiographic data on the mouse retina reveal a more distributed topography of mitotic activity. Cell divisions occur throughout the retinal sheet, although as the eye matures this activity is more prominent toward the retinal periphery (Sidman, 1961; in cat, Rapaport and Stone, 1983). The pattern in neural retina has been recently investigated in detail with a monoclonal antibody to neurofilament protein which cross-reacts strongly with mitotic cells (Dräger, 1985). Mitotic activity is homogeneously distributed throughout the neural retina prenatally but becomes progressively more biased toward the periphery after birth (U. Dräger, personal communication). Using the same probe, we have demonstrated an analogous pattern in the PE (Bodenstein and Sidman, 1986a). In the late-stage embryo and early postnatal animal, mitotic activity is confined to a broad circumferential band of retina abutting the ora serrata. At the earliest embryonic stages examined, the breadth of this band approximates the retinal radius, and mitotic activity is noted throughout the retina. By early postnatal stages, the retinal radius has increased considerably, and the width of the mitotic zone decreases somewhat; mitotic activity is then more prominent toward the periphery (anterior) of the eye than toward the center (posterior) of the eye. We describe this mitotic pattern as a form of *edge-biased interstitial growth* to indicate that (1) mitoses occur within the tissue sheet (interstitial), but (2) the activity is greater toward the ora serrata (edge biased). This pattern contrasts with that seen in the frog, where all activity is confined to a very narrow growth zone at the periphery of the PE (strict *edge growth*).

In the frog, a variety of clonal geometries are generated when

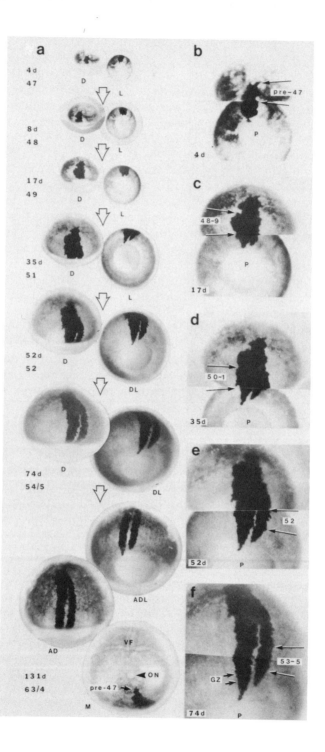

small pieces of marked tissue are implanted surgically into an unmarked host eye. The fine structure of these polyclones is dependent on the experimental details and has been reviewed (Conway *et al.*, 1980). In general, grafts generate large patches which appear as pie slice-shaped sectors extending from the optic disk to the ora serrata (Fig. 4).

In the mouse, retinal cell patches in experimental chimeras and mosaics are highly disrupted but seem to follow a general radial plan (Mintz and Sanyal, 1970; Sanyal and Zeilmaker, 1977) reminiscent of that seen in frogs. Using whole-mount preparations, we have investigated the PE cell patterns in detail; in accord with other investigators (West, 1976a; Schmidt *et al.*, 1986) we note that the radial design is most obvious at the retinal periphery where patches are strongly oriented perpendicular to the ora serrata (Fig. 5; Bodenstein and Sidman, 1986b).

FIG. 4. Development of a wedge-graft chimera in the frog. Excerpts from the photographic record of case SI2631 (×19), an orthotopic graft chimera of *Xenopus,* kindly provided by Dr. R. K. Hunt, Salk Institute, La Jolla, California. (a) Growth of the eye proceeds from top to bottom; for each photographic timepoint the age of the tadpole (in days postoperation) and its developmental stage (Nieuwkoop and Faber, 1967) are given to the left, beginning at 4 days postoperation when the tadpole had progressed to stage 47 (onset of feeding). At each stage two or more views of the eye are shown and are labeled as dorsal (D), lateral (L), dorsolateral (DL), etc. All three photomicrographs from the last timepoint (131 days, stage 63/4) were photographed in buffer after immersion fixation of the eye. In addition to providing an unobstructed medial view of the eye, the eye can be rotated on its axis to provide oblique views such as anterodorsal (AD) and anterodorsolateral (ADL) to visualize optimally the black sector which the transplant had formed in the pigmented retinal epithelium (pre). (b–f) Montages formed from ×2.5–3 blow-ups of the D view and the DL or L view at the left for the corresponding timepoint. Boundary characters established since the previous photographic session are bracketed on each montage. Note that "new" boundary characters are added to the distal edge of the PRE and accordingly provide a cumulative spatial record of the developmental growth of the transplant. For ready orientation, the pupil (P) is labeled, and the germinal zone (GZ) is also labeled on the last montage. This chimera was prepared by orthotopic transplant of a small patch of dorsal germinal cells from the right eye of a stage 34 pigmented donor embryo into the dorsal germinal zone of a stage 34 albino host embryo. Photomicrographs were taken at ×37.5 through a Ziess dissecting microscope with the anesthetized tadpole lying on a bed of agar and transilluminated at low angles with two Dolan–Jenner fiber optic illuminators. This case is somewhat idiosyncratic in that one or more albino germinal cells, at a later stage of development, "invaded" the black arc of germinal cells descended from the transplant. Whereas, before that event, the pigmented germinal cells added coherent arcs of black cells to the pre, subsequent to that event, arcs of newly added material were broken by one or more albino pre cells arising from the invading germinal cell. This idiosyncratic event is common for *dorsal* transplants but not typical of germinal cell transplants at other positions around the eye rim.

Fɪɢ. 5. Mouse chimera PE whole mounts. Chimeras were formed by fusion of 8-cell morulas of pigmented and albino strains of mice (C57BL/6J-c^{2J}-Gpi-1^a ↔ C57BL/-6J-c^+-Gpi-1^b). Whole mounts were prepared with the neural retina "face down" and the PE exposed by removal of the overlying sclera and choroid (see Bodenstein and Sidman, 1986a,b). In these preparations, pigmented PE cells are dark and unpigmented PE cells occupy the clear areas. The optic nerve head appears as a small, sharply bounded clear spot in the approximate center of each specimen. At the edge of the PE sheet, in the region of the ora serrata, the ciliary body and iris are evident. A notch has been placed in the iris at the dorsal pole for orientation (top of each figure). The specimens shown range from about 10% pigmented (left), through 50% (middle), to 90% pigmented (right). Inspection reveals that (1) both parental cell types (pigmented and albino) are well distributed throughout each specimen, (2) whereas isolated patches are clearly discernible in the extreme cases, most patches become linked in a large network covering the entire specimen when the ratio of pigmented to unpigmented cells approaches unity (middle), (3) patch size increases with increasing distance from the optic nerve head, and (4) peripheral patches show a consistent radial orientation which is not present in the central region. Bar, 500 μm.

Frog PE, then, grows by accretion of cells at the periphery (edge growth), while mouse PE grows by addition of cells in a more distributed fashion (edge-biased interstitial growth). In addition, polyclones in the frog PE are easily defined, smoothly bounded, and clearly radial; patches in the mouse PE are poorly defined and "ragged" but suggestive of a radial organization (especially toward the periphery). Our growth model is consistent with these findings: when we limit the positions of dividing cells to the tissue edge, the model generates "frog-like" patches, and when we allow interstitial growth, it produces "mouse-like" patches (Bodenstein, 1986).

MODELING THE PIGMENT EPITHELIUM

We have modeled the PE as a spherical sheet of cells. A small founder population is divided into two subpopulations which are identical except for a marker. Tissue growth then proceeds either at the tissue edge (frog-like conditions; Fig. 7) or in edge-biased interstitial fashion (mouse-like conditions; Fig. 8).

The initial cell pattern in each case is chosen to represent the experimental method of chimera construction. In the frog case, a small wedge of cells which represents a surgical graft is marked; the remaining cells in the model PE are unmarked. In the mouse case, the early patterns have not been described in detail but are known to be highly variegated (West, 1976a). We have therefore begun these simulations with a random mix of marked and unmarked cells. In the face of limited data, this crude approximation is viable since elsewhere we have demonstrated that the relevant features of the final pattern are less dependent on the initial conditions and more dependent on the mode of growth (Bodenstein, 1986).

In the frog case, only cells abutting the ora serrata are eligible for division. In the mouse case, the spatiotemporal distribution of mitotic activity is derived from recent observations on cell division in this tissue (Bodenstein and Sidman, 1986a). These distributions are approximated in the simulation model in *de facto* fashion. The *in vivo* mechanisms by which such distributions may be generated are not addressed here, although it is possible to mimic these distributions using putative cell lineages in the absence of additional (ad hoc) mechanisms (unpublished simulation data).

A simulated frog eye reproduces several features of the experimental case (compare Figs. 4 and 7). The marked polyclone extends in a coherent and radial fashion from the site of implantation to the ora serrata. Pattern features at the lateral edges of the polyclone are generated at the ora serrata, evolve somewhat as they pass out of the growth ring, and then stabilize in form; they become displaced successively further from the growth ring as new cells are generated anteriorly. Although the polyclone is very coherent, the edge is prone to some slight disruption which causes both a jaggedness and the occasional isolation of individual cells.

The mouse eye simulation is radically different from that of the frog eye and corresponds well to the experimental data (compare Figs. 5 and 8). Comparable longitudinal data are not available for the mouse. However, the experimental and simulation-derived final state patterns can be compared. Pattern similarities include a high degree of variegation with a patch size distribution that is skewed toward the small end of the scale, increasing average patch size with increasing distance from the optic nerve head, and more radial orientation of patches toward the periphery (see Bodenstein and Sidman, 1986b). Also note that although the patches appear well distributed throughout the tissue, when the proportion of marked cells is low, several large, peripheral regions are devoid of marked patches. This is most evident in the simulation with 5% marked cells (Fig. 5) and can be compared with the chimeric specimen with a very high proportion of pigmented cells (Fig. 8); in the latter case, the scattered albino patches should be considered the "marked" population for comparison.

FIG. 6. Cell mixing in a model tissue. In (a), a model tissue consisting of two abutting polyclones grows from the 36-cell stage to the 2500-cell stage. When cell divisions are not spatially restricted, considerable cell mixing and disruption of the clonal boundary occurs (top); when cell division is restricted to the tissue edge, little cell mixing occurs and the boundary remains largely intact (bottom). The degree of cell mixing is dependent on the amount of cell movement associated with each division event. The final state pattern (2500-cell stage) of an analogous simulation using an algorithm chosen to minimize this movement is also shown (b). Less mixing is apparent, but there continues to be a greater disruption of the clonal boundary in the interstitial growth case (top) than in the edge growth case (bottom).

FIG. 7. Simulation of frog PE chimera. A frog PE simulation has been run from the 143-cell stage to the 2500-cell stage. Seven contiguous cells in the initial grouping have been color coded (blue) to mimic a marked wedge graft. Only cells at the edge of the tissue are eligible for division, and all new cells are generated in this "germinal zone." As the total number of cells increases, the eye shell enlarges to maintain a curved, close-packed, one-cell-thick array of cells. Notice that the "graft" grows out a contiguous sector as cells are added in annuli at the germinal zone. Prominent features at the sector boundary stabilize in form and are passively displaced away from the germinal zone as the eye grows. The elaboration of the contiguous sector, and the trapping of sector boundary features or characters, is characteristic of experimental material (see Fig. 4). In addition, some disruption of the boundary with occasional isolation of single cells occurs in both the model and *in vivo*. We have not attempted to model the idiosyncratic "invading" albino PE cells as captured in Fig. 4; the model does produce (1) radial propagation of clones from the germinal zone, and (2) occasional cell displacement events along sector boundaries, which can combine to produce this idiosyncratic type of pattern feature.

FIG. 8. Simulation of mouse PE chimeras. Shown in (a) is the growth of a simulated mouse eye from the 598-cell stage to the 4000-cell stage (spherical images). The 4000-cell case is then displayed as a "whole mount" for comparison with experimental data. The low cell density in the center of each whole-mount display is a reflection of the geometric distortion introduced when a spherical surface is mapped onto a plane; this low density is not present in the original spherical representation. The geometric conversion causes minimal distortion of patch geometry as can be seen by comparing the spherical and flattened images. In the initial pattern (598 cells), cells were assigned to one (gray) or the other (blue) population on a random basis (see text). Cell division was biased toward the edge of the tissue, with increasing bias as the number of cells increased. Thus, at the 598-cell stage, all cells are eligible for division, but cells in the periphery have a 2- to 3-fold higher likelihood of division than cells in the center of the eye. By the 4000-cell stage, cell division in the central region is virtually absent, and cell division is increasingly confined to the tissue edge. In this simulation, the proportion of "marked" cells (blue) is 0.25. Shown in (b) are 4000-cell whole-mount images of simulations in which the proportions of marked cells are 0.05, 0.10, 0.15, and 0.20.

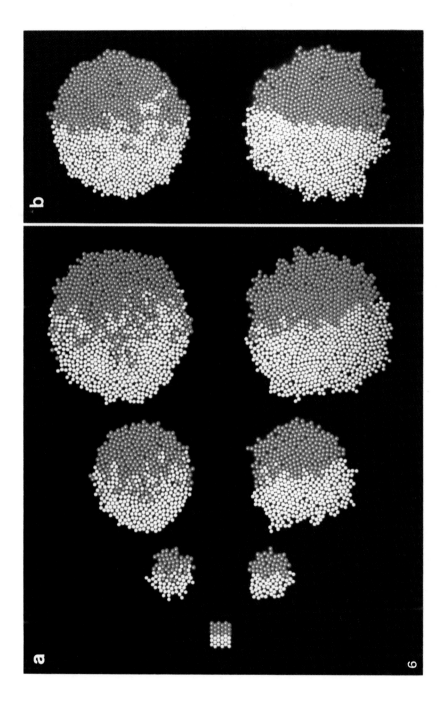

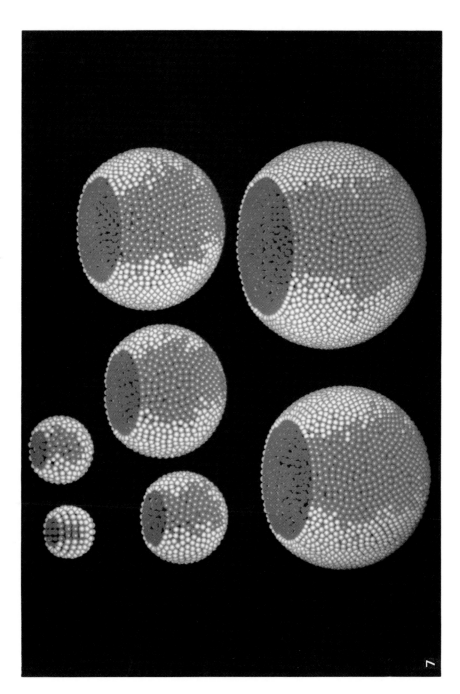

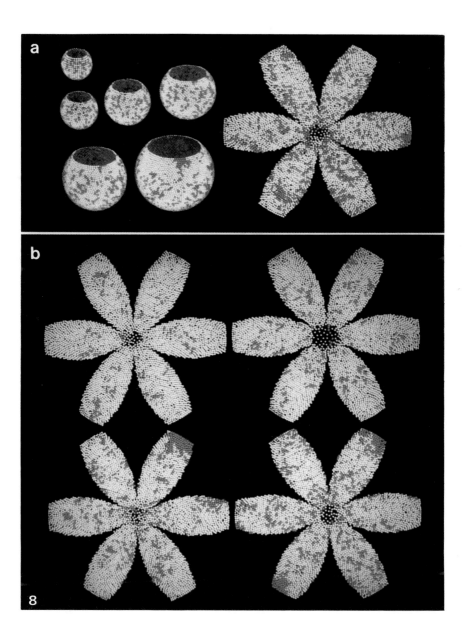

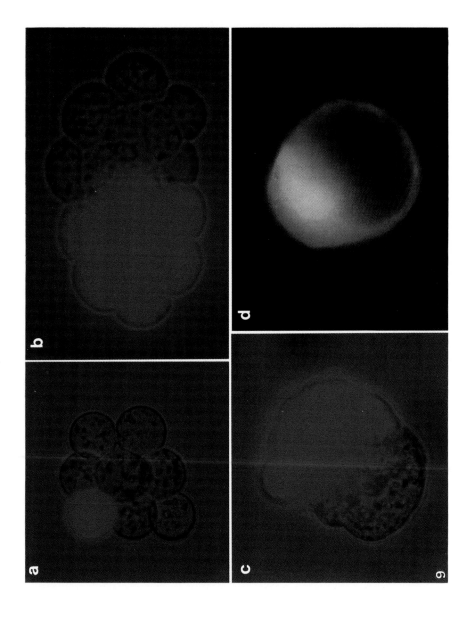

IV. A Three-Dimensional System

Although numerous studies in the mouse have demonstrated a high degree of cell mixing, detectable in late-embryonic and postnatal cell patterns, studies of the very early embryo (through blastulation) do not detect significant cell mixing (Garner and McLaren, 1974; Kelly, 1979; Balakier and Pedersen, 1982). We have examined the cell movements associated with the fusion of morulas to produce chimeric embryos and the formation of the blastocyst. Methods have been developed in which morulas (in part or whole) are vitally stained and followed for up to 30 hours postfusion. During this time, fusion, compaction, and blastulation occur. In agreement with previous studies, very little cell mixing is seen at these early stages (Fig. 9).

Although during cleavage stages the embryo grows by interstitial cell division (all cells are mitotically active), daughter cells do not grow to parent cell size and the embryo does not appreciably enlarge. Because daughter cells occupy about the same space as the parent cell, little cell movement should be associated with these division events. We have done simulation modeling of these early stages. In the absence of postdivision cell growth, mixing in the model is neglible (data not shown); this agrees with the experimental observations.

V. Concluding Remarks

Developing systems involve large numbers of cells, each carrying out a sequence of behaviors mediated by the complex interplay of genetic and environmental factors. Understanding the means by which these behaviors meld to evolve multicellular structures has remained a vexing problem. Introduction of cell autonomous lineage markers into developing embryos provides insight into the clonal

---

FIG. 9. Vital staining of mouse early embryo chimeras. In (a), a single rhodamine-labeled blastomere dissociated from an 8-cell embryo adheres to an unlabeled 8-cell embryo. Such cells incorporate and produce labeled clones visible through blastulation (not shown). In (b), a rhodamine-labeled 8-cell embryo is aggregating with a corresponding unlabeled 8-cell embryo. In (c), a labeled ↔ unlabeled aggregate embryo, formed at the 8-cell stage, has undergone compaction to produce a late morula. The component embryos are clearly distinguishable as is a single labeled cell which has not compacted. In (d), a rhodamine-labeled 8-cell embryo was aggregated with a fluorescein-labeled 8-cell embryo and the pair allowed to develop to the blastocyst stage in culture. A medium-sized blastocyst cavity is evident and imparts a signet ring appearance to the embryo. Notice that the two components can be clearly distinguished and that little cell mixing has occurred. The yellow boundary region reflects overlap of cells since the plane of fusion is not exactly parallel to the angle of view. More detail can be realized by "optically sectioning" through the intact specimen.

nature of embryo construction. However, there has been little success in extracting anything but rudimentary information from these cell patterns, especially in mammals. As another approach, we have devised a simulation model which can (1) reveal and evaluate the mechanisms underlying pattern generation, and (2) mimic cell patterns seen in experimental material. Although highly schematized, this model is capable of reproducing the very different amphibian and mammalian cell patterns. With a more robust tissue representation (e.g., tensile as well as compressive forces), the model may be applied to other, more complex developmental systems.

APPROACHES TO MODELING

The use of modeling to understand the evolution and function of complex systems has a rich history in biology and the other sciences. The model which we have described is of a particular genre of relatively recent vintage.

Models can be broadly classified into two major categories, those of hypothesis and those of synthesis. Models of hypothesis, an extension of Baconian inductive reasoning, are generally the more daring. Built on an established base of data, they engender hypotheses which are then tested against newly acquired data. The term "model" rather than "hypothesis" is used when the intent is to encompass a large data base, derived from multiple sources, under a single rubric. In biology, the more celebrated models of hypothesis include the *lac operon* model of prokaryotic gene expression (Jacob and Monod, 1961) and the sliding filament model of muscle contraction (A. F. Huxley and Niedergerke, 1954; H. E. Huxley and Hanson, 1954). The field of morphogenesis or pattern formation has seen a wealth of such models, including strict diffusion models (Crick, 1970; Conway, 1979), reaction–diffusion models (Turing, 1952; Gierer and Meinhardt, 1972; Shoaf *et al.*, 1984), a phase-shift model (Goodwin and Cohen, 1969), an incremental transduction model (Bodenstein and Hunt, 1979) and a metastable, binary state model based on reaction–diffusion interactions (Kauffman *et al.*, 1978). As with other hypotheses, these models can never be verified with certainty, and their long-term value is often dependent on continued concordance with newly acquired data. Yet the utility of such models lies not only in their accuracy, but in their serviceability to (1) review the data base and clearly define the outstanding, unresolved issues in a field, (2) pull together similar themes in different fields which may indicate a common mechanistic underpinning, (3) initiate and rejuvenate discussion in areas which have remained dormant through oversight or neglect, and (4) generate concrete, testable predictions which call forth experimental inquiry. An excellent example of such a model is the regeneration-reduplication model of French *et al.* (1976; also see Bryant *et al.*, 1981). The model provides a simple series of rules which summarize a myriad of results obtained when developmental anlagen are disrupted and reassembled in unorthodox configurations. The power of the model rests in its simplicity (two rules) and its broad applicability (it has been applied to developmental systems from the cockroach leg to the frog eye). Yet the most salient feature of the model is not its novelty (see Bateson, 1894, and Harrison, 1921, for qualitative and less ambitious attempts to devise similar schemes), but rather that this robust model served to rejuvenate scientific inquiry. A flurry of excitement followed introduction of the model, and dozens of thought-provoking experiments, in several different experimental systems (review Stocum, 1980; Iten, 1982; Walthall and Murphey, 1984), were devised (ostensibly to "test" the model). That the model, or portions of the model, failed direct test in some instances (Stocum, 1978; Hunt *et al.*, 1982; Shelton *et al.*, 1983) in no way lessens the contribution made to experimental science by such a timely bit of theory.

Models of synthesis are less flashy than those of hypothesis, but they may become indispensible to the study of complex biological systems. Because of the complexity of these systems, most synthetic models feature computer simulations to test the degree to which putative mechanisms can account for empirical observations. A particularly successful example of such a model is the Hodgkin–Huxley model of nerve conduction (Hodgkin and A. F. Huxley, 1952d) which established that the physicochemical properties of the axonal membrane, coupled with the known intracellular and extracellular ion concentrations, could account in detail for the resting membrane potential and the form and time course of the action potential.[4] Synthetic models which address aspects of embryonic development are of more recent vintage (Ede and Law, 1969; Ransom, 1975, 1977; Matela and Fletterick, 1980; Ransom and Matela, 1984; Kurnit *et al.*, 1985; Bodenstein, 1986). Development features the interaction of many cells and

---

[4]If the Hodgkin–Huxley model is defined broadly to include the wealth of experimental work leading to an understanding of nerve conduction [and culminating with the now famous series of papers in the *Journal of Physiology* (Hodgkin and A. F. Huxley, 1952a–d)], then it is largely a model of hypothesis. If, however, we narrow our focus to the final mathematical threatment (Hodgkin and A. F. Huxley, 1952d), then the model is aptly described as one of synthesis. In this case, the primary role of the model is to demonstrate that the accumulated data are sufficient to "account for conduction and excitation in quantitative terms;" no new data or explanations are offered.

cell types, each with diverse developmental programs, over the relatively long time periods of organ formation. In such a setting, subtle mechanisms can, over time and space, generate large perturbations in the system. The random motions of a single cell may appear an insignificant factor in organ formation, but summed over all cells and over reasonable time periods, these motions may drive sorting phenomena distinguished on a large scale (Steinberg, 1962a,b; Moscona, 1968). In homogeneous populations, where individual cells obey the same and simple rules, mathematics borrowed from statistical mechanics are of use. In a grossly inhomogeneous population, with complex cell–cell interactions, the calculations become hopelessly involved.[5] In such systems it is appropriate to generate a detailed simulation, filling in the appropriate parameters and interactions, and allowing the model system to evolve over time. Such a simulation model synthesizes all known and hypothesized parameters and relationships into a single composite. The model provides only the framework, the details are provided by experiment and theory.[6] A recent example of such an approach applied to a developmental system is a model of nerve cell specificity and plasticity provided by Fraser (1980; Fraser and Hunt, 1980). By combining previously hypothesized cell–cell interactions with a novel homophilic chemistry in a quantitative simulation model, the authors were able to predict accurately the pattern of connections which are generated when nerve fibers from the frog eye innervate the optic tectum.

To be useful, models of synthesis must (1) possess a degree of complexity and detail equal to the task, (2) allow for the introduction of all

[5]This is particularly true if the final state is sensitive to initial conditions. The robust ability of developing systems to overcome perturbations and produce a "normal" final product (especially in vertebrates) suggests that complex regulative mechanisms exist. Such mechanisms, which may include redundant parameters (Bodenstein and Hunt, 1979; Hunt *et al.*, 1982), may serve to complicate these calculations but tend to stabilize final results.

[6]The long-range weather forecast model used by the National Weather Bureau is an excellent example of a computer-implemented synthetic model. For global patterns the atmosphere is divided into cubes of fixed size (a three-dimensional grid). Each cube functions as a homogeneous unit (a "black box") and interacts with its neighboring cubes and global energy systems (e.g., sunlight). Forecasting is improved by (1) making the grid finer, (2) using more refined algorithms to generate the "behavior" of individual cubes, and (3) modifying the set of global parameters (e.g., older models did not account for the presence of sunlight during the day and its absence at night). Accuracy is limited by the ability to collect the huge amount of data and the ability of even the fastest digital computers to perform the necessary calculations in a reasonable period of time. In addition, this system is particularly sensitive to initial conditions, a modeling problem which has only recently attracted serious attention (see footnote 5).

physically reasonable entities and interactions, and (3) not generate artifacts which are specific to the model or its implementation. These must be viewed only as goals since even with high-speed computers and sophisticated model-building it is doubtful that one could provide a perfect framework for anything but the simplest of systems. Conversion from the real system to a theoretical model and then to a practical implementation of this model (usually a computer program) entails considerable risk for the generation of artifact, and special attention must be given to this conversion process.[7] In spite of these difficulties, this form of synthetic modeling is at least useful, and may become indispensible, to the study of complex biological systems.

VI. Appendix: The Mathematics of Biological Patches and Clones *Revisited*

The most widely used method of analysis of the cell patterns in mammalian chimeras and mosaics consists of determining, by a combination of experimental and theoretical means, the size of the groupings of clonally related and spatially contiguous cells, termed *coherent clones* (Nesbitt, 1974). Application of these methods by a number of investigators has revealed significant limitations (West, 1976a,b; Mullen, 1978; Schmidt *et al.*, 1986). First, the analysis requires a highly restrictive (and rarely, if ever, valid) set of assumptions about the biological material. These assumptions include size and spatial homogeneity of clones, tissue isotropy, limited cell mixing, and little cell death (see main text). Some of these assumptions are prerequisites for an accurate assessment while others are related to the amount of information one is likely to glean from the analysis; no attempt has been made to ascertain the degree to which departures

---

[7]Synthetic models of complex systems usually require computer implementation. In this scheme, a *theoretical model* is designed from the *experimental system,* and then a *computer implementation* is built from the theoretical model. Experimental systems are generally analog while computers are digital. The analog to digital conversion may be left to the final stages so that the theoretical model is unencumbered by conversion artifacts and can parallel the real system as closely as possible, i.e., the theoretical model is analog. The analog to digital conversion is often a laborious process; even the simplest of analog processes may be maddeningly difficult to implement in a digital format (Dewdney, 1984) and the risk of introducing artifacts is high. One method to reduce this problem is to construct the theoretical model in such a way that the implementation is more feasible. In this approach the model is still analog, but the potentially intractable analog process in the real system is modeled as another (different) analog process (or a combination of other processes) more amenable to digital conversion. Finally, to help exclude implementation artifacts, it is useful to devise and test several implementation strategies. In this case, concordance of results is reassuring (Bodenstein, 1986).

from these assumptions will bias results. Second, patch size analysis represents an untested extrapolation (West, 1975) of a study of the mathematics of random clumping (Roach, 1968); no attempt has been made to validate this extrapolation or to explore its use in relation to experimental data.

As a theoretical exercise, we have investigated the mathematics of conversion from patch size data to coherent clone size information. As in previous studies (West, 1975; Whitten, 1978) we have used computer analysis to solve what is an analytically intractable problem. These previous studies provided "normal tables" for use in analysis of experimental material and furnished some guidelines for the selection of this material (e.g., the ratio of marked and unmarked cells in the tissue should depart markedly from unity).

Our method of simulation analysis is an extension of that described by West (1975) and Whitten (1978). A model tissue is constructed as a regular array of cells. Cells are then assigned as either marked or unmarked on a random basis; contiguous like-marked cells form patches.

We have investigated the one-dimensional case, three forms of the two-dimensional case (triangular cells, square cells, and hexagonal cells) and two forms of the three-dimensional case (cubic cells and tetrakaidodecahedral cells). In each instance, we have varied the size of the cellular array (100 to 1000 cells in one-dimension; 10×10 to 100×100 in two-dimensions; $20 \times 20 \times 20$ to $60 \times 60 \times 60$ in three-dimensions) and the proportion of marked and unmarked cells (where p is the proportion of marked cells and can vary from 0.0 to 1.0). In each simulation, average patch size for the marked population was calculated; simulations were run 100 times, and the means and standard deviations for representative cases are given in Table I.

In addition to calculating average patch size (West, 1975; Whitten, 1978), we have calculated a *weighted average patch size* for each case. This parameter is generated by weighting each patch by its size. The usefulness of this approach is clear at high p values. For example, given a 100×100 hexagonal array of cells with $p = 0.9$, most simulations consist of a single large patch of 9000 cells. Rarely, one or two cells "escape" from this single large patch, and patches of 8999 and one, or 8998 and one and one are seen. Patch size as averaged over all simulations is significantly lowered by these rare cases (and standard deviations are high): 8535.01 ± 1402.24; a better description of the nature of the pattern is provided by the weighted average patch size (which is very much less sensitive to the "escapees"): 8999.79 ± 0.91. In essence, whereas average patch size describes the distribution of

TABLE I

AVERAGE PATCH SIZE AND WEIGHTED AVERAGE PATCH SIZE
FOR SELECTED CELLULAR GEOMETRIES[a]

p^b	One dimension (1000)	Two dimensions (100 × 100)		Three dimensions (20 × 20 × 20)
		Triangular	Hexagonal	Cubic
0.1[c]	1.1 ± 0.2	1.2 ± 0.01	1.4 ± 0.03	1.4 ± 0.03
	1.2 ± 0.3	1.4 ± 0.04	1.8 ± 0.1	2.0 ± 0.1
0.2	1.2 ± 0.1	1.4 ± 0.02	2.1 ± 0.04	2.2 ± 0.1
	1.4 ± 0.3	2.0 ± 0.1	3.6 ± 0.2	5.7 ± 0.7
0.3	1.4 ± 0.2	1.8 ± 0.03	3.4 ± 0.1	4.5 ± 0.2
	1.8 ± 0.3	3.1 ± 0.1	8.7 ± 0.6	66.0 ± 32.4
0.4	1.7 ± 0.2	2.4 ± 0.04	7.2 ± 0.2	12.5 ± 0.8
	2.3 ± 0.4	5.4 ± 0.3	37.6 ± 6.8	2193.5 ± 147.2
0.5	2.0 ± 0.2	3.7 ± 0.1	23.8 ± 1.6	39.1 ± 3.8
	2.9 ± 0.5	12.3 ± 1.1	1392.2 ± 636.5	3720.9 ± 33.7
0.6	2.4 ± 0.2	6.7 ± 0.2	109.7 ± 13.4	150.5 ± 36.1
	3.7 ± 0.6	46.9 ± 8.1	5697.5 ± 101.3	4725.9 ± 15.8
0.7	3.3 ± 0.3	18.2 ± 1.0	624.0 ± 166.4	699.7 ± 285.6
	5.2 ± 1.0	1948.4 ± 931.9	6963.4 ± 14.7	5582.9 ± 6.2
0.8	4.9 ± 0.5	75.8 ± 6.7	3661.4 ± 2023.9	3986.1 ± 1838.8
	8.6 ± 1.9	7596.8 ± 74.2	7995.3 ± 4.2	6398.0 ± 2.0
0.9	9.2 ± 1.0	616.9 ± 174.8	8535.0 ± 1402.2	6984.0 ± 854.9
	15.7 ± 3.7	8960.1 ± 13.6	8999.8 ± 0.9	7199.9 ± 0.5

[a]One-dimensional (1000 cells), two-dimensional (100 × 100 = 10,000 cells) and three-dimensional (20 × 20 × 20 = 8000 cells) tissues are presented. Cell geometry is either triangular or hexagonal in the two-dimensional case, and cubic in the three-dimensional case; other forms (square in two dimensions and tetrakaidodecahedron in three dimensions) are not shown.

[b]p, The proportion of marked cells in the tissue.

[c]In each case and in each cell proportion, the average patch size (mean ± SD) is given (above) along with the weighted average patch size (below).

patches, weighted average patch size describes the distribution of cells which form patches, i.e., the average size patch with which a given cell is associated. The use of other patch pattern parameters has been suggested in different contexts (Schmidt et al., 1986; Bodenstein, 1986), but these have not been applied to the problem of extracting coherent clone information from patch size data.

These simulations serve to emphasize several points. The geometrical form chosen to represent a cell determines the number of near neighbors each cell will have, and this in turn dramatically affects the average patch size at given values of p. The hexagon in two dimensions and the tetrakaidodecahedron in three-dimensions have the lowest surface area to volume ratio of the possible semiregular polygons and polyhedrons (Thomson, 1887) and are reasonable models of close-packed cellular arrays (Thompson, 1961). In real tissue systems, however, the number of near neighbors varies on a cell-by-cell basis. Since average patch size is particularly sensitive to this parameter, consideration must be given to determining the distribution of cell forms in an experimental tissue if patch size analysis is to be valid.

As pointed out by West (1975), at low values of p patch size is largely independent of array size, but at high value of p all cells tend to become linked in a single large patch, and patch size is strongly dependent on array size. Therefore, "any attempt to infer clone size from patch size will be most reliable when the ratio of the two cell populations departs markedly from unity" and the smaller population is used in the analysis (i.e., p is small). This handicap may considerably limit the selection of experimental material, since small values of p are necessary to assure accuracy (especially when the number of cell near neighbors is high as in three-dimensions). Indeed, if p is very small then coherent clones will rarely abut and patches are single clones (Whitten, 1978); in this case no analysis is needed, but only a small percentage of experimental cases fulfill this strict criterion (Schmidt et al., 1986).

Patch size is sensitive to the value of p since edge effects dominate when the ratio of patch size to tissue size is not kept very small. If the tissue under investigation is not naturally bounded, then some patches will be cut at the artificially imposed edges (truncation). In addition, large patches will be less likely to be included within the circumscribed piece of tissue under investigation (exclusion); in the limiting case, the tissue can contain no patch larger than itself, although in an infinite tissue all patch sizes would be present. Both of these effects act to remove large patches differentially from the analysis and therefore lessen average patch size. We can correct for truncation (see below), but exclusion is inescapable and becomes significant if some patch sizes approach tissue size. For this reason, it is preferable to analyze tissue which is complete with natural boundaries. In this case, the set of patches is also complete and neither truncation nor exclusion occurs.

The rationale for calculating average patch size in simulated tissues is to provide a baseline for the degree of clumping associated with the state of random assortment, i.e., complete disruption of the spatial integrity of clones by cell mixing. Clumping in addition to this level is then thought to represent nonrandom mechanisms of cell asso-

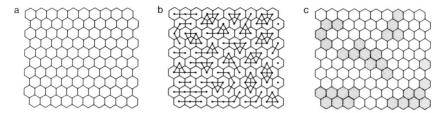

FIG. 10. Creation of model tissue array. In (a), a 10×10 array (tissue) of hexagonal cells is constructed. In (b), this array is segregated into randomly arranged 4-cell clones. Smaller clones at the boundaries represent 4-cell clones "cut off" by the edge of the tissue. In (c), clones in this tissue have been randomly assigned as either marked (shaded) or unmarked (blank) such that the proportion of marked cells is about 0.3. The patch pattern consists of six patches of average size 5.17. Patch size analysis is designed to deduce the clonal pattern (b) from the patch pattern (c).

ciation, the most important of which is coherent clonal growth by cell division. For example, if average patch size in an experimental tissue is 5-fold greater than that calculated in a matched simulated tissue in which individual cells are randomly assorted, then we infer that the experimental tissue is formed as a composite of 5-cell coherent clones. We have devised a theoretical study to test the approach. A model tissue array is subdivided into coherent clones of equal size (Fig. 10). Each clone is then randomly assigned, as marked or unmarked, to yield a patch pattern. This pattern is then analyzed to determine if the expected coherent clone size can be recovered: the average patch size in this tissue is divided by the average patch size in a theoretical tissue, in which clone size is unity, to yield the estimated clone size for the tissue being analyzed (West, 1975). The study demonstrates that patch size analysis tends to underestimate coherent clone size (Table II). This underestimation increases with increasing value of p and increasing coherent clone size, and is related to the edge effects of truncation and exclusion. We can eliminate the edge effect of truncation by discounting from the analysis patches which abut the tissue edge; the edge effect of exclusion cannot be eliminated but should be small at low p values. Additional error results from accumulated small effects; in part, these are due to variability in the number of near neighbors abutting each coherent clone or patch.

    Patch size analysis requires a strict selection of experimental material (e.g., small value of p), a detailed representation of the tissue (e.g., number of near neighbors on a cell-by-cell basis), and a method of treating edge effects (e.g., a tissue with natural boundaries). In addition to these mathematical limitations, the analysis has significant

TABLE II

EXTRACTION OF CLONE SIZE FROM PATCH SIZE DATA IN SIMULATED TISSUES[a]

	All patches		Edge-patches removed	
p^b	Clone size 1 (a)	Clone size 5 (b)	Clone size 1 (c)	Clone size 5 (d)
0.1	1.34 ± 0.11	5.53 ± 1.24 (4.13) [−17%, −1%]	1.35 ± 0.15	6.37 ± 1.76 (4.72) [−5%, +14%]
0.2	1.96 ± 0.18	7.37 ± 1.50 (3.76) [−25%, −10%]	1.95 ± 0.27	8.42 ± 2.88 (4.32) [−14%, +4%]
0.3	3.10 ± 0.34	11.19 ± 2.42 (3.61) [−28%, −13%]	2.87 ± 0.60	11.10 ± 7.17 (3.86) [−23%, −7%]
0.4	5.76 ± 0.77	17.20 ± 4.58 (2.99) [−40%, −28%]	3.81 ± 1.10	7.67 ± 6.51 (2.01) [−60%, −52%]
0.5	14.16 ± 3.59	31.57 ± 11.25 (2.23) [−55%, −47%]	2.95 ± 2.09	4.48 ± 6.09 (1.52) [−70%, −64%]

[a]In column (a), average patch size is given for randomly assorted hexagonal cells in a 50 × 50 = 2500 array; the array is then divided into five cell clones (see Fig. 10) and the calculation repeated in column (b). Note that some clones are "cut off" at the tissue boundary, so that the actual average clone size for the array is 4.17. The number in parentheses is the division quotient of the patch size in (b) and the patch size in (a); this represents the clone size as determined from the patch size. Below this number are two numbers in brackets—the first represents the error from the expected value of five, the second represents the error from the actual value of 4.17; the former is the correct error if the tissue has been cut (i.e., the actual tissue contains only five cell clones, some of which have been cut off prior to analysis); the latter is the correct error if the tissue is naturally bounded (i.e., there are smaller clones at the natural tissue edge). In (c) and (d) the same analysis has been repeated, except that patches which abut the tissue edge have been excluded. Note that even at low p values (the only region where the analysis might work) both methods usually underestimate clone size. This error worsens with larger clone size (not shown).

[b]p, The proportion of marked cells.

biological limitations since it is based on a highly schematized, and almost certainly inaccurate, developmental scenario (see above and main text). For these reasons, such analyses are not easily managed in an experimental setting, and we conclude that other, more powerful, methods are required to decode the elusive clonal history of mammalian tissue.

ACKNOWLEDGMENTS

This work was supported by NIH Grants NS20820 and NS20822 (R. L. S.), and Public Health Service, National Service Award 2T 32 GM07753 from the National Institute of General Medical Science (L. B.).

REFERENCES

Balakier, H., and Pedersen, R. A. (1982). *Dev. Biol.* **90**, 352–362.
Bateson, W. (1894). "Materials for the Study of Variation." Macmillan, London.
Beach, D. H., and Jacobson, M. (1979). *J. Comp. Neurol.* **183**, 603–614.
Benirschke, K. (1970). *Curr. Top. Pathol.* **51**, 1–61.
Bodenstein, L. (1986). *Cell Differ.* **19**, 19–33.
Bodenstein, L., and Hunt, R. K. (1979). *Biophys. J.* **25**, 84a (Abstr.).
Bodenstein, L., and Sidman, R. L. (1986a). *Dev. Biol.* (in press).
Bodenstein, L., and Sidman, R. L. (1986b). *Dev. Biol.* (in press).
Bryant, P. J. (1970). *Dev. Biol.* **22**, 389–411.
Bryant, P. J., and Schneiderman, H. A. (1969). *Dev. Biol.* **20**, 263–290.
Bryant, S. V., French, V., and Bryant, P. J. (1981). *Science* **212**, 993–1002.
Conway, K. (1979). *J. Biophys.* **25**, 159a (Abstr.).
Conway, K., Feiock, K., and Hunt, R. K. (1980). *Curr. Top. Dev. Biol.* **15**, 217–317.
Crick, F. H. C. (1970). *Nature (London)* **225**, 420–422.
Crick, F. H. C., and Lawrence, P. A. (1975). *Science* **189**, 340–347.
Deol, M. S., and Whitten, W. K. (1972). *Nature (London) New Biol.* **238**, 159–160.
Dewdney, A. K. (1984). *Sci. Am.* **250**, 19–26.
Dräger, U. (1985). *In* "Cell and Developmental Biology of the Eye: Hereditary and Visual Development" (J. B. Sheffield and S. R. Hilfer, eds.), pp. 43–62. Springer-Verlag, New York.
Ede, D. A., and Law, J. T. (1969). *Nature (London)* **221**, 244–248.
Ford, C. E. (1969). *Br. Med. Bull.* **25**, 104–109.
Fraser, S. (1980). *Dev. Biol.* **79**, 453–464.
Fraser, S., and Hunt, R. K. (1980). *Annu. Rev. Neurosci.* **3**, 319–352.
French, V., Bryant, P. J., and Bryant, S. V. (1976). *Science* **193**, 969–981.
Garcia-Bellido, A. (1975). *Ciba Found. Symp.* **29**, 161–178.
Garcia-Bellido, A., Ripoll, P., and Morata, G. (1973). *Nature (London) New Biol.* **245**, 251–253.
Gardner, R. L. (1984). *J. Embryol. Exp. Morphol.* **80**, 251–288.
Garner, W., and McLaren, A. (1974). *J. Embryol. Exp. Morphol.* **32**, 495–503.
Gierer, A., and Meinhardt, H. (1972). *Kybernetik* **12**, 153–178.
Gimlich, R. L., and Cooke, J. (1983). *Nature (London)* **306**, 471–473.
Goodwin, B. C., and Cohen, M. H. (1969). *J. Theor. Biol.* **25**, 49–107.
Green, M. C., Durham, D., Mayer, T. C., and Hoppe, P. C. (1977). *Genet. Res.* **29**, 279–284.
Harrison, R. G. (1921). *J. Exp. Zool.* **32**, 1–136.
Harrison, R. G. (1933). *Am. Nat.* **67**, 306–321.
Hirose, G., and Jacobson, M. (1979). *Dev. Biol.* **71**, 191–202.
Hodgkin, A. L., and Huxley, A. F. (1952a). *J. Physiol. (London)* **116**, 449–472.
Hodgkin, A. L., and Huxley, A. F. (1952b). *J. Physiol. (London)* **116**, 473–496.
Hodgkin, A. L., and Huxley, A. F. (1952c). *J. Physiol. (London)* **116**, 497–506.
Hodgkin, A. L., and Huxley, A. F. (1952d). *J. Physiol. (London)* **117**, 500–544.
Hollyfield, J. G. (1971). *Dev. Biol.* **24**, 264–286.

Hunt, R. K., and Cohen, J. S. (1984). *Soc. Neurosci. Abstr.* **10,** 788.
Hunt, R. K., Tompkins, R., Reinschmidt, D., Bodenstein, L., and Murphey, R. K. (1982). *Am. Zool.* **22,** 185–204.
Huxley, H. E., and Hanson, J. (1954). *Nature (London)* **173,** 973–976.
Huxley, A. F., and Niedergerke, R. (1954). *Nature (London)* **173,** 971–973.
Iten, L. E. (1982). *Am. Zool.* **22,** 117–129.
Jacob, F., and Monod, J. (1961). *Cold Spring Harbor Symp. Quant. Biol.* **26,** 193–209.
Jacobson, M. (1978). "Developmental Neurobiology," 2nd ed. Plenum, New York.
Jacobson, M. (1985). *Annu. Rev. Neurosci.* **8,** 71–102.
Jacobson, M., and Klein, S. L. (1985). *Philos. Trans. R. Soc. London Ser. B* **312,** 57–65.
Kauffman, S. A., Shymko, R. M., and Trabert, K. (1978). *Science* **199,** 259–270.
Kelly, S. J. (1979). *J. Exp. Zool.* **207,** 121–130.
Kimmel, C. B., and Warga, R. M. (1986). *Science* **231,** 365–368.
Kurnit, D. M., Aldridge, J. F., Matsuoka, R., and Matthysse, S. (1985). *Am. J. Med. Genet.* **20,** 385–399.
Le Douarin, N. M. (1980). *Nature (London)* **286,** 663–669.
Le Douarin, N. M., and McLaren, A., eds. (1984). "Chimeras in Developmental Biology." Academic Press, New York.
Lewis, J. (1973). *J. Theor. Biol.* **39,** 47–54.
McLaren, A. (1976). "Mammalian Chimaeras." Cambridge Univ. Press, London.
Martinez-Arias, A., and Lawrence, P. A. (1985). *Nature (London)* **313,** 639–642.
Matela, R. J., and Fletterick, R. J. (1980). *J. Theor. Biol.* **84,** 673–690.
Mintz, B. (1967). *Proc. Natl. Acad. Sci. U.S.A.* **58,** 344–351.
Mintz, B., and Sanyal, S. (1970). *Genetics Suppl.* **64,** 43–44.
Moore, W. J., and Mintz, B. (1972). *Dev. Biol.* **27,** 55–70.
Moscona, A. A. (1968). *Dev. Biol.* **18,** 250–277.
Mullen, R. J. (1977). *Nature (London)* **270,** 245–247.
Mullen, R. J. (1978). *In* "The Clonal Basis of Development," pp. 83–101. Academic Press, New York.
Nesbitt, M. N. (1974). *Dev. Biol.* **38,** 202–207.
Nieuwkoop, P. D., and Faber, J. (1967). "Normal Table of *Xenopus laevis* (Daubin)." North-Holland Publ., Amsterdam.
O'Brochta, D. A., and Bryant, P. J. (1985). *Nature (London)* **313,** 138–141.
Oster-Granite, M. L., and Gearhart, J. (1981). *Dev. Biol.* **85,** 199–208.
Pearlstein, R. A., and Sidman, R. L. (1986). *Anal. Quant. Cytol.* **8,** 89–95.
Pearlstein, R. A., Kirschner, L., Simons, J., Machell, S., White, W. F., and Sidman, R. L. (1986). *Anal. Quant. Cytol.* **8,** 108–115.
Peterson, A. C., and Bray, G. M. (1984). *J. Comp. Neurol.* **227,** 348–356.
Ponder, B. A. J., Wilkinson, M. M., Wood, M., and Westwood, J. H. (1983). *J. Histochem. Cytochem.* **31,** 911–919.
Ponder, B. A. J., Festing, M. F. W., and Wilkinson, M. M. (1985). *J. Embryol. Exp. Morphol.* **87,** 229–239.
Ransom, R. (1975). *J. Theor. Biol.* **53,** 445–462.
Ransom, R. (1977). *J. Theor. Biol.* **66,** 361–377.
Ransom, R., and Matela, R. J. (1984). *J. Embryol. Exp. Morphol.* **83** (Suppl.), 233–259.
Rapaport, D. H., and Stone, J. (1983). *J. Neurosci.* **3,** 1824–1834.
Roach, S. A. (1968). "The Theory of Random Clumping." Spottiswoode, Ballantyne, Ltd., London.
Rossant, J., Vijh, M., Siracusa, L. D., and Chapman, V. M. (1983). *J. Embryol. Exp. Morphol.* **73,** 179–191.

Sanyal, S., and Zeilmaker, G. H. (1976). *J. Embryol. Exp. Morphol.* **36,** 425–430.
Sanyal, S., and Zeilmaker, G. H. (1977). *Nature (London)* **265,** 731–733.
Schmidt, G. H., Garbutt, D. J., Wilkinson, M. M., and Ponder, B. A. J. (1985a). *J. Embryol. Exp. Morphol.* **85,** 121–130.
Schmidt, G. H., Wilkinson, M. M., and Ponder, B. A. J. (1985b). *Cell* **40,** 425–429.
Schmidt, G. H., Wilkinson, M. M., and Ponder, B. A. J. (1985c). *J. Embryol. Exp. Morphol.* **88,** 219–230.
Schmidt, G. H., Wilkinson, M. M., and Ponder, B. A. J. (1986). *J. Embryol. Exp. Morphol.* **91,** 197–208.
Shelton, P. M. J., Pfannenstiel, H.-D., and Wachmann, E. (1983). *J. Embryol. Exp. Morphol.* **76,** 9–25.
Shoaf, S. A., Conway, K., and Hunt, R. K. (1984). *J. Theor. Biol.* **109,** 299–339.
Sidman, R. L. (1961). *In* "The Structure of the Eye" (G. K. Smelser, ed.), pp. 487–505. Academic Press, New York.
Steinberg, M. S. (1962a). *Proc. Natl. Acad. Sci. U.S.A.* **48,** 1577–1582.
Steinberg, M. S. (1962b). *Proc. Natl. Acad. Sci. U.S.A.* **48,** 1769–1776.
Steinberg, M. S. (1963). *Science* **141,** 401–408.
Stern, C. (1968). "Genetic Mosaics and Other Essays." Harvard Univ. Press, Cambridge, Massachusetts.
Stocum, D. L. (1978). *Science* **200,** 790–793.
Stocum, D. L. (1980). *Dev. Biol.* **79,** 276–295.
Straznicky, K., and Gaze, R. M. (1971). *J. Embryol. Exp. Morphol.* **26,** 67–79.
Thompson, D'A. (1961). "On Growth and Form." Cambridge Univ. Press, London.
Thomson, W. (1887). *Philos. Mag.* **5,** 503–514.
Turing, A. M. (1952). *Philos. Trans. R. Soc. London Ser. B* **237,** 37–72.
Walthall, W. W., and Murphey, R. K. (1984). *Nature (London)* **311,** 57–59.
Weisblat, D. A., Sawyer, R. T., and Stent, G. S. (1978). *Science* **202,** 1295–1298.
West, J. D. (1975). *J. Theor. Biol.* **50,** 153–160.
West, J. D. (1976a). *J. Embryol. Exp. Morphol.* **35,** 445–461.
West, J. D. (1976b). *J. Embryol. Exp. Morphol.* **36,** 151–161.
West, J. D. (1978). *In* "Development in Mammals" (M. H. Johnson, ed.), Vol. 3, pp. 413–460. North Holland Publ., Amsterdam.
West, J. (1984). *In* "Chimeras in Development" (N. Le Douarin and A. McLaren, eds.), pp. 39–67. Academic Press, London.
West, J. D., and McLaren, A. (1976). *J. Embryol. Exp. Morphol.* **35,** 87–93.
Wetts, R., and Herrup, K. (1982). *J. Neurosci.* **2,** 1494–1498.
Whitten, W. K. (1978). *In* "Genetic Mosaics and Chimeras in Mammals" (L. B. Russell, ed.), pp. 445–463. Plenum, New York.
Wolpert, L. (1971). *Curr. Top. Dev. Biol.* **6,** 183–224.

CHAPTER 2

POSITION-DEPENDENT CELL INTERACTIONS AND COMMITMENTS IN THE FORMATION OF THE LEECH NERVOUS SYSTEM

Marty Shankland

DEPARTMENT OF ANATOMY AND CELLULAR BIOLOGY
HARVARD MEDICAL SCHOOL
BOSTON, MASSACHUSETTS 02115

I. Introduction

In order to fully appreciate the development of the nervous system, it is necessary to understand the means by which neurons acquire their distinctive identities. This differentiative process occurs during the development of any multicellular assembly, but the nervous system is an apt choice for study because it offers a vast array of unique cell types, each of which is characterized by a constellation of highly specialized structural and chemical properties and makes a precise set of functional connections with other cells. One approach to the problem is to study factors in the cellular environment which modulate the terminal differentiation of the postmitotic neuron. However, many aspects of a neuron's mature identity are likely to be determined at the time of its birth and to devolve from the developmental constraints of its progenitors. The nature of these constraints must be sought among those cell lineages and cell interactions from which the final components of the nervous system arise.

Invertebrates such as nematodes (Sulston, 1976; Sulston *et al.*, 1983), leeches (Stent *et al.*, 1982), and insects (Goodman, 1982) have proven especially favorable for the analysis of neural development at the level of defined cell lineages. Their adult nervous systems are composed for the most part of uniquely identifiable neurons, and these neurons arise from a sequence of equally identifiable progenitors via nearly invariant patterns of cell division. In theory, a cell could obtain a unique identity either from its own developmental history or its interaction with other cells. An invariant sequence of cell divisions suggests that lineage history is the primary determinant of cell fate in organisms such as these. Nonetheless, there are several documented instances in which pairs or groups of cells—often cells which exhibit

31

some ambiguity in lineal identity—become specified to follow divergent developmental pathways on the basis of position and, therefore, as a result of interaction with the cellular environs (Sulston and White, 1980; Kimble, 1981; Weisblat and Blair, 1984; Shankland, 1984). Indeed, insect neurogenesis seems to be composed of discrete phases in which one or the other mechanism predominates (Doe *et al.*, 1985).

This chapter will focus on the cellular basis of a particular instance of positional specification—the divergence of the O and P cell lines—that occurs during the embryonic development of the leech. Positional specification requires that cells are able to detect and respond to positional cues by choosing one of two or more alternative developmental pathways. The means by which cells detect and respond to position is a central issue in the field of pattern formation, and by studying this process within a context of defined cell lineages it may be possible to resolve positional cues in terms of specific interactions occurring among a manageable number of cells. Particular attention will be given to the way in which position-dependent cell interactions become converted into cell-intrinsic states of commitment, leading to a discussion of ways in which developmental commitment of an embryonic progenitor cell may influence the composition of its descendant clone. Position-dependent cell interactions are a fundamental part of vertebrate neurogenesis (Spemann and Mangold, 1924), and an understanding of pattern formation in invertebrate nervous systems may lend insight there as well.

II. Embryonic Development of the Leech

A. Cell Lineage Analysis

A few hours after being deposited within the cocoon by its maternal parent, the fertilized leech egg initiates a stereotyped sequence of embryonic cleavages. These cleavages have been described in a wide variety of species, with the most complete descriptions coming from leeches of the family Glossiphoniidae, whose large, yolky eggs develop directly into the mature form without passing through an intervening larval stage (Anderson, 1973). The sequence of early cleavages is so regular that individual blastomeres can be uniquely identified, either by their relative sizes and positions within the egg or by their differing lines of cellular descent (Whitman, 1878; Fernandez, 1980; Stent *et al.*, 1982). However, such identifications become infeasible as the leech embryo gains cellular complexity, and the actual sequence of embryonic cell divisions is not known past the stage of gastrulation.

Intracellular injection of histologically identifiable cell lineage tracers has afforded a means of ascertaining cell lineage relationships at later stages of leech development. The cleavages of the leech embryo are holoblastic in nature, and hence sister blastomeres—though they may communicate via gap junctions—do not retain cytoplasmic continuity (Weisblat et al., 1978). High-molecular-weight compounds introduced into one blastomere will remain confined to that cell and its progeny, and in this way one can selectively label a single embryonic cell line and follow its development even after the labeled cells have become intermingled with other unlabeled cell lines. The enzyme horseradish peroxidase (HRP) was first used as an injectable cell lineage tracer in a study which characterized the embryonic origin of the leech's central nervous system (Weisblat et al., 1978), and since that time a number of fluorescently labeled peptides and dextrans have been developed and put to similar use (Weisblat et al., 1980b; Gimlich and Braun, 1985).

A cell's *developmental fate* is that ensemble of cellular phenotypes manifested by it or its descendants. The technique of lineage-tracer injection is of considerable value in constructing embryonic fate maps because it provides an extremely accurate method for the identification of the cells within a descendant clone. Cell fates can be arbitrarily defined with respect to any single developmental stage, and the complete sequence of descendant fates which comprise the subsequent development of a given cell line will here be referred to as its *developmental pathway*. Leech blastomeres tend to follow stereotyped pathways which lead to the formation of specific complements of descendants (Weisblat et al., 1978, 1980b, 1984; Zackson, 1984), and the fate of a given lineally defined blastomere in one leech embryo is, with few exceptions, characteristic of that same blastomere throughout the species as a whole. This is not the case for the embryos of many other organisms, such as vertebrates, in which there is considerable variability in the developmental origin of mature pattern elements (Hirose and Jacobson, 1979; Kimmel and Warga, 1986) and the fate of each lineally identified blastomere must be defined at the species level by a series of probability functions covering a variety of descendant cell types.

B. Formation of the Basic Body Pattern

The main, segmented portion of the leech nervous system—as well as the skin, musculature, and a number of internal organs—arises from specialized blastomeres known as *teloblasts* which themselves arise via a bilaterally symmetrical sequence of cleavages from the largest blastomere of the 4-cell embryo (Stent et al., 1982). In the glossiphoniid leech embryo, the teloblasts are arrayed about the animal pole of the egg in two bilateral sets of five. Three of the teloblasts in each set—designated as cells M, N, and Q—have unique ancestries and can be reliably identified by their size and position within the egg, but the two remaining teloblasts are sister cells of roughly equal size which share the common designation of O/P (Weisblat and Blair, 1984). Being sisters, the O/P teloblasts have the same mother cell and hence cannot be distinguished on the basis of ancestry. In some leech species the O/P teloblasts arise from a cleavage of variable orientation (Shankland and Stent, 1986) and also cannot be meaningfully distinguished on the basis of position. The sister O/P teloblasts are therefore deemed to be lineally equivalent. Experimental results indicate that these two cells are equivalent in developmental potential as well (see Section III,C).

Each teloblast undergoes an iterated series of highly asymmetrical divisions to generate several dozen *primary blast cell* daughters which depart from the teloblast in a coherent column called a *bandlet* (Fig. 1A). Throughout this process the teloblast retains its original designation, with the primary blast cells being designated by the corresponding lowercase letter. The M, N, and Q teloblasts begin the production of m, n, and q blast cells soon after their respective births, but the OP blastomere, mother of the O/P teloblasts, produces a few op blast cells before undergoing its symmetrical cleavage (Fernandez and Stent, 1980). The sister O/P teloblasts produced by this cleavage then resume blast cell production, and thus give rise to two lengthy o/p bandlets which are joined at their leading ends.

The five ipsilateral bandlets come together to form a bilayered *germinal band* (Fig. 1A). The merger of bandlets into the germinal band begins with the oldest blast cells, which lie furthest from the teloblast, coming together to form primordia for the anteriormost body segments. The germinal band elongates by the addition of progressively younger blast cells posteriorly, and thus the spatial order of blast cells within a given bandlet both reflects the temporal order of birth rank and presages the longitudinal placement of their descendant clones along the body axis.

The n, q, and two o/p bandlets form the germinal band's outer, ectodermal cell monolayer, while the larger m bandlet forms the deeper, mesodermal cell layer (Fig. 1B). The q bandlet lies at the edge of the germinal band adjacent to the micromere cap, an ephemeral structure located at the animal pole of the egg which will give rise at later stages of development to a provisional embryonic integument (Anderson, 1973; Weisblat *et al.*, 1984). The n bandlet lies at the opposite or anticapward edge of the germinal band, and the two o/p bandlets lie between n and q. Once within the germinal band, the two o/p bandlets

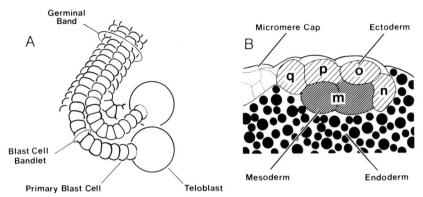

Fig. 1. Formation of the right germinal band in a glossiphoniid leech embryo. (A) Each teloblast undergoes a sequence of highly asymmetrical divisions to generate a continuous bandlet of much smaller primary blast cell daughters. The individual bandlets come together to form a unified germinal band which ultimately gives rise to the nervous system and other segmental tissues on one side of the mature leech. (B) Transverse section of the right germinal band displaying the organization of the individual bandlets. The n, o, p, and q bandlets form an ectodermal cell monolayer and lie over the surface of a larger, mesodermal m bandlet. The o and p bandlets arise from lineally equivalent O/P teloblasts and are defined by their position within the band. The germinal band is supported by yolk-filled endodermal blastomeres and is covered by an epithelium which extends out over its surface from the micromere cap. Taken from Shankland and Weisblat (1984).

are distinguished on the basis of position, with the anticapward o/p bandlet being designated as the o bandlet and the capward o/p bandlet being designated as the p bandlet.

Injection of lineage tracers has revealed that the five ipsilateral teloblasts give rise to distinct and interdigitated patterns of descendants in the mature leech (Weisblat et al., 1980a, 1984). These descendant patterns are segmentally iterated, show little variation between segments, and are confined to one side of the body midline. The M, N, and Q patterns are named after their teloblasts of origin, while the O and P patterns derive, respectively, from those O/P teloblasts whose daughter cell bandlets occupy the o and p positions within the germinal band (Weisblat and Blair, 1984). There is no apparent correlation of the O and P fates with relative positions of the two O/P teloblasts.

The primary blast cells within either o/p bandlet give rise to virtually identical clones of descendants in different body segments, and it is from this arrangement that the segmentally iterated character of the O and P patterns derives (Weisblat and Shankland, 1985). Thus, the divergence of the O and P cell lines can be reduced to a question of how individual o and p blast cells give rise to their distinct types of descendant clone. The M pattern also arises from a single m blast cell per segment (Zackson, 1982), but each segmental complement of the N and Q patterns arises from a consecutive pair of primary blast cells within the ancestral bandlet (Weisblat and Shankland, 1985).

The two hemilateral germinal bands migrate circumferentially over the egg's surface and fuse at the future ventral midline to form a bilaterally symmetrical structure known as the *germinal plate*. Prior to this fusion the blast cells in different bandlets come into their final segmental register by sliding past one another along the length of the germinal band (Weisblat and Shankland, 1985). Thus, it is within the germinal plate that the individual primary blast cell clones first exhibit the segmental organization found in the adult leech. Throughout this time the blast cells and their progeny are actively dividing, and it is shortly after the formation of the germinal plate that the ectoderm separates into neural and epidermal components and organogenesis begins.

C. FORMATION OF THE NERVOUS SYSTEM

Lineage tracers have been used to ascertain the genealogical relationship of a large number of individual neurons and glia in the leech nervous system (Weisblat et al., 1980a, 1984; Shankland and Weisblat, 1984; Kramer and Weisblat, 1985; Weisblat and Shankland, 1985). Many of these neurons had been previously identified in adult leeches by physiological or pharmacological criteria (for an overview, see Muller et al., 1981), and a host of previously unknown neurons and neural progenitors were discovered as a result of these investigations. Nearly identical cell lineage relationships have been observed in the several different glossiphoniid leech species examined (Kramer and Weisblat, 1985; Torrence and Stuart, 1986), and there is also considerable homology in the cellular composition of the nervous system in glossiphoniid leeches and the more widely studied hirudinid leeches (Kramer and Goldman, 1981; Kramer, 1981; Cline, 1983; Glover et al., 1987). However, the small size of the hirudinid egg has so far precluded a comparable analysis of cell lineage.

All five bandlets contribute descendants to the nervous system (Weisblat et al., 1984). The blast cells of the n, o, p, and q bandlets give rise to both central and peripheral neurons as well as to patches of epidermis, while the m blast cells give rise—in addition to musculature, nephridia, and other internal organs—to a small number of central neurons (Kramer and Weisblat, 1985). Thus, the segregation of purely neuronal lineages from epithelial or muscle cell lineages occurs secondary to the segregation of the major ectodermal and mesodermal cell lines. Within the O and P cell lines this histotypic segregation is not complete until the primary blast cell has undergone several additional divisions (Shankland, 1987a,b).

The two n bandlets make the major contribution to the leech's central nervous system. When the right and left germinal bands fuse, the n bandlets come to lie adjacent to one another along the ventral midline, and it is there that their progeny establish the primordia of the 32 segmental ganglia which comprise the ventral nerve cord of the adult leech. The n blast cell progeny account for approximately two-thirds of the 400 neurons in an adult leech ganglion, while the remaining central neurons arise from other, more lateral bandlets (Weisblat et al., 1984) from where they or their immediate precursors migrate into the ganglionic primordium (Torrence and Stuart, 1986). Most adult leech neurons are bilaterally paired in accordance with their dual origin on the two sides of the germinal plate. However, there are also a few central neurons which arise bilaterally but become unpaired as the result of the selective loss in each segmental ganglion of either the right or left homologue during the time of postmitotic differentiation (Stuart et al., 1983; Kramer and Weisblat, 1985). The segmental ganglia also contain five bilateral pairs of identified glial cells which arise from diverse ectodermal bandlets (Kramer and Weisblat, 1985).

Blast cells in the four ectodermal bandlets also give rise to a number of peripheral neurons located at stereotyped positions within each body segment (Weisblat and Shankland, 1985). These peripheral neurons extend afferent axons which fasciculate with the efferent axons of central neurons to establish a stereotyped array of peripheral nerves (Kramer and Kuwada, 1983; Braun, 1985). A considerable amount has been learned about the morphogenesis and biochemical differentiation of embryonic leech neurons (Kuwada and Kramer, 1983; Kramer and Stent, 1985; Glover and Mason, 1986; Glover et al., 1987), but such studies lie outside the scope of the present review.

III. Positional Specification of the O and P Cell Lines

A. NORMAL DEVELOPMENTAL PATHWAYS

The developmental fate of the o and p bandlets has been most thoroughly studied in the glossiphoniid species *Hellobdella triserialis*. The primary blast cells in these two bandlets arise from lineally equivalent teloblasts, and are themselves morphologically similar at the time when they first assume their definitive positions within the germinal

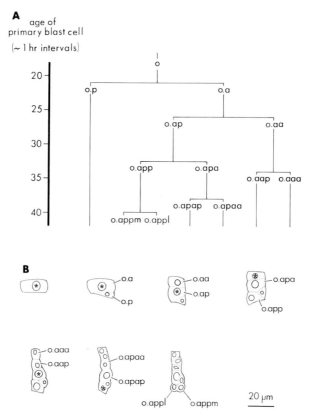

Fig. 2. Descendant lineage of the primary o blast cell. (A) Sequence of blast cell divisions. (B) Camera lucida tracings of the primary o blast cell clone following sequential divisions, with developmental stages reading from left-to-right in rows. The asterisk marks the nucleus of the cell which will undergo the next division at each stage, and the daughters of that division are designated by name at the succeeding stage. Anterior is at the top in this and all subsequent tracings. Taken from Shankland (1987a).

band. However, soon after entering into the band the o and p blast cells undertake divergent patterns of cell division and begin to follow distinct O and P developmental pathways. Blast cells in both bandlets experience their first division at an age of approximately 20 hours, but the geometry of their first division is strikingly different (Zackson, 1984). The primary o blast cell divides in a highly asymmetrical fashion (Fig. 2) to yield a large anterior daughter, cell o.a, and a much smaller posterior daughter, cell o.p, while the primary p blast cell divides in a more nearly symmetrical fashion (Fig. 3). The designations given to these secondary blast cells, as well as their subsequent

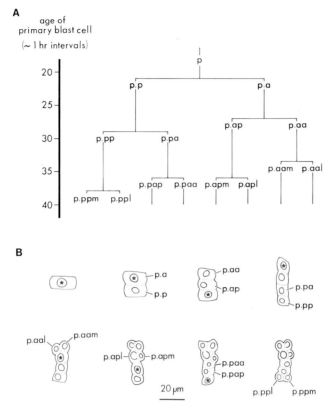

FIG. 3. Descendant lineage of the primary p blast cell, with the same conventions as Fig. 2. Taken from Shankland (1987b).

progeny, consist of the name of the mother cell plus a lowercase letter representing the position of the daughter cell relative to the cleavage furrow. Thus, the anterior and posterior daughters of cell o.a are designated as o.aa and o.ap, respectively.

The o and p blast cell lineages have been traced through several mitoses (Zackson, 1984; Shankland, 1987a,b) and manifest numerous differences in the timing, symmetry, and spatial orientation of cell divisions. Cell counts suggest that there are 3–4 as yet uncharacterized rounds of mitosis separating the latest of these lineally identified blast cell progeny from their postmitotic descendants (Shankland and Weisblat, 1984), but the fate of the o and p blast cell clones at yet later stages of development has been ascertained with the aid of intracellularly injected lineage tracers (Weisblat and Shankland, 1985;

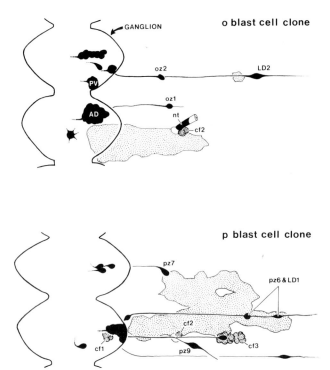

FIG. 4. Cellular composition of the o and p blast cell clones at stage 10 of embryonic development. Clones are drawn in dorsal view with the segmental ganglia of the ventral nerve cord shown in outline at the left; drawings are based on camera lucida tracings taken from embryos in which the ancestral blast cell had been labeled with a rhodamine-conjugated dextran. Contributions to the epidermis are stippled, with the cell florets (cf) being shown darker than the rest. Other cells, including neurons, glia, and the cells of the nephridial tubule (nt) are shown in black. The o and p clones contain approximately the same number of cells and are both distributed on one side of the body midline over two consecutive segments, but exhibit a number of reliable differences in cellular composition. Pattern elements discussed in either the text or Table I have been marked. Taken from Shankland (1987a,b).

Shankland, 1987a,b). At embryonic stage 10, when organogenesis nears completion, the o and p blast cell clones contain 70–100 cells apiece and are for the most part composed of the same general histotypes, including postmitotic neurons and glia as well as cells of the definitive epidermis (Fig. 4). However, despite these similarities the o and p blast cell clones exhibit a number of specific differences in cellular composition which have proven useful as diagnostic criteria for distinguishing the O and P fates:

1. The o blast cell contributes more than twice as many descendants as the p blast cell to the ganglia of the ventral nerve cord. Both blast cells give rise to neurons with somata located in the middle region of the ganglion, but the o blast cell also gives rise to distinctive anterodorsal (AD) and posteroventral (PV) cell clusters. These central contributions show little or no variation between different segments or different individuals, and a number of the constituent neurons and glia have been uniquely identified by either morphological or physiological criteria in embryos of the giant leech *Haementeria ghilianii* (Kramer and Weisblat, 1985).

2. The o and p blast cells give rise at stage 10 of embryogenesis to distinctive complements of three and seven peripheral neurons respectively. Each neuron can be identified by its cell body location and the route taken by its axon to reach the CNS. Moreover, the p blast cell derivative LD1 and the o blast cell derivative LD2, whose cell bodies lie in different peripheral nerves, contain the neurotransmitter dopamine and can be distinguished from other neurons by induced histofluorescence (Blair, 1983; Glover et al., 1987). None of the neurons identified at embryonic stage 10 corresponds to the domelike sensilla that adorn the integument of the adult leech (Mann, 1962), and it seems likely that additional peripheral neurons are born and undergo axonogenesis at some later stage in development, possibly arising from the embryonic structures known as cell florets.

3. The p blast cell gives rise in the epidermis of the stage 10 embryo to two structures—cell florets 1 and 3—which contain no o blast cell derivatives. The cell florets are islands of rounded cells located at specific sites within the sheet of squamous cells that comprise the definitive epidermis (Shankland and Weisblat, 1984). Both the o blast cell and the p blast cell contribute descendants to cell floret 2, which is present in all abdominal segments and marks the future site of the excretory nephridiopore in those 15 segments that house definitive nephridia (Weisblat and Shankland, 1985).

4. In those segments that house nephridia, the o blast cell contributes two cells to that end of the nephridial tubule which is connected to the inner surface of cell floret 2. The p blast cell makes no contribution to this tubule.

5. The p blast gives rise to approximately twice as many squamous epidermal cells as the o blast cell. In a normal embryo these two epidermal contributions are distributed in a fairly predictable pattern on

the ventrolateral body surface (Fig. 4). However, the positioning of squamous domains depends upon morphogenetic interactions within the nascent epidermis (Blair and Weisblat, 1984) and is not a reliable diagnostic indicator of O or P fate in embryos which have been subjected to embryonic lesion.

Lineage tracers have not only proven useful in characterizing normal development, but also provide a means of monitoring experimentally induced changes in blast cell fate. The following section discusses studies in which patterns of blast cell division and the cellular composition of blast cell clones are used as diagnostic criteria in order to demonstrate that the cells in one o/p bandlet can be experimentally induced to forsake their normal developmental pathway for that pathway usually followed by cells in the other o/p bandlet.

B. EXPERIMENTALLY INDUCED CHANGES IN BLAST CELL FATE

Blast cells in the o bandlet can be experimentally diverted from the O pathway into the P pathway by elimination of the neighboring p bandlet, either through the prior ablation of its parental teloblast by the injection of toxic enzymes (Weisblat and Blair, 1984; Zackson, 1984) or by directly photoablating fluorescently labeled p blast cells within the germinal band (Shankland and Weisblat, 1984; Shankland, 1987c). An o blast cell which is deprived of p blast cell neighbors by either technique will undergo the cell division pattern typical of a p blast cell and will give rise to a descendant clone containing no unambiguous O pattern elements and all of the normal P pattern elements (Fig. 5). These findings indicate that o blast cells are initially competent to follow either the O or the P developmental pathway and choose the O pathway as the result of an interaction with the p bandlet. This interaction affects the morphological characteristics of particular o blast cell divisions and also the ensemble of terminally differentiated descendants to which the o blast cell will give rise. Elimination of the o bandlet has little or no effect on the development of blast cells in the neighboring p bandlet (Weisblat and Blair, 1984; Zackson, 1984; Shankland and Weisblat, 1984), and thus it would appear that if either o/p bandlet is eliminated, the blast cells in the remaining o/p bandlet follow the P pathway regardless of their original fate.

Recent studies suggest that the primary blast cells of the p bandlet can be diverted from the P pathway into the O pathway by transposition into the o bandlet position and, hence, are also competent to follow either of these two developmental pathways. These findings have not been published elsewhere and, hence, will be covered in some detail.

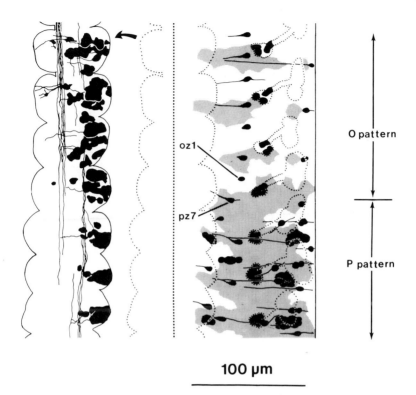

100 μm

FIG. 5. Blast cells of the o bandlet give rise to descendant clones typical of p blast cells when deprived of their normal p blast cell neighbors. Left, Nerve cord; right, body wall. In this embryo the o bandlet was labeled with HRP and the posterior half of the p bandlet eliminated by injecting its parental teloblast with a toxic enzyme midway through blast cell production. The pattern of HRP-labeled descendants reveals that o blast cells located in the anterior portion of the germinal band gave rise to their normal O descendant pattern, but that o blast cells located in the posterior portion of the germinal band—and therefore lacking p blast cell neighbors—gave rise to a P descendant pattern. This P pattern contains all of the P pattern elements (e.g., neuron pz7) and none of the unambiguous O pattern elements (e.g., neuron oz1). Labeled squamous epidermal cells are shown by stippling; all other labeled cells—including the cell florets—are drawn in black. Taken from Shankland and Weisblat (1984).

The technique for bandlet transposition relies on the observation that an ectodermal bandlet which breaks while entering the germinal band will, under some conditions, switch to the other side of the embryo and become incorporated into the contralateral germinal band (Blair and Weisblat, 1982). The likelihood of bandlet switching can be enhanced by first labeling the bandlet with fluorescein-conjugated dextran—

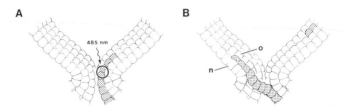

FIG. 6. Switching of bandlets into the contralateral germinal band by means of a localized photolesion. (A) The right and left germinal bands are adjacent to one another at the point where they emerge onto the surface of the egg. Switching can be induced by first labeling a bandlet with a fluorescein-conjugated lineage tracer (stippling) by injection of the parental teloblast and then focally irradiating a small portion of the labeled bandlet with intense 485-nm illumination. This illumination excites the fluorescein within the living cells and brings about a rapid and selective degeneration (Shankland, 1984). (B) Contralateral switching of a focally lesioned p bandlet. The leading fragment of the sundered bandlet has continued into the right germinal band, but the trailing fragment has switched sides and become incorporated into the left germinal band. As shown here, a switched p bandlet usually became insinuated between the native n and o bandlets.

through the prior injection of its parental teloblast—and then irradiating 2 or 3 of the labeled blast cells with the fluorescein excitation wavelength at the point where they enter the germinal band (Fig. 6). This irradiation leads to a rapid and selective degeneration of the labeled cells (Shankland, 1984) and thereby sunders the labeled bandlet into leading and trailing fragments. Lesions performed at the point of entry often block the incorporation of the trailing fragment into the ipsilateral germinal band, and in 9 of the more than 100 embryos in which a p bandlet was so lesioned, the trailing fragment switched sides and became incorporated into the contralateral band.

Seven such embryos were examined shortly after switching had occurred. The switched bandlet could be identified by its fluorescein label, and the nuclear morphology of the blast cells was ascertained with the DNA stain Hoechst 33258 (Zackson, 1984). In four of these embryos the switched bandlet had insinuated itself into the ectodermal cell layer between the native n and o bandlets. In one embryo the native o bandlet had already been eliminated by teloblast ablation, and the switched bandlet insinuated itself between the n bandlet and the original p bandlet, i.e., in the normal o bandlet position. In the two remaining embryos the switched bandlet had not insinuated itself into the ectodermal cell layer, but rather came to lie immediately superficial to the other ectodermal bandlets. In every case the leading portion of the switched bandlet contained alternating pairs of cells with large lightly staining nuclei and small intensely staining nuclei, sug-

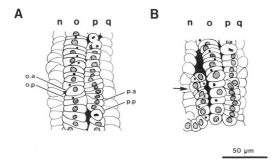

Fig. 7. Blast cells of the p bandlet manifest an o-type first division if made to lie on the anticapward side of another o/p bandlet. (A) Camera lucida tracing of a left germinal band. Nuclei were stained with Hoechst 33258, and have been drawn as stippled or black to distinguish lightly and intensely staining cells. Primary blast cells located in the posterior portion of the band have not yet divided and display large, lightly staining nuclei, but more anterior blast cells have divided and given rise to daughter cell pairs. Blast cells in the o bandlet divide asymmetrically to yield an anterior daughter with a large, lightly staining nucleus, cell o.a, and a posterior daughter with a small, intensely staining nucleus, cell o.p. In contrast, blast cells in the p bandlet divide to yield two nearly symmetrical daughters, cells p.a and p.p. Also note the presence of two actively mitosing cells in the p bandlet, one of which lies immediately behind the marked cell pair. (B) Tracing of a left germinal band which has incorporated a supernumerary o/p bandlet (arrow). Although this bandlet was destined to occupy the p bandlet position in the right germinal band, it has come to lie on the anticapward side of the native o bandlet in the left germinal band. Several of its most anterior blast cells have produced the highly asymmetrical daughter cell pairs indicative of an o-type first division. Blast cells in the native o and p bandlets seem to be undergoing their normal patterns of division.

gesting that the primary blast cells in the switched bandlet had divided in the highly asymmetrical manner normally characteristic of o blast cells (Fig. 7). The presence of the switched bandlets did not induce any apparent changes in the division pattern of the native o or p bandlets and thus, in cases where there were three o/p bandlets sharing the same germinal band, blast cells in both of the more anticapward bandlets were seen to follow the O pathway as judged by cleavage pattern.

I have also characterized the pattern of neurons generated by a switched p bandlet in collaboration with D. A. Weisblat. In one of two embryos examined the switched bandlet had insinuated itself between the native n and o bandlets; in the other, the position of the switched bandlet was unknown. In both cases the switched bandlet gave rise to a pattern of descendant neurons and epidermal cells suggestive of the normal O fate. This pattern contained many more central descendants than would normally be expected from a p bandlet, including groups of neuronal somata resembling the AD and PV clusters that are characteristic of the O pattern. Unfortunately, the switched bandlets only contributed descendants to the extensively fused and poorly studied segments of the tail sucker, and it was not clear whether the pattern of peripheral descendants was characteristic of an O or a P fate.

These experiments indicate that primary blast cells in both o/p bandlets can be induced to follow either the O or the P developmental pathway. The interchangeable character of the two o/p bandlets does not extend to the other three bandlets, which arise from lineally distinct teloblasts and show little capacity to undergo compensatory regulation following embryonic ablations (Blair, 1982, 1983; Zackson, 1984).

C. The Role of Blast Cell Position in the Choice of Developmental Pathway

Weisblat and Blair (1984) were the first to suggest that the o and p blast cells are developmentally equivalent at the time of their birth and proposed that these cells follow divergent developmental pathways as a result of the differing positions that they occupy within the germinal band. This interpretation has received additional support from a number of observations. For instance, the blast cells in the o bandlet do not lose their competence to follow both pathways until after they have acquired their definitive positions (Shankland and Weisblat, 1984). In addition, there is a strict correlation between a bandlet's position in the germinal band and its developmental fate following a variety of bandlet rearrangements (Fig. 8). If a germinal band contains a single o/p bandlet, that bandlet will follow the P pathway. If a germinal band contains two o/p bandlets—as occurs during normal development—the capward bandlet follows the P pathway and the anticapward bandlet follows the O pathway. Finally, if a germinal band contains three o/p bandlets, the most capward bandlet follows the P pathway and the two more anticapward bandlets follow the O pathway.

The fact that a single o/p bandlet follows the P pathway regardless of its original fate is revealing about the way in which o/p blast cells choose between their two alternative fates. This finding implies that the P fate is dominant to the O fate for an o/p blast cell that has no chance to interact with other o/p bandlets. Moreover, adjoining o/p bandlets must interact in such a way that the blast cells in the more anticapward bandlet are excluded from the P pathway and thereby diverted into an alternative O pattern of development (Fig. 9). Thus,

Total Number of o/p Bandlets	Experiment	Number of Observations	(Source)	Fate of bandlet(s)

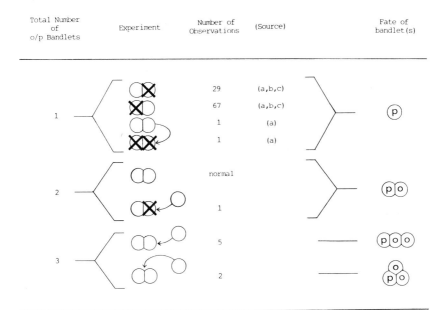

FIG. 8. Correlation of bandlet position and developmental fate following a variety of bandlet rearrangements. A single o/p bandlet can be experimentally isolated either by ablating its neighbor or by inducing one of a pair of o/p bandlets to switch into a germinal band whose native o/p bandlets have been ablated. In all cases the isolated o/p bandlet follows the P developmental pathway regardless of its original fate. A normal germinal band contains two o/p bandlets, and a similar condition can be obtained by ablating one of the native o/p bandlets and switching in an additional o/p bandlet from the contralateral side. Two adjacent o/p bandlets interact in such a way that cells in the capward bandlet follow the P pathway and cells in the anticapward bandlet follow the O pathway. Capward is here shown to the left. Three o/p bandlets can be made to occupy the same germinal band by the contralateral switching paradigm shown in Fig. 6. The most capward o/p bandlet follows the P pathway, and both of the more anticapward bandlets follow the O pathway. a, Weisblat and Blair, 1984; b, Zackson, 1984; c, Shankland and Weisblat, 1984.

the change in o blast cell fate observed following p bandlet ablation is thought to result from the elimination of a normally occurring cell interaction. This interaction appears to be unidirectional in nature, since the fate of an o blast cell which is deprived of interaction with the p bandlet is virtually indistinguishable from that of a normal p blast cell (Shankland and Weisblat, 1984).

Although the bandlets clearly do interact in an operational sense, the cellular basis of this interaction is poorly understood. It could be that the p bandlet excludes the o bandlet from the P pathway simply

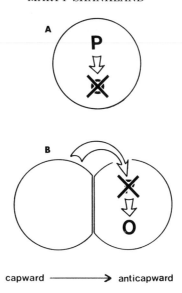

Fɪɢ. 9. Rules governing the O and P fates of o/p blast cells within the germinal band of the leech. A, P fate is dominant to O fate; B, capward blast cell suppresses P fate in anticapward blast cell.

by occupying a critical position. For instance, there might be a signal which is restricted to the capward side of the germinal band and induces only the most capward o/p bandlet to follow the P pathway. Alternatively, the p blast cells could play a more active role and themselves generate some intercellular signal which induces neighboring o/p blast cells to follow the O pathway. This issue is further complicated by the recent finding that the micromere epithelium which extends out over the germinal band seems to play an obligatory role in the interaction of these two bandlets (Ho and Weisblat, 1987). Fate changes are very precisely restricted to zones of bandlet rearrangement (Zackson, 1984; Shankland and Weisblat, 1984), and whatever the nature of the interaction it would appear that the p blast cells are only able to exert an influence on o blast cell development over relatively short distances. It may be that blast cells in the two bandlets require direct physical contact in order for the interaction to occur.

If one accepts that blast cells in the two o/p bandlets are equivalent and bear no inherent proclivity for one or the other pathway, then there must be some additional cue which governs the spatial orientation of their interaction. This line of reasoning suggests that the o/p blast cells have access to information regarding the capward–anticapward polarity of the germinal band, and that this polarity information biases the direction of interaction (Fig. 9). Information regarding polarity cannot be originating solely from the surrounding bandlets because two o/p bandlets will follow their normal, position-specific pathways even after ablation of all other ipsilateral bandlets (Zackson, 1984). If polarity information arises extrinsic to the o/p bandlets, it must come either from the micromere cap epithelium or from some other, more distant structure (cf. Fig. 1). One cannot completely discount the possibility that o/p blast cells—though indeterminate with respect to the O and P pathways—are born with an intrinsic polarity which they carry into the germinal band and which subsequently serves to bias the direction of their interaction. However, the latter alternative is more difficult to reconcile with the finding that a contralaterally switched o/p bandlet seems capable of recognizing and responding to its ectopic position within a germinal band of inverse polarity.

Experiments in which three o/p bandlets share the same germinal band may also be informative about the way in which o/p bandlets interact (Fig. 8). The observed pattern of positional specification can be explained by two different mechanisms. One possibility is that the most capward of the three o/p bandlets diverts *both* of the more anticapward o/p bandlets into the O pathway. A second possibility is that the most capward bandlet only influences its immediate neighbor, the middle o/p bandlet, and that it is this middle bandlet which diverts the most anticapward o/p bandlet into the O pathway. The latter hypothesis would require that an o/p bandlet can serve as a diverting influence independent of its own choice of pathway. This idea deserves further consideration, since it has been shown that blast cells in the p bandlet are capable of diverting neighboring o blast cells into the O pathway even though their own development has been blocked prior to first division and they display no overt evidence of following the P pathway at the time of the interaction (C. L. Wyman and M. Shankland, unpublished).

IV. Commitment Events

A. DEVELOPMENTAL POTENTIAL AND COMMITMENT

The factors which constrain an embryonic cell to its normal developmental pathway can originate either from properties intrinsic to the cell or from the environmental circumstance in which it develops. One concept which is useful in addressing cell intrinsic factors is that of *developmental potential*. Developmental potential is ideally defined as that range of fates a cell is competent to manifest when exposed to an

exhaustive domain of developmental situations, but from a more practical standpoint it must be equated to the fates manifested in response
to an arbitrary set of experimental conditions. A second useful concept
is that of *developmental commitment*, which will here be defined as any
significant change in the cellular partitioning of developmental potential. By this definition, commitment can be said to take place if (1) a
cell line undergoes a reduction in developmental potential—either at
or between cell divisions—and thereby loses the capacity to manifest
certain fates, or (2) a cell division partitions developmental potential
into a novel cellular array. It is best to view commitment not as being a
property of the cell—whose identity is terminated when it divides—
but rather as a property of a developmental cell line.

B. COMMITMENT TO THE O PATHWAY

The blast cells which comprise the o bandlet in *H. triserialis* are
initially competent to follow either the O or the P developmental pathway. However, soon after they enter the germinal band these blast
cells undergo a commitment to the O pathway which is manifested by
a loss of competence to follow the P pathway in response to photoabla-
tion of the neighboring p bandlet (Shankland and Weisblat, 1984;
Shankland, 1987a).

One defining feature of the O pathway is the set of differentiated
pattern elements to which it gives rise. The potential of an o blast cell
clone can be measured by its ability to produce various O and P pattern
elements following p bandlet ablation and has been studied as a function of the primary blast cell's developmental age (Shankland and
Weisblat, 1984). That study revealed that the o blast cell loses its
potential to follow the P pathway after it has entered the germinal
band and showed that the primary o blast cell clone becomes committed to the O pathway in a sequence of at least three discrete steps (Fig.
10). Each of these commitment events independently determines the
fate of a different subset of o blast cell descendants and, thus, brings
about only a partial commitment of the primary blast cell clone as a
whole (Table I).

If an o blast cell is deprived of the neighboring p bandlet around the
time that it enters the band, it will give rise to the full complement of
P pattern elements but to no unambiguous O pattern elements. However, if an o blast cell has already undergone a partial commitment to
the O pathway at the time of the ablation, it will give rise to a subset of
P pattern elements and a subset of O pattern elements as well. Only
when the clone has undergone all three commitments to the O path-

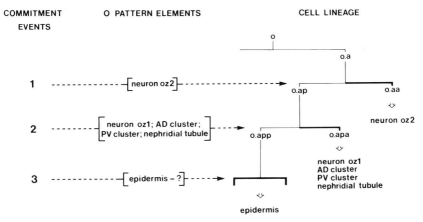

Fɪɢ. 10. Model depicting the stepwise commitment of the o blast cell clone as the sequential commitment to the O developmental pathway of different branches of its descendant lineage. The first two commitment events insure that the primary blast cell clone will produce particular subsets of O pattern elements regardless of any subsequent p bandlet ablation. These same subsets normally arise from particular o blast cell sublineages, and it is thought that both of these two commitment events involve the selective commitment of only one of the two daughters (bold face) of a particular blast cell division, leaving the other daughter cell competent to follow either the O or the P developmental pathway. The third commitment event seems to bring about a commitment to produce epidermal descendants, and may, as shown, be involved in the commitment of that sublineage which arises from cell o.app.

way is the fate of its descendant clone no longer dependent upon the continued presence of the neighboring p bandlet.

These findings suggest that an o blast cell clone is sensitive to the presence of the p bandlet over a period of many hours, and that each commitment event represents only one of a sequence of critical periods for bandlet interaction. If the p bandlet is present during a given critical period, then the o blast cell clone will undergo a partial commitment to the O pathway and, as a result, give rise to a certain subset of O pattern elements even without the continued presence of the p bandlet. If the p bandlet is not present during the critical period, then the o blast cell clone will not produce that subset of O pattern elements and will produce a particular subset of P pattern elements instead (Table I).

A second feature distinguishing the O and P pathways is the sequence of cell divisions by which the primary blast cells give rise to their descendants. The first division of an o blast cell can be readily

TABLE I

SET OF UNAMBIGUOUS O AND P PATTERN ELEMENTS PRODUCED BY A PRIMARY o
BLAST CELL CLONE FOLLOWING ABLATION OF THE ADJACENT p BANDLET[a]

	Age of clone at time of ablation			
	0–20 hours	21–25 hours	26–31 hours	>32 hours
O pattern elements				
Nephridial tubule	−	−	+	+
AD neuron cluster	−	−	+	+
PV neuron cluster	−	−	+	+
Neuron oz1	−	−	+	+
Neuron oz2	−	+	+	+
P pattern elements				
Neuron pz7	+	+	+	−
Cell floret 3	+	+	+	−
Cell floret 1	+	+	−	−
Neuron pz9	+	+	−	−
Neuron pz6/LD1	+	+	−	−
Neuron pz6/LD1	+	−	−	−
	▲	▲	▲	
Commitment events	1st	2nd	3rd	

[a]Each vertical column represents the typical set of unambiguous O and P pattern elements produced by a primary o blast cell following ablation of the neighboring p bandlet. Neurons pz6 and LD1 were not distinguished in these experiments, and it is not known which of these cells disappears in association with the first commitment event. Adapted from Shankland and Weisblat (1984).

distinguished from that of a p blast cell on morphological grounds, and if the p bandlet is experimentally eliminated, neighboring o blast cells manifest a p-type first division (Zackson, 1984; Shankland, 1987c). This finding indicates that the interaction of the o and p bandlets not only governs the descendant fate of the o blast cell but also exerts an influence on the positioning of the cleavage furrow at the first o blast cell division (Fig. 11A).

C. ASSOCIATION OF COMMITMENT EVENTS WITH PARTICULAR BLAST CELL DIVISIONS

There is a stereotyped segregation of descendant pattern elements between different o blast cell sublineages (Shankland and Stent, 1986; Shankland, 1987a), and comparison reveals that it is those groups of pattern elements which arise from different sublineages that become independently committed to the O pathway at three commitment

A   GEOMETRY OF DIVISION

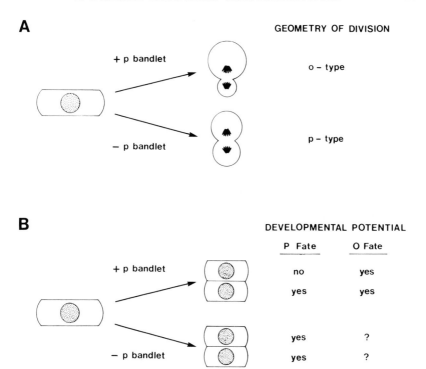

B   DEVELOPMENTAL POTENTIAL

		P Fate	O Fate
+ p bandlet		no	yes
		yes	yes
− p bandlet		yes	?
		yes	?

FIG. 11. Summary of known influences of the p bandlet on o blast cell development. (A) The presence or absence of the p bandlet determines whether a primary o blast cell will undergo a highly asymmetrical (o-type) or a more symmetrical (p-type) first division. (B) At other divisions in the o blast cell lineage, the presence or absence of the p bandlet determines the pattern of daughter cell commitment. If the p bandlet is present, one o blast cell daughter will be committed to the O pathway while the other daughter remains uncommitted with respect to these two pathways. Thus, the presence of the p bandlet alters the state of O or P commitment of at least one of the two daughters of the division. It is not known whether either of the daughters produced in the absence of the p bandlet are competent to manifest O fates.

events described above. The first commitment event is thought to occur in association with the second o blast cell division, at which cell o.a cleaves to yield an anterior daughter, cell o.aa, and a posterior daughter, cell o.ap. During normal development cell o.aa gives rise to a clone of 10–15 descendant neurons—including the peripheral neuron oz2—and one isolated epidermal cell, while cell o.ap gives rise to a descendant clone which contains 25–30 neurons—including the peripheral neuron oz1 and the AD and PV clusters of central neurons—as well as two cells of the nephridial tubule and a large patch of epidermal cells (Fig. 10).

The first commitment to the O pathway insures that the o blast cell clone will give rise to neuron oz2 but does not insure the formation of the other unambiguous O pattern elements (Table I). If these readily identifiable pattern elements are indicative of the clone as a whole, then it would appear that the first commitment event involves a selective commitment to the O pathway of the o.aa sublineage (Fig. 10). The first commitment event apparently does not restrict the fate of the sublineage arising from the sister cell o.ap, as if this cell were still competent to generate either O or P pattern elements depending upon the outcome of the next commitment event.

The second commitment event is thought to occur in association with the next division of the uncommitted o.ap sublineage (Shankland, 1987a). This cell normally divides to yield an anterior daughter, cell o.apa, which exclusively gives rise to the neuronal and nephridial descendants, and a posterior daughter, cell o.app, which gives rise to squamous epidermal cells. In a manner analogous to the first commitment event, the second commitment to the O pathway appears to divide the o.ap clone into committed and uncommitted subunits which, respectively, correspond to the o.apa and o.app sublineages (Fig. 10). The third commitment event is thought to involve the o.app sublineage, and considerations of timing suggest that the third event may be associated with the next division of that cell, although there is no clear indication whether that commitment would affect one or both of its daughters.

This model envisions the stepwise commitment of the o blast cell clone as a sequence of restrictions on the fate of different branches within its lineage tree. There is circumstantial evidence which suggests that at least two of the early divisions in this lineage yield one daughter cell which is committed to the O pathway and one daughter cell which is still uncommitted. A cell which is committed to the O pathway would give rise to its normal descendant clone regardless of any subsequent ablation of the neighboring p bandlet, and hence its fate would not depend upon the continued presence of bandlet interaction. In this way position-dependent intercellular signals would be converted into cell-intrinsic states of commitment.

During certain periods of development the primary blast cell clone would contain both committed and uncommitted cells, and itself would be only partially committed. The available evidence suggests that each commitment event determines the O or P fate of an entire o blast cell sublineage and, if so, would by definition be a heritable cellular property. However, the present data is insufficient to be conclusive on this point.

The first and second commitment events have been estimated to take place several hours prior to those cell divisions which segregate the committed and uncommitted sublineages (Shankland and Weisblat, 1984; but see also Shankland, 1987a). It is not yet known which phase of the cell cycle is associated with the commitment event, but it would appear that the p bandlet is acting on a premitotic o blast cell in such a way as to alter the outcome of its next division. If the p bandlet is present, only one of the two daughter cells will remain competent to produce P pattern elements; but if the p bandlet is eliminated prior to this interaction, then both daughter cells retain this potential (Fig. 11B). Thus, the developmental potential of at least one of the two daughter cells is manifestly dependent on the foregoing bandlet interaction.

It is not known whether daughter cells that are produced in the absence of the p bandlet are still competent to follow the O pathway, but one could imagine that under these circumstances the cell in question would divide in a manner analogous to the normally occurring O commitment, producing one daughter which is uncommitted and one daughter which is committed to the alternative P pathway.

There is no known commitment of descendant cell fates associated with the primary blast cell division, whose morphology differs so dramatically between the two bandlets. This puzzling finding could be explained by assuming that, as occurs at the two subsequent divisions, only one of the two daughter cells is committed to the O pathway. Cell o.a is uncommitted according to the model presented in Fig. 10, and experiments in which this cell was selectively labeled and the neighboring p bandlet photoablated lend support to this idea (unpublished results). However, the smaller daughter, cell o.p, gives rise to none of the unambiguous O pattern elements and few, if any, other descendants (Shankland, 1987a). Even if this cell were committed to the O pathway, the consequences of such a commitment would not have been detectable in the experiments performed to date. This situation calls for further study, and it may turn out that a similar pattern of daughter cell commitment is manifested throughout the first several o blast cell divisions.

D. Relationship of Blast Cell Commitments to Neuronal Differentiation

Commitment of an o blast cell clone to the O developmental pathway occurs several divisions prior to the terminal differentiation of its descendants, but this commitment nonetheless places constraints on the pattern of differentiation that those descendants will manifest. In theory one could imagine that the O and P pathways lead to equivalent sets of postmitotic descendants that are secondarily diversified by selective cell loss or phenotypic modulation. However, such ideas are completely untenable in view of the available evidence: the early divergence of o and p blast cell lineages (Zackson, 1984; Shankland, 1987a,b), the fact that these two cell lines manifest radically different patterns of migration prior to their terminal mitoses (Braun, 1985; see also Torrence and Stuart, 1986), and the early stage at which the o blast cell clone loses its competence to produce P pattern elements (Shankland and Weisblat, 1984). Rather, one must assume that the O and P developmental pathways lead to what are fundamentally distinct sets of terminally differentiated neuronal and nonneuronal descendants and grant that the commitment of a blast cell to the O or P pathway has a long-term influence on the pattern of differentiation that its descendants will manifest.

There are a few pairs of o and p blast cell derivatives which exhibit a striking phenotypic similarity. For example, the dorsal and ventral pressure-sensitive neurons are derived from the o and p blast cells, respectively (Kramer and Weisblat, 1985). These two neurons display the same sensory modality and electrical properties, as well as making many of the same synaptic connections (Nicholls and Baylor, 1968; Nicholls and Purves, 1970). The o and p blast cells also give rise to a pair of peripheral neurons—LD1 and LD2, which are morphologically similar and contain the same neurotransmitter (Blair, 1983)—as well as functionally interchangeable contributions to the epidermis (Blair and Weisblat, 1984). Thus, some branches of the o and p blast cell lineages lead to very similar differentiative pathways. However, even the functionally related pressure-sensitive neurons exhibit unique traits that correlate with their ancestral origins, and further study will be required to ascertain whether these two neurons are in fact equivalent cells which differentiate somewhat differently according to the environs into which they are born, or if they inherit from their ancestral blast cells a differing state of O or P commitment which evokes subtle modulations in the expression of an otherwise common pattern of differentiation.

E. OTHER ASPECTS OF BLAST CELL COMMITMENT

There is a second aspect of blast cell commitment which has not yet been considered. The various o blast cell progeny give rise to distinct subsets of O pattern elements, and there must be some developmental constraint which insures that a given cell will—once committed to the O pathway—give rise to the appropriate subset. For example, the sister cells o.aa and o.ap differ not only with respect to the stage at which they become committed to the O pathway, but also in terms of which O pattern elements they produce (Fig. 10). It would be improper at the present time to refer to this constraint as a commitment, since there is no evidence to distinguish whether the various o blast cell progeny are, in fact, restricted in potential or if their fates are differentiated by their relative positions within the primary blast cell clone.

Because it is possible to experimentally dissociate the commitment of the various o blast cell sublineages to the O pathway, one might imagine that these commitments involve some common cellular mechanism which has the same effect—loss of potential to follow the P pathway—regardless of which subset of O pattern elements is being affected. Other findings also lend support to the idea that there may be distinct mechanisms governing a cell's constraint to produce a particular subset of pattern elements and its commitment to the O or P pathway (Shankland, 1987b). For instance, an o blast cell clone which has undergone one partial commitment to the O pathway will, in response to p bandlet ablation, give rise to some of the P pattern elements normally produced by each of the two daughters of the primary p blast cell division but not to the entire set. A clone which has undergone two partial O commitments gives rise to P pattern elements whose normal lines of descent are segregated at the first p blast cell division but not to other pattern elements that are more closely related in the normal p blast cell lineage (Fig. 12). These findings indicate that a partially committed o blast cell clone which is induced to generate P pattern elements will produce a subset that overlaps with but is not equivalent to any normal p blast cell sublineage, and implies that the individual cells in the partially committed o blast cell clone must be giving rise to subsets of mature pattern elements that could not be clonally related in normal development (Shankland, 1987b). Thus, it would appear that even those o progeny which are uncommitted with respect to the production of O or P pattern elements may have developmental identities that are unique to the O cell line.

One explanation for these findings is to assume that the presence of the p bandlet has two quite distinct effects on the developmental potential of the daughter cells produced at an o blast cell division. Bandlet interaction would not only commit one of the daughters to the exclusive formation of O pattern elements but would also govern the segregation between the two o blast cell daughters of the capacity to produce different P pattern elements. Thus, that o blast cell daughter which is still competent to produce P pattern elements following the interaction (i.e., the uncommitted daughter) would give rise to a subset

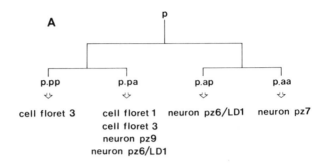

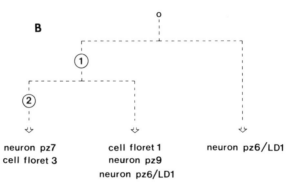

FIG. 12. Segregation of P fates in different blast cell lineages. (A) Individual P pattern elements normally originate from different p blast cell sublineages. Two of the six P pattern elements listed—peripheral neurons pz6 and LD1—were not distinguished in these studies and their origins are ambiguous. Each cell floret is composed of several cells, and cell floret 3 was found to include descendants from two separate sublineages. (B) Probable pattern of segregation of the same six P pattern elements during the o blast cell lineage. The primary o blast cell can generate all of these same P pattern elements, but its clone loses the capacity to generate certain elements when it undergoes partial commitments to the O pathway. Comparison with A reveals that an o blast cell clone which has undergone either 1 or 2 partial commitments to the O pathway will generate subsets of P pattern elements which do not represent clonally related sublineages during the normal development of the p blast cell.

of P pattern elements which is different from that subset which would have arisen from either of the two daughter cells if the interaction had not taken place. The presence of the p bandlet has been shown to alter the segregation process at the first o blast cell division (Shankland, 1987a,b), and it may be that a similar sort of control is also exerted at later divisions where the presence of the p bandlet is known to bring about a partial O commitment as well.

One could construe O and P commitment as two different states of a developmental switch which may be akin—from an operational but not necessarily mechanistic standpoint—to the action of "selector genes" such as those described in the bithorax and antennapedia gene complexes of the fruit fly *Drosophila melanogaster* (Garcia-Bellido, 1977; Lewis, 1978). It is thought that positional signals within the fly egg specify states of gene expression in these two complexes and that differential states of gene expression confer alternative, segment-specific states of commitment upon the cells of the fly embryo blastoderm. Segmental commitments are then passed along from the blastoderm cells to their descendants and eventually bias the way in which different segments pattern their mature tissues in response to a common, segmentally iterated array of positional information (Bryant, 1979).

It may be in a similar fashion that the o and p blast cell clones become positionally specified to take on alternative, bandlet-specific states of commitment early in leech development and as a result give rise to descendant clones which, like different segments, are distinct in overall pattern but composed of a similar range of cellular histotypes. If there is in fact an analogy in the mechanism of pattern formation in these two diverse systems, then the o and p blast cell lineages may share a common means of distinguishing the different pattern elements in their descendant clones. The alternative states of O and P commitment would then lead to the differential interpretation of a common determinant by homologous pattern elements within the two clones.

V. Concluding Remarks

The leech embryo develops into its mature form via a stereotyped sequence of cell divisions. Most embryonic cells can be uniquely identified on the basis of lineage, and it appears that a cell's lineage history is often a major determinant of its developmental fate (Blair, 1982, 1983; Zackson, 1984). However, the present review has focused not on cases of lineal determination but rather on one of several known instances in which equipotent cells become secondarily committed to divergent developmental pathways as a result of position-dependent cell interactions (see also Shankland, 1984; Stuart *et al.*, 1983). Experimental studies indicate that the o and p blast cells of the leech embryo are equally capable of following the O and P developmental pathways at the time when they take on their definitive positions within the germinal band. Subsequently, they interact in such a way that the p blast cells—which follow the P developmental pathway—exclude the o blast cells from this pathway and thereby divert them into an alternative O pattern of development. It remains for future studies to learn why it is the blast cells in the p bandlet position that assume the dominant role in this interaction and to describe the way in which these p blast cells communicate this influence to their o blast cell neighbors.

There are other reports of positional specification among groups of lineally identified cells which have led to similar conclusions. The somatic cell lineages of the nematode *Caenorhabditis elegans* have been described in their entirety (Sulston *et al.*, 1983), and several instances have been found in which cells follow different descendant lineages as a result of a position-dependent interaction (Sulston and White, 1980; Kimble, 1981). Many of these instances involve pairs of cells which behave much like the o and p blast cells of the leech. If either cell is eliminated, the survivor follows its so-called primary pathway; if both cells are present, one follows the primary pathway and the other is diverted into a secondary pathway. As with the o and p bandlets, it is the relative position of cells, not their lineage, that determines which one will take on the dominant role in the interaction. Instances of pathway exclusion involving pairs of sister neurons have been reported in the insect nervous system, but in those cases the exclusion seems to occur without any apparent positional bias (Kuwada and Goodman, 1985). Positional specification may represent a specialized case of what is a more generalized mechanism for the secondary diversification of cells which are not provided with differential states of commitment by their lineage.

Pathway exclusions of the sort discussed here can also play a role in more complicated instances of pattern formation. The vulva of the *C. elegans* hermaphrodite arises from a set of founder cells located in the ventral hypodermis. Experimental studies indicate that there are five contiguous hypodermal cells which have the same developmental potential for generating elements of the vulva and that these five cells normally take on different fates in accordance with their positions (Sulston and White, 1980). Vulva formation is induced by a signal emanating from the gonad (Kimble, 1981), and those hypodermal cells with the greatest access to this signal effectively exclude their neighbors from their own differentiative pathway (Sternberg and Horvitz, 1986). A slightly different picture has emerged from studies of insect neurogenesis. The insect ectoderm contains neurogenic cells that are competent to differentiate into multiple types of neuronal progenitor or nonneuronal support cells (Lawrence, 1973; Doe and Goodman, 1985; Doe *et al.*, 1985). Those ectodermal cells which differentiate into neuronal progenitors exclude their potentially neurogenic neighbors from following the same pathway and thereby divert them into one of the support cell pathways. There also appears to be a set of positional cues which specify not only the location at which the neuronal progenitors arise but also their position-specific identities. Even in cases where a positional signal has a graded or diffuse distribution and acts over a large number of competent cells, local pathway exclusions can help to insure that only the appropriate number of cells will respond.

Insight has also been gained into the means by which a primary o/p blast cell clone becomes irreversibly committed to the O pathway, i.e., one of its two original developmental alternatives. This commitment is not a single event, but rather a sequence of at least three separate events which appear to be associated with different blast cell divisions. Each commitment selectively influences a different branch of the primary blast cell's descendant lineage and may place restrictions upon the fate of all of the cells within that sublineage. These committed cells are no longer competent to produce pattern elements that would normally derive from the other developmental pathway, and at least to a first approximation their development is no longer dependent upon interaction with cells of the other bandlet. It would be interesting to compare the way in which cells in the p bandlet become committed to the P pathway, but unfortunately little is known about that process.

One of the most intriguing aspects of blast cell commitment is the finding that there are multiple pathways which can lead to the formation of a particular descendant pattern element. The early commitments to the O pathway only seem to affect one of the two daughters of a blast cell division, and through the first several divisions of the o lineage the primary blast cell clone contains cells which are still competent to produce P pattern elements. Because the early o and p blast cell divisions are morphologically distinct, P pattern elements that have arisen from a secondary or tertiary o blast cell are generated by a sequence of cell divisions which is distinct from their normal line of descent. There is also evidence which indicates that those o blast cell progeny which are diverted into the P pathway generate subsets of P pattern elements that are not clonally related in the normal p blast cell lineage. The fact that the O and P cell lines exhibit the regulative capacity to generate the same pattern elements via different lines of descent may prove to be significant for a transphyletic analysis of development, for it suggests that the leech embryo—with its normally invariant sequence of cell lineages—may share some of the same patterning mechanisms as those animals in which mature pattern elements do not normally exhibit a fixed genealogical relationship (Shankland and Stent, 1986).

For the most part it remains a mystery what characteristics of an embryonic cell commit it to the formation of a particular ensemble of phenotypically diverse descendants. However, studies such as those described here help to resolve the sequence of determinative events which link the progenitor to its descendants, and the discovery of multiple commitment events in the o blast cell lineage is a useful first step in unraveling this process. Each time a portion of the o blast cell clone becomes restricted to the O pathway, it is converting position-dependent intercellular signals into a cell-intrinsic commitment. This conversion process is an essential part of the way in which the early events of development limit and define the arena in which later interactions and commitments will take place.

ACKNOWLEDGMENTS

I would like to thank Gunther Stent for insightful discussion and laboratory facilities. I am also indebted to David Weisblat, Seth Blair, and Saul Zackson for their contributions to this field of study, to Jochen Braun for the design and synthesis of a sublime repertoire of fluorescent lineage tracers, and to Robert Ho for introducing me to the technique of blast cell injection. The work reported here was supported by National Research Service Award 5 F32 NS06814-02, as well as research grants NS12828 and HD17088 from the National Institutes of Health, BNS79-12400 from the National Science Foundation, 1-738 from the March of Dimes Birth Defects Foundation, and the Rowland Foundation.

REFERENCES

Anderson, D. T. (1973). "Embryology and Phylogeny in Annelids and Arthropods." Pergamon, Oxford.
Blair, S. S. (1982). *Dev. Biol.* **89,** 389–396.
Blair, S. S. (1983). *Dev. Biol.* **95,** 65–72.
Blair, S. S., and Weisblat, D. A. (1982). *Dev. Biol.* **91,** 64–72.
Blair, S. S., and Weisblat, D. A. (1984). *Dev. Biol.* **101,** 318–325.
Braun, J. (1985). Doctoral thesis, University of California, Berkeley.
Bryant, P. J. (1979). *In* "Determinants of Spatial Organization" (S. Subtelny and I. R. Konigsberg, eds.), pp. 295–316. Academic Press, New York.
Cline, H. T. (1983). *J. Comp. Neurol.* **215,** 351–358.
Doe, C. Q., and Goodman, C. S. (1985). *Dev. Biol.* **111,** 206–219.
Doe, C. Q., Kuwada, J. Y., and Goodman, C. S. (1985). *Philos. Trans. R. Soc. London* **312,** 67–81.
Fernandez, J. (1980). *Dev. Biol.* **76,** 245–262.
Fernandez, J., and Stent, G. S. (1980). *Dev. Biol.* **78,** 407–434.
Garcia-Bellido, A. (1977). *Am. Zool.* **17,** 613–629.
Gimlich, R. L., and Braun, J. (1985). *Dev. Biol.* **109,** 509–514.
Glover, J. C., and Mason, A. (1986). *Dev. Biol.* **115,** 256–260.
Glover, J. C., Stuart, D. K., Cline, H. T., McCaman, R. E., and Magill, C. (1987). *J. Neurosci.* (in press).
Goodman, C. S. (1982). *In* "Neuronal Development" (N. C. Spitzer, ed.), pp. 171–212. Plenum, New York.
Hirose, G., and Jacobson, M. (1979). *Dev. Biol.* **71,** 191–202.

Ho, R. K., and Weisblat, D. A. (1987). *Dev. Biol.,* in press.

Kimble, J. (1981). *Dev. Biol.* **87,** 286–300.

Kimmel, C. B., and Warga, R. (1986). *Science* **231,** 365–368.

Kramer, A. P. (1981). *J. Comp. Physiol.* **144,** 449–457

Kramer, A. P., and Goldman, J. R. (1981). *J. Comp. Physiol.* **144,** 435–448.

Kramer, A. P., and Kuwada, J. Y. (1983). *J. Neurosci.* **3,** 2474–2486.

Kramer, A. P., and Stent, G. (1985). *J. Neurosci.* **5,** 768–775.

Kramer, A. P., and Weisblat, D. A. (1985). *J. Neurosci.* **5,** 388–407.

Kuwada, J. Y., and Goodman, C. S. (1985). *Dev. Biol.* **110,** 114–126.

Kuwada, J. Y., and Kramer, A. P. (1983). *J. Neurosci.* **3,** 2098–2111.

Lawrence, P. A. (1973). *In* "Developmental Systems: Insects" (S. J. Counce and C. H. Waddington, eds.), Vol. 2, pp. 157–209. Academic Press, New York.

Lewis, E. B. (1978). *Nature (London)* **276,** 565–570.

Mann, K. H. (1962). "Leeches (Hirudinea). Their Structure, Physiology, Ecology and Embryology." Pergamon, New York.

Muller, K. J., Nicholls, J. G., and Stent, G. S. (1981). "Neurobiology of the Leech." Cold Spring Harbor Laboratory, Cold Spring Harbor, New York.

Nicholls, J. G., and Baylor, D. A. (1968). *J. Neurophysiol.* **31,** 740–756.

Nicholls, J. G., and Purves, D. (1970). *J. Physiol. (London)* **209,** 647–667.

Shankland, M. (1984). *Nature (London)* **307,** 541–543.

Shankland, M. (1987a). *Dev. Biol.,* in press.

Shankland, M. (1987b). *Dev. Biol.,* in press.

Shankland, M. (1987c). *Dev. Biol.,* in press.

Shankland, M., and Stent, G. S. (1986). *In* "Genetics, Development and Evolution" (J. P. Gustafson, G. L. Stebbins, and F. J. Ayala, eds.), pp. 211–233. Plenum, New York.

Shankland, M., and Weisblat, D. A. (1984). *Dev. Biol.* **106,** 326–342.

Spemann, H., and Mangold, H. (1924). *Arch. Mikrosk. Anat. Entwicklungsmech.* **100,** 599–638.

Stent, G. S., Weisblat, D. A., Blair, S. S., and Zackson, S. L. (1982). *In* "Neuronal Development" (N. C. Spitzer, ed.), pp. 1–44. Plenum, New York.

Sternberg, P. W., and Horvitz, H. R. (1986). *Cell* **44,** 761–772.

Sulston, J. E. (1976). *Philos. Trans. R. Soc. London* **275,** 287–297.

Sulston, J. E., and White, J. G. (1980). *Dev. Biol.* **78,** 577–597.

Sulston, J. E., Schierenberg, E., White, J. G., and Thomson, J. N. (1983). *Dev. Biol.* **100,** 64–119.

Torrence, S. A., and Stuart, D. K. (1986). *J. Neurosci.* **6,** 2736–2746.

Weisblat, D. A., and Blair, S. S. (1984). *Dev. Biol.* **101,** 326–335.

Weisblat, D. A., and Shankland, M. (1985). *Philos. Trans. R. Soc. London B* **312,** 39–56.

Weisblat, D. A., Sawyer, R. T., and Stent, G. S. (1978). *Science* **202,** 1295–1298.

Weisblat, D. A., Harper, G., Stent, G. S., and Sawyer, R. T. (1980a). *Dev. Biol.* **76,** 58–78.

Weisblat, D. A., Zackson, S. L., Blair, S. S., and Young, J. D. (1980b). *Science* **209,** 1538–1541.

Weisblat, D. A., Kim, S. Y., and Stent, G. S. (1984). *Dev. Biol.* **104,** 65–85.

Whitman, C. O. (1878). *Q. J. Microsc. Sci.* **18,** 215–315.

Zackson, S. L. (1982). *Cell* **31,** 761–770.

Zackson, S. L. (1984). *Dev. Biol.* **104,** 143–160.

CHAPTER 3

ROLES OF CELL LINEAGE IN THE DEVELOPING MAMMALIAN BRAIN

Karl Herrup

DEPARTMENT OF HUMAN GENETICS
YALE UNIVERSITY
SCHOOL OF MEDICINE
NEW HAVEN, CONNECTICUT 06510

I. Introduction

The goal of this chapter is to explore the role or roles that cell lineage plays in the development of the mammalian CNS. Since the discussion will be confined to mammals, this means, of necessity, a focus on the mouse as the experimental animal and on genetic mosaics as the analytical tool. The specific reasons for this are outlined below, but the restriction to the genetic mosaic requires the inclusion of a substantial section on the techniques currently available for marking cells in mosaics—a chapter within a chapter. This is done both to inform the reader of the technical procedures by which lineage relationships are approached in this field as well as to illustrate the limitations of these marking systems and, hence, the limitations on the questions that can be answered in the mammal.

The first section is a discussion of concepts and definitions. This is followed by my chapter within a chapter and finally an overview of what has been gleaned thus far from the combination of concepts and methodology.

II. Concepts and Definitions

It is axiomatic that cell lineage relationships exist during the development of any multicellular organism. All creatures are essentially a single clone of cells by virtue of the fact that all of the cells of an organism are descended from a single founder cell: the fertilized egg. In the mammal, this can be extended since all cells in the adult are (poly)clonal descendants of the few cells that form the inner cell mass of the preimplantation embryo and, later, from the cells of the embryonic ectoderm of the early postimplantation embryo (Gardner and Pa-

CURRENT TOPICS IN
DEVELOPMENTAL BIOLOGY, VOL. 21

paioannou, 1975; Gardner and Rossant, 1979; Gardner, 1982; Slack, 1983). Further, since each tissue in the adult contains many more cells than it did at the time in embryogenesis when it was first established, the adult tissue can be considered as a collection of cellular clones, the cells of each having descended from the ancestral cells that founded the tissue. Taken to the extreme, it would be theoretically possible to describe the ancestral family tree of every cell in an organism such as a mouse. This, in itself, is not the goal; such a lineage diagram is merely the statement of the problem. The study of cell lineage is immensely valuable, not to document its own existence, but rather to probe the question of whether the developing organism makes use of the information available in cell lineage relationships to guide and regulate the various developmental processes. We would like to know which of the many morphogenetic and histogenetic events that we observe during embryogenesis are based on information about lineage and which are not.

We can ask this question at a cellular level as well and, in many ways, this is more productive. As we wish to confine ourselves to the nervous system, we can begin by asking what is it that defines a neuron and then ask whether these properties are expressions of a cell's lineage history or of other, extrinsic, influences. One must be careful in posing the question this way, i.e., intrinsic vs extrinsic influences. A postmitotic neuroblast just leaving the ventricular zone generally has a great deal of growth and differentiation to accomplish, yet many aspects of its final shape and form are already intrinsic to that cell (e.g., Caviness and Rakic, 1978; Jensen and Killackey, 1984). Experimental manipulations of final position often have only minor effects on the form or function of a neuron. One would not want to conclude from this evidence alone, however, that the final shape and form were dependent on cell lineage. One would instead say that the cell had begun a program of final differentiation that included a certain size and shape. By contrast, if one considers that as early as the neural plate stage (in axolotl) certain cells in the anterior region are already specified to produce eye structures, even if transplanted to a developmentally neutral site (Boterenbrood, 1970), one would conclude that the developing eye probably relies on information unique to the descendants of those cells to form a proper adult structure. In this sense, cell lineage information is developmentally useful to the cells of the eye. The key here is that a given neuronal phenotype must not only be cellularly autonomous, it must also be heritable over several cell divisions in order to ascribe an importance to cell lineage in the establishment of that phenotype.

Let us return to the first part of our original question: what it is that characterizes a neuron. In simple terms, a neuron can be defined by (1) what it looks like, (2) what it acts like, (3) to whom it talks (and to whom it listens), and (4) how many other neurons look, act, and talk in the same way. The first item is meant to represent the cytostructural features that characterize a given neuronal type. For example, an adult cerebellar Purkinje cell is recognizable by shape alone. The second item represents those distinguishing features of a neuron's physiology, for example, its transmitter type(s). The third item represents the neuroanatomical relationship of the neuron to the other cells of the brain. This is meant to include both the pattern of efferents and afferents of a cell as well as its microanatomical position within a cell group. The final item, cell number, is a characteristic that is less often discussed as a part of a cell's identity. Yet as much as shape, or transmitter type, or connectivity, the impact that a neuronal type will have on the development and function of the CNS will be a direct function of the balance that this population strikes with the rest of the brain.

There are many studies that illustrate how changes in the developmental environment (cellularly *extrinsic* factors) can perturb these four basic neuronal properties. For example, how a neuron looks can be altered dramatically by the pattern of inputs to that cell (e.g., Rakic, 1972; Caviness and Rakic, 1978). An extrinsic environmental change, such as removing or reducing inputs, usually results in an atrophic pattern of dendrites as well as a shrinkage in the size of the cell body (e.g., Roffler-Tarlov and Herrup, 1981; Berry *et al.*, 1980). How a neuron acts can also be altered by extrinsic influences. Hormones (e.g., Jonakait *et al.*, 1981) as well as diffusible proteinaceous substances (Black and Patterson, 1980) can alter the transmitter choice of a neural crest cell. In fact, the cells of neural crest, as a rule, seem to be highly dependent on their developmental environment for selecting their adult properties (LeDouarin, 1980). Illustrations of extrinsic factors that can influence to whom a neuron talks come from lesion experiments that deprive cells of their normal target. Many cell types will respond by making secondary choices and producing what are known as heterologous synapses. Finally, the number of cells in a population is also sensitive to extrinsic influences. Known variously as target-related cell death or numerical matching (Cowan, 1973; Oppenheim, 1981; Katz and Grenader, 1982), there is a developmental neuronal imperative such that, during a sensitive period of a neuron's development, it must interact with its target or be removed by cell death.

The above examples of neuronal plasticity (in the broadest sense of the word) are well known by most students of neurobiology. For the purposes of our discussion, these experiments illustrate several neuronal properties or phenotypes that are not intrinsic to the cell and hence do not (at least in their final details) depend on cell lineage for their expression. Along with these examples, however, come a number of observations that suggest that certain features of a neuron's identity are intrinsic to the cell over many mitoses and are not sensitive to environmental changes. For example, while the final form of a neuronal dendrite is altered by deafferentation, it remains recognizable as a dendrite and often, as in the case of the cerebellar Purkinje cell, a distinct type of dendrite. Following afferent removal (by drugs, X rays, or mutation), the Purkinje cell remains identifiable and even the small characteristic tertiary branchlet spines form and are maintained (e.g., Sotelo, 1978; Crepel et al., 1980; Berry and Bradley, 1976; Bradley and Berry, 1978). The neural plate transplantation experiments discussed above (as well as others performed by many authors) are classic examples of a cellular autonomy of fate. The experiments of Landmesser and colleagues (reviewed in Landmesser, 1984) suggest that even patterns of axon growth may be governed in part by factors intrinsic to the cells. Examples of intrinsically controlled cell number are known in invertebrate embryology (see Sulston et al., 1983; Horvitz et al., 1982). However, apart from the data discussed below, there are no clear-cut examples of autonomous cell number regulation in the CNS of vertebrates.

The point of these examples is that, while the cells of the developing nervous system are clearly highly malleable and able to respond in various ways to perturbations in the developmental terrain, they nonetheless retain certain strong intrinsic imperatives which do not respond to environmental changes and which are stably passed from one cell to its descendents. In this sense these cellular progeny all share something by virtue of their common ancestry and, while they may still look, talk, or act differently, they are related by common "imperatives." To what extent and in what ways these imperatives shape the developing CNS are the questions examined in the last section of the chapter.

III. How to Study Cell Lineage

There are basically four ways to study cell lineage in an organism. The simplest, and historically oldest, is simply to observe the developing organism. The second method, which represents an extension of the first, is to use locally applied dye or tracer substances. By following the cells that inherit the marker, one extends the powers of direct observation to situations where extensive cell movement or optical inaccessibility are significant problems. These two methods of study only permit the construction of a descriptive "fate map" of the cells observed. While relevant to lineage studies, they provide no information on the various states of commitment or potencies of the cells involved. The third method of study is the use of embryonic surgical manipulations. This can take the form of procedures such as lesion (removal), rotation, transplantation, or growth in tissue culture. While suffering from the disadvantage that the organism must be disrupted and manipulated at fragile stages of development, the data so obtained provide insight into intrinsicly vs extrinsicly determined features, and hence a better perspective which traits of the developing tissue(s) are related to cell position and which are related to cell lineage.

These three techniques are extremely powerful tools, especially when used in combination, to explore the roles that cell lineage plays in a developing system. Unfortunately, all three are difficult to apply to the mammalian nervous system. The major reason, although not the only one, is that nearly the entire developmental process occurs in the uterus. This makes direct observation impossible. Further, while some surgery has been performed on the nervous systems of mid- to late-gestation fetuses in species such as mouse (e.g., Silver and Ogawa, 1983), cat (e.g., Shatz, 1983), and monkey (e.g., Goldman and Rakic, 1980), attempts at fetal manipulation often lead to absorbtion of the fetus or spontaneous abortion. Precision manipulations of mammalian embryos at the neural plate to neural groove stage are virtually unknown.

In addition to the difficulties imposed by the uterine location of the fetus, the cells of the embryo are so tiny that injection of tracer substances is difficult and the quantities that can be injected are small. There is also a large number of cell divisions between the stages during which one would wish to apply tracers and stages when morphogenesis is sufficiently advanced that the fate of the cells can be determined. Further, unlike the development of amphibian embryos, tremendous growth occurs during this time period. Together, these qualities conspire to produce a situation where any tracer substance would quickly be diluted below the level of detection. Progress in the die tracing of early embryonic lineages is reviewed in Rossant (1987). The use of such techniques in the CNS, however, seems a long way off.

The fourth method by which cell lineage can be studied is through the use of the genetic mosaics or chimeras. It is the use of this last technique that allowed the first approaches to lineage relationships in a mammal to be made during the last 10–15 years. There are three basic types of mosaics. The first is a true mosaic; genetically distinct cells are analyzed among the cellular descendants of a single zygote. In

Drosophila, this situation can be created by the technique of X-ray-induced somatic crossing over or genetically induced gynandromorphs. In eutherian mammals, it occurs naturally for all X-linked genes in every organism with two X chromosomes. Early in embryogenesis, in normal females, one of the X chromosomes becomes condensed and late replicating, and the genes on this chromosome become inactive (in the human, some genes on the short arm of the X chromosome are believed to escape inactivation). This process, known as X inactivation (see review, Gartler and Riggs, 1983), occurs when there are relatively few cells in the embryo proper—22 to 40 by one estimate (Nesbitt, 1971). Which one of the two X chromosomes remains active is apparently random in any one embryonic cell, although in extraembryonic tissues the maternal X is preferentially active. This means that a female mammal is a mosaic for any gene for which the alleles of the two X chromosomes differ. Unfortunately, no genetic variants of any histological usefulness in the CNS are currently known on the X chromosome of any experimental mammal. One useful genetic translocation has been studied by Deol and Whitten (1972) and by West (1976), but other than this effort, no use has been made of this marker in the CNS.

The second and third types of genetic mosaics have been termed chimeras, as they result from the experimental mixing of cells of distinct genotypes that are descendants of different zygotes (see discussion in McLaren, 1976). The chick/quail transplantation studies as well as the placement of mouse blastomere cells of one embryo into the blastoceol cavity of a second embryo (e.g., Gardner and Lyon, 1971) both represent the class of transplantation chimeras. In the CNS of the mammal, this technique has not yet been used to study lineage relationships.

By far the most widely used method of cell lineage analysis in the mammal is the experimental aggregation chimera. These animals are formed by techniques developed basically by Mintz (1962, 1965) and by Tarkowski (1961). Eight cell embryos are harvested from the oviducts at 2.5 days of gestation and the zonae pellucidae are removed. The embryos are naturally sticky at this stage if kept at 37°C. Two embryos of different genotypes are placed in contact and cultured overnight. During this time, they form a single, double-sized blastocyst. This double-sized embryo is transplanted to the uterus of a pseudopregnant host mother who, as she has mated with a vasectomized male, will have no fertilized eggs of her own. The transplanted embryos soon implant in the uterine wall, adjust in size, and finish development normally. They are born and develop subsequently as mice that are normal in proportion but whose bodies are mosaics of cells of two distinct genotypes.

If the two embryos used to make the chimera were chosen in such a way that they differ at some genetic locus that controls coat pigmentation, then the resulting animals can be visually quite striking (see, for example, Mullen and Whitten, 1971). To study the CNS, however, such differences are, with a few exceptions, useless; most nerve and glial cells are not pigmented. Different marking strategies are needed to reveal the genotype of an individual nerve cell in a chimera. It is this need that brings us to the "chapter within a chapter." To do any sort of clonal analysis in a mosaic, one must have a reliable cell marking system. The following section deals with the progress that has been made to date in this area.

IV. Cell Markers and Strategies for Their Use

The cells of a genetic mosaic contain, in their numbers, ratios and positions, a vast storehouse of information on the role(s) of cell lineage during the development of the organism. To unlock this storehouse, however, one must be able to know, consistently and unambiguously, which cell is of what genotype. Stated differently, one must have a way of marking each cell so as to reveal its genotype independently of its phenotype. Although much thought has gone into what the properties of an ideal marking system would be, the *ideal* system does not exist. Nonetheless, a statement of the goal may be useful as a yardstick by which to measure the various marking systems described below.

As outlined by Gearhart and Oster-Granite (1978), by McLaren (1976), and by others, the perfect marker would have a number of important qualities. It should be easily visualized in routinely prepared histological material. It should be present in all cells of all tissues. The presence or absence of the marker should have no effect on either the function or development of the tissue. This property must be further extended such that cells carrying the marker are not at a competitive advantage or disadvantage during either embryogenesis or adult life. The marker must be cell autonomous, that is, its presence or absence in a cell should be a function only of the genotype of that cell. As a corollary to this, its expression should not be up or down regulated in such a way as to obscure its usefulness by the presence of cells of diverse genotype in the mosaic. The strains that carry the marker should be fully viable and fertile. Finally, a perfect marker would be detectable at all developmental stages.

The marking systems used to date (with one recent exception) can be broken down into three basic categories: (1) systems that rely on genetic differences in the activity of a specific enzyme, (2) systems that rely on genetic differences in an immunocytochemically detectable antigen, and (3) systems based on genetic differences in a cyto-structural feature. While none of these systems has yet proved to be ideal, each is highly useful when applied to a specific cell type (or types) at a particular developmental age.

Historically, the first marking systems used relied on enzyme activity differences between cells of the two genotypes making up the mosaic. The absence of tyrosinase activity in the albino mutant made this a logical first choice as a cell marker. The appearance of the "patchwork" coat of albino ↔ pigmented chimeras was the first indication of the successful production of these mosaic animals (Mintz, 1965; Tarkowski, 1961). In the nervous system, however, these pigmentation differences have a limited usefulness. They are restricted to normally pigmented tissues, which include only the pigment epithelium and iris of the eye and cells in the inner ear. Despite this limited range, this marker has been used extensively in cell lineage analyses as we shall see in Section V.

A more widely applicable marking system was introduced by Dewey et $al.$ (1976) and was based on a genetic difference at the locus of the β-galactosidase structural gene (Bgs). In mouse strains with the Bgs^h/Bgs^h genotype, the specific activity of the enzyme is twice that in strains of the Bgs^d/Bgs^d genotype (Felton et $al.,$ 1974). Use of a histochemical stain for enzyme activity allowed the authors to visualize the genotypes of the cells of many neuronal populations in C57BL/6 ↔ AKR/J chimeras (Bgs^h/Bgs^h ↔ Bgs^d/Bgs^d).

A similar marking system was developed to exploit differences in β-glucuronidase activity due to differences at the structural locus for this enzyme (Gus). Mice which are Gus^b/Gus^b in genotype have much higher levels of enzyme activity per cell than mice which are Gus^h/Gus^h (Morrow et $al.,$ 1949). This difference was first successfully used to detect cellular genotypes in mosaic liver tissue by Condamine et $al.$ (1971). In addition to differences in activity, the enzyme produced by the Gus^h/Gus^h allele is significantly more heat labile than that produced by the Gus^b/Gus^b allele. Herrup and Mullen (1979) tried to use this feature to amplify the activity differences seen in histological sections. Focusing on the cells of the liver, they confirmed the observations of Condamine et $al.$ (1971)—a clear pattern of high- and low-staining mosaicism was seen in the cells of the chimeric tissue. The boundaries between the two types of cells were unambiguous, but when they pretreated the histological sections with heat to destroy the Gus^h/Gus^h enzyme, they discovered that low-staining Gus^h/Gus^h cells contained small amounts of heat-stable Gus^b/Gus^b enzyme. These results suggest that low levels of enzyme exchange are occurring among the cells of the chimera. The amounts of exchange are small and do not affect the ability to reliably distinguish Gus^h/Gus^h from Gus^b/Gus^b cells (this is of great importance in interpreting some of the results presented below). To date, this marker system has been successfully used only in larger cells since the enzyme is cytoplasmic and small neurons have very little cytoplasm in their perikarya. This same size restriction also applies to the β-galactosidase marker.

In many ways it seems odd that genetically programmed immunocytological differences are not a widely used marking system in mosaics. Gearhart and Oster-Granite (1978) have argued eloquently for this approach. The existence of many dozens, if not hundreds, of enzyme structural polymorphisms coupled with the exquisite sensitivity of the immune system to subtle differences in epitopes plus the ability to obtain large quantities of antibodies via monoclonal antibody technology would argue that this marker sysem should be the one of choice. Despite this, only two systems have been developed to date that make use of this potentially valuable method.

In an early report, Dewey et al. (1976) illustrated H-2 haplotype mosaicism in the cerebellum of a C3H ↔ C57BL/6 (H-$2^b/H$-2^b ↔ H-$2^k/H$-2^k) chimera. The antisera used to detect these differences were raised by reciprocal immunizations using "normal lymphoid tissues" as immunogen. Gardner and Johnson (1973) also used this marker to examine mouse ↔ rat chimeras early in gestation. More recently, Ponder et al. (1983), using highly characterized monoclonal antibodies, have reexamined this method of cell marking. They constructed two types of H-2 chimeras: H-$2^k/H$-2^k ↔ H-$2^b/H$-2^b and H-$2^a/H$-2^a ↔ H-$2^b/H$-2^b. In unfixed cryostat sections they obtained convincing demonstration of mosaicism in those tissues in which they could detect H-2 immunocytochemically. A tremendous advantage of this marking system is that adjacent sections, stained with complementary antisera, reveal complementary staining patterns. Two disadvantages, discussed by the authors themselves, are that "cellular detail is lost in frozen sections" and "adjacent cells labelled by a surface membrane marker are difficult to distinguish." For studies of cell lineage in tightly packed populations, this last problem is particularly worrisome. Of greater immediate concern for the use of this marker in the study of cell lineage in the CNS is the additional comment that "Ironically, we were unable to demonstrate H-2 antigens in the central nervous system: the one tissue in which they have previously been used as histological cellular markers in chimaeric tissue. . . . The discrepancy is unexplained."

A more widely applicable marking system, based on antigenic differences between cytoplasmic allozymes, has also been described. Developed by Gearhart and Oster-Granite (1978), this marking system potentially meets many of the requirements of an ideal system. The system is based on antigenic differences between the gene products of two alleles of the structural gene for the enzyme glucose phosphate ismoerase (GPI). Originally described as electrophoretic variants (De-Lorenzo and Ruddle, 1969), the $Gpi-1^a$ allele codes for a slower migrating form of the enzyme while the $Gpi-1^b$ allele codes for a faster migrating form. Gearhart and Oster-Granite found that they could isolate the enzyme from inbred mouse strains homozygous for one or another of the two alleles, immunize (using each purified enzyme separately), and then perform reciprocal absorptions (e.g., absorb the antiserum to the $Gpi-1^a/Gpi-1^a$ enzyme with enzyme isolated from $Gpi-1^b/Gpi-1^b$ mice and vice versa). As a result of these efforts they produced an antiserum to the $Gpi-1^a$ enzyme that, unfortunately, recognized both $Gpi-1^a/Gpi-1^a$ and $Gpi-1^b/Gpi-1^b$ GPI without any apparent distinction. Nonetheless, their antiserum to the $Gpi-1^b$ enzyme appeared to contain antibodies that were specific for the $Gpi-1^b/Gpi-1^b$ form of the protein.

The advantage of using a marking system based on GPI is that the enzyme is ubiquitous. GPI is synthesized early in embryogenesis (Chapman et al., 1971) and, since it is a basic enzyme for both glycolysis and gluconeogenesis, it is found in nearly all cell types. Using the "allozyme specific antibody," Oster-Granite and Gearhart (1981) were able to illustrate mosaicism in the cerebella of BALB/cJ ↔ C57BL/6J chimeras ($Gpi-1^a/Gpi-1^a$ ↔ $Gpi-1^b/Gpi-1^b$). The mosaicism was apparent in both the Purkinje and granule cell population, i.e., a large and a small cell population.

The third basic group of cell marking systems is based on differences in cytostructural features of the cells of one genotype compared to a second. The first of such markers to be used, albeit indirectly, was the beige gene (bg). This mutation was originally discovered as a coat color mutation as it lightens a black pelt to dark grey. This lightening is caused by a clumping of the lysosomes and melanosomes of the cells of the affected animals (Oliver and Essner, 1973). Dewey et al. (1976) made use of the lysosomal aggregation to amplify the staining differences between Bgs^h and Bgs^d strains of mice. This was possible because β-galactosidase is a lysosomal enzyme, and thus in Bgs^h/Bgs^h bg/bg ↔ Bgs^d/Bgs^h +/+ chimeras they could compare high clumped to low unclumped enzyme activity.

A second, more direct use of this type of marker was introduced by Goldowitz and Mullen (1982a). In 1975, Green et al. discovered that the heterochromatin of the lymphoid cells of the ichthyosis (*ic*) mutant was clumped and more centrally located compared to normal animals. This phenotype was retained when the cells were transferred to the circulation of wild-type histocompatible hosts and is therefore assumed to be intrinsic to the cells themselves. Goldowitz and Mullen showed that the clumping phenotype was also visible in some of the small cell populations of the CNS, and that the number of cells exhibiting clumping in the brains of *ic/ic* ↔ +/+ mice was intermediate to that observed in either mutant or wild type. Unfortunately, the clumping is a "weak" phenotype that demonstrates incomplete penetrance at a cellular level. When viewed in 1 μm sections, the authors report that only 24% of the granule cells of the cerebellar cortex (a favorable cell population) exhibits an obvious clumping phenotype. In thicker sections (~8 μm) where any one cell is far more likely to be totally contained in a single section, this percentage increases to 65% (Herrup, 1983) or 90% (Wetts and Herrup, 1982a) depending on the stringency with which one scores. Compounding this problem is the reciprocal observation that a few of the cells in a wild-type animal score as having clumped heterochromatin. This percentage ranges from 6 to 11% in the literature on cerebellar granule cells. These considerations mean that the ichthyosis mutation is not valuable for determining the genotype of a single cell in a tissue section. On the other hand, the differences in the percentages of clumped phenotype are consistent and reproducible. This allows the fairly accurate determination of the genotype ratios in the entire population. For many studies this is an important piece of information that is not available through the use of other marking systems. The percentage of ichthyosis cells in a chimera is estimated by the following formula devised by Goldowitz and Mullen:

$$\frac{(\%\text{IC cells in chimera}) - (\%\text{IC cells in } +/+)}{(\%\text{IC cells in } ic/ic) - (\%\text{IC cells in } +/+)}$$

where IC cells are cells with the ichthyosis phenotype.

Using this system, three cerebellar mutants that suffer granule cell loss have been examined. In one of these, the weaver (gene symbol, *wv*), the ectopia of *wv*/+ granule cells, was shown to be an intrinsic property of the heterozygous cells (Goldowitz and Mullen, 1982b). In two others, lurcher (gene symbol, *Lc*) and staggerer, the phenotype of postnatal death of the cells was shown to be caused by factors extrinsic to the cells themselves (Wetts and Herrup, 1982a; Herrup, 1983).

No current discussion of cell marking systems in the mammal

would be complete without mentioning some of the newer strategies that the techniques of molecular genetics are making available to workers in this field. Once they are fully established, these will represent the fourth, and possibly the most powerful, system available. The best developed to date was introduced from the laboratories of J. Rossant and V. Chapman (Siracusa *et al.*, 1983; Rossant *et al.*, 1983; Rossant and Chapman, 1983). It is based on the observation of significant sequence nonhomology between the satellite DNAs of *Mus musculus,* the common house mouse, the *Mus caroli,* a close relative. Satellite DNA refers to a short sequence of nucleotides (a few hundred base pairs) which is repeated approximately one million times in the mouse genome. These highly repetitive sequences have diverged among the related species of the genus *Mus* (Sutton and McCallum, 1972), and the differences can be detected by the relative efficiencies of DNA–DNA hybridization between intra- and interspecific mixing experiments. The key to the strategy lies in the fact that these hybridization differences can also be detected using the DNA present in the cell nuclei of a tissue section. A radioactive probe, complementary to the *M. caroli* satellite sequences, will not bind to nuclei in sections from *M. musculus.* In the interspecific chimeras, tissue mosaicism can be seen even in embryos as early as a few days after implantation (Rossant *et al.,* 1983). As of this writing, this marking system has not been used to examine adult tissue, but there is no reason to believe such reports will not be surfacing soon.

There are only two potential drawbacks that apply to this system. The first is a practical one. The physical conditions needed to obtain efficient DNA–DNA or DNA–RNA hybridization (high temperature and strong alkali) are precisely those that are most actively avoided to preserve basic morphology and vice versa. The quality of the *in situ* hybridization methodology has been steadily improving, however, and one hopes that this problem will soon be overcome. The second drawback is theoretical. To what extent are the lineage patterns in an *inter*specific mosaic representative of the situation in normal, non-mosaic development? Surely the situation is less artificial than the mouse ↔ rat chimeras (Gardner and Johnson, 1973) since *M. musculus* and *M. caroli* will interbreed (although the offspring are sterile "mouse mules"). Nonetheless, caution seems advised initially.

Finally, under the heading of long-term prospects, the technique of microinjection of specific DNA fragments into the pronuclei of newly fertilized mouse oocytes offers the opportunity to attempt to literally create new markers. The theoretical advantage of this system is that one is free to choose any gene product for which a gene clone is avail-

TABLE I

CELL MARKERS REPORTED USEFUL IN MAMMALIAN CHIMERAS

Marker	Reported cell types	Used by other than original developer(s)	Use reported in embryo
I. Enzyme Activity Differences			
β-Galactosidase	Some large neurons	No	No
β-Glucuronidase	Liver	Yes	No
	Many large neurons		
Tyrosinase	Retinal PE	Yes	Yes
	Coat melanocytes		
	Iris		
II. Immunocytochemical Differences			
H-2	Some large neurons?	Yes	No
	(Thymus)		
	(Kidney)		
	(Testes)		
GPI	Large and small neurons	No	No
III. Cytostructural Differences			
Beige	Large neurons	No	No
Ichthyosis	Some small neurons	Yes	No
	(Lymphocytes)		
Lurcher	Purkinje cells only	No	No
Retinal degeneration	Retinal photoreceptors?	Yes	No
IV. In situ Hybridization			
Satellite DNA	All cells	No	Yes

able. This includes a variety of prokaryotic and eukaryotic proteins as well as viral proteins. The major disadvantage, currently, is the paucity of our knowledge of what molecular characteristics are required to allow a given cell at a given time and place to express the gene. Thus, there is no certainty that the marker of our choice will be expressed by the mouse when and where we would like. Our molecular understanding of tissue-specific gene expression is growing rapidly, however, and this concern may not be as significant in a few years.

Table I presents the markers thus far discussed in a tabular form. Two markers that have not been discussed yet have been placed into category III. Their description and use will appear below. Data are included *only* in regard to published work on mosaic tissue. Thus, while GPI presence has been successfuly revealed by immunocytochemistry in early embryos, chimeric tissue has never been formally examined. The marker thus receives a "No" under "Use reported in embryo."

V. Cell Lineages in the Mammalian CNS

Using the markers described in Table I, various workers over the past 20–25 years have examined the patterns of mosaicism obtained in the mouse in an attempt to discover the existence of cell lineage patterns. While some of this work was directed at structures outside of the nervous system, the focus of this chapter does not permit a detailed examination of these studies. The reader is referred, instead, to several excellent reviews (West, 1978; McLaren, 1976; Mintz, 1974; Rossant, 1987).

In the mosaic nervous system, the placements of nerve cells have been scanned, using a variety of cell markers, for hints of clonal relationships. In retrospect, this search was probably biased somewhat by the clonal analyses performed in *Drosophila melanogaster*. In these elegant studies (e.g., Hotta and Benzer, 1972; Garcia-Bellido, 1972; Crick and Lawrence, 1975; Kankel and Hall, 1976) the clones of cells (whether induced by X rays or created by nondisjunction) appear as coherent spatial groups. The compound eye shown in Fig. 1A is an example of the striking visual remembrance of the clonal organization of the compound eye in invertebrates.

When this type of analysis is applied to the mouse, the phylogenetic distance between the two species becomes quite apparent. The eye of the albino ↔ pigmented chimera shown in Fig. 1b illustrates this point. Observed from a distance, a crude organization of radiating sectors can be seen. On closer examination, however, it becomes apparent that the genotypes of these iris melanocytes are not obeying obvious clonal boundaries in anything approaching the near crystalline order of the fruit fly. In virtually every region, cells of both genotypes are found.

This pattern of intermingling of the genotypes is reflected in the retina in both the photoreceptor layer and in the pigment epithelium of the retina. The latter has been studied in both X-inactivation mosaics (Deol and Whitten, 1972; West, 1976) and in albino ↔ pigmented chimeras (Mystkowska and Tarkowski, 1968; Mintz and Sanyal, 1970; Sanyal and Zeilmaker, 1976, 1977; LaVail and Mullen, 1976; West, 1976, 1978). The literature contains two views on the meaning of the relationship of the two genotypes. In one view, the broad sector organization is emphasized (Mintz and Sanyal, 1970; Mintz, 1974; Sanyal and Zeilmaker, 1977) and from a composite pattern it is argued that,

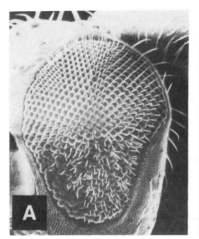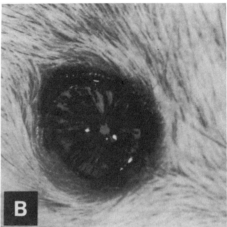

FIG. 1. Comparison of the cell lineage relationships in the formation of the eye in fly and mouse. (A) This figure is a scanning electron micrograph of the compound eye of a Glued — wild type mosaic created by X ray-induced somatic crossing over. Cells of like genotype are clearly associated and a near razor-edged boundary can be drawn which defines the two clonal domains. (Reprinted with the kind permission of the author, Douglas Kankel, to whom I would like to express my gratitude.) (B) The melanocytes of the iris of an albino ↔ pigmented chimera are visible in this photograph. In contrast to the segregation by genotype observed in A, these melanocytes demonstrate a rough sector organization but each sector is clearly "invaded" by cells of the opposite genotype. Mixing is clearly a factor here, and this pattern is reflected in bona fide CNS derivatives as well.

in SAM (the *s*tatistical *a*llophenic *mo*use), the retina consists of 10 sectors, each sector representing a clone of cells. The intermingling observed, it is argued, represents "developmental noise" (Mintz, 1974). In the other view, the mixing is emphasized. Analysis of sections of retina from albino ↔ pigmented chimeras (West, 1976) at embryonic day 13 (the first age at which pigment is reliably detected in melanocytes) reveals a near random dispersion of the cells of the two genotypes. A similar analysis, performed in the adult, suggests the existence of small coherent clones of cells numbering six to eight cells each. The suggestion is that, initially, the cells of the pigment epithelium are extremely well mixed. There follows, however, a period of limited coherent clonal growth with little or no mixing.

A similar controversy exists over the clonal relationships of the photoreceptor cells using the retinal degeneration mutation (*rd*) as a cell marker for photoreceptor cells (Mintz and Sanyal, 1970; West, 1976; Mintz, 1974). In this author's view, however, there are too many ambiguities associated with this particular marker to warrant an extended discussion here. Essentially, the site of *rd* gene action has been narrowed to neural retina; pigment epithelium is specifically excluded (LaVail and Mullen, 1976). There is no proof, however, that the gene acts autonomously within the photoreceptor. If the photoreceptor dies as an indirect consequence of primary gene action elsewhere, then the cell death in the *rd*/*rd* ↔ +/+ chimeras provides little or no information on cell lineage relationships [for a more detailed discussion, see Herrup and Silver (1985) and Mullen (1984)].

Other CNS regions have been studied using primarily histochemical markers (such as β-galactosidase and β-glucuronidase). These regions include the hypoglossal nucleus (Mullen, 1977b), the hippocampus and several peripheral ganglia (Dewey *et al.,* 1976), the inferior olive (Wetts and Herrup, 1982b), and the facial nerve nucleus (Herrup *et al.,* 1984a,b). In all regions except for hippocampus, the positions of the cells reveal a pattern of extensive intermingling of the two genotypes with no obvious clustering of cells of like genotype.[1] To these observations can be added the few additional brain regions found in Figs. 2 and 3. In none of these is there any hint of genotype-specific aggregation of cells in examining single sections.

The region of the brain that has been analyzed in most detail in regard to the spatial arrangement of its cells is the cerebellar cortex, specifically, the cerebellar Purkinje cells. Three different analyses have been performed. The first was by Mullen (1977a) using the cell autonomous action of the Purkinje cell degeneration (*pcd*) mutation. Reconstruction of nearly 3×10^5 μm^2 of Purkinje cell layer from a *pcd*/*pcd* ↔ +/+ chimera revealed a seemingly random pattern of cell survival (i.e., +/+ neurons). Histochemical markers [β-glucuronidase,

---

[1]The pyramidal cells of the hippocampus may prove to be somewhat unique. As reported by Dewey *et al.* (1976) this structure tends to be made up of cells descended primarily from only one of the two embryos used to make the chimera. This would argue that the hippocampus might represent one or a small number of clones. The authors were cautious in their interpretation, however, and my own limited experience is that this caution is justified. The montage in Fig. 2 suggests that in favorable preparations, mosaicism can be seen. It may be that cytoplasmic markers are giving a false picture in this region.

FIG. 2. Staining pattern of β-glucuronidase in several *Gus^h*/*Gus^h* ↔ *Gus^b*/*Gus^b* chimeras. In this and the following figure, the brain centers illustrated have all been stained for β-glucuronidase activity which appears red in the original material and black in these photographs. The sections have been lightly counter stained with methyl green so as not to obscure the reaction product. In both Figs. 2 and 3, the positions of the stained cells have been indicated with arrows. Note the absence of spatial segregation by genotype. A, Hippocampus; B, locus coeruleus.

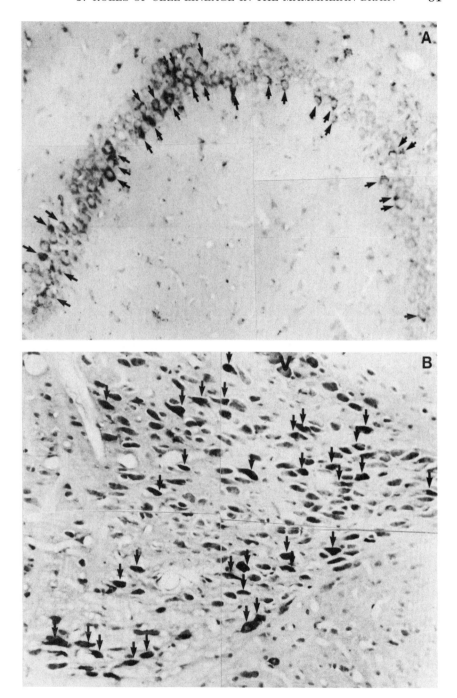

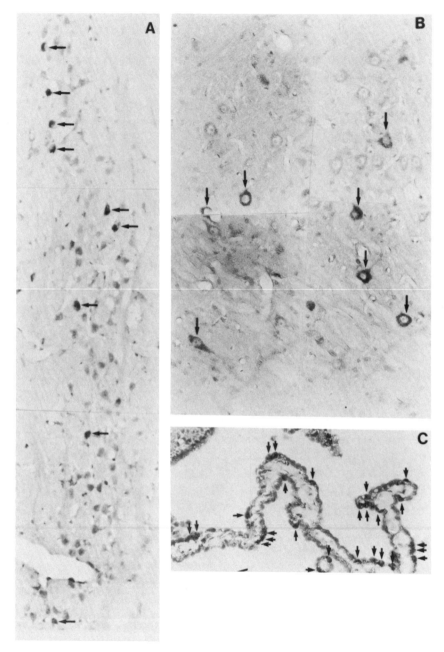

FIG. 3. Staining pattern of β-glucuronidase in several $Gus^h/Gus^h \leftrightarrow Gus^b/Gus^b$ chimeras. A, Trapezoid nucleus; B, motor nucleus of the fifth cranial nerve; C, choroid plexus in the third ventrical.

(Mullen, 1977b)] or immunocytochemical markers [anti-GPI allozyme-specific antibodies (Oster-Granite and Gearhart, 1981)] reveal a similar pattern of intermingling, but statistical analysis suggests there may be some order among the cells. While Mullen (1977b) has estimated a nearly random arrangement of the cells of the two genotypes, Oster-Granite and Gearhart (1981) have found evidence for the existence of coherent clones of cells with approximately four cells in each clone. This difference is far from trivial. On the one hand, if the distribution of cells is truly random, one must ask what mechanism or mechanisms ensure such perfect "anti-order." On the other hand, if there are, regularly, clones consisting of a small number of cells, one must ask whether there is a true functional significance of these cohorts. Recent studies suggest that there may indeed by a physiological relevance to these smaller aggregates of cells (Herrup and Bower, 1986).

If one considers the cerebellum as a single organ, however, both results argue that clonal relationships do not make a significant contribution to the construction of the cerebellum as a whole. Given that there are anywhere from 75,000 to 100,000 Purkinje cells per cerebellar half (depending on genotype), even if a clone contained as many as 8 cells this would still mean that there were 9000 to 12,000 such clones per half. In particular, when these results are viewed in the context of the clonal organization of *Drosophila,* they lead to the conclusion that cells do not sort by lineage in the mammalian CNS, rather they intermingle extensively in all brain regions. When analyzed closely there are suggestions of a final short period of coherent clonal growth, for 1–3 cell divisions in some organs. These later cohorts tend to be quite small, however, compared to the entire population.

The preceding discussion, it must be emphasized, has considered only spatially contiguous cells as evidence of clonal relationships. As pointed out in the introduction, however, larger clonal relationships exist; the only question is whether the developing system makes use of the information inherent in these relationships. The absence of spatial evidence for clonal histories strongly suggests that these histories are not relevant in guiding cells to their final, precise three-dimensional position (see, however, below). But position, despite the aphorism, is *not* everything in life. As argued above, the number of cells present is also an important property of any cell group.

Numerical clones are abstract concepts, not "visible" in any one animal. They are revealed only in the comparisons of data from many mosaic populations. The method by which these clones can be studied is best illustrated by the baking of a cherry pie. Imagine that your favorite recipe calls for seven cans of cherries. You have some cans of light and some cans of dark colored cherries in the cupboard. Every time you bake a pie, the cupboard has different numbers of light and dark cans but, since they taste pretty much the same, it does not matter what the ratio of light to dark is. The recipe also calls for other ingredients that must be well mixed with the cherries before putting the filling in the crust and baking it. When you finally remove the pie from the oven, you will not find all of the light cherries on one side and all of the dark on the other. If you let the pie cool, however, you can count the number of light and dark cherries separately. If you do this every time you bake a pie you will find that the numbers of dark cherries all turn out to be multiples of the number of cherries in one can of the dark variety.

Keeping this analogy in mind, consider the cerebellar Purkinje cell population in a chimera. We count the number of Purkinje cells of one genotype separately and do so in a number of chimeric animals. If numerical clones exist they would show up as "quantal" changes in the number of such cells. A study of this nature requires an accurate cell marker since, depending on the size of the clone, rather small differences would have to be reliably detected. It turns out that the ideal marker for the study of cerebellar Purkinje cells is a neurological mutation, lurcher (Lc), that uniquely and specifically destroys postnatal cerebellar Purkinje cells (Philips, 1960; Caddy and Biscoe, 1979; Swisher and Wilson, 1977). The gene acts intrinsically to the Purkinje cell—only genotypically wild-type cells survive in the adult chimera (Wetts and Herrup, 1982b).

The numbers of cells surviving in several lurcher ↔ wild-type chimeras are shown in Table II. The wild-type cells are from embryos that are either C3H/HeJ or AKR/J × C3H/HeJ hybrids. The cherry pie analogy is followed rather well. The numbers of Purkinje cells in the 12 cerebellar halves examined do not distribute randomly across the spectrum of values from 0 to wild-type. Instead they increase in quanta of 10,200 Purkinje cells. These quanta represent numerical clones, each of which has descended from a single progenitor cell. Note that the nonchimeric, wild-type values also obey this quantal system (the C3H/HeJ mouse has 8 clones per cerebellar half), suggesting that this organization is not some artifact of the chimera or the lurcher gene.

Fig. 4 illustrates the spatial distribution of the cells in one animal that had only one clone of Purkinje cells remaining per cerebellar half. Viewed either spatially in the sagittal plane (Fig. 4A) or numerically along the mediolateral axis (Fig. 4B), there is no tendency for the cells to cluster into regions or form large patches of like genotype. This

TABLE II

CLONAL RELATIONSHIPS OF C3H/HeJ AND AKR × C3H HYBRIDS AS
REVEALED BY PURKINJE CELL COUNTS IN LURCHER CHIMERIC MICE[a]

Animal	Purkinje cells remaining	Assumed quantal size			
		9200		10,200	
C57BL-*Lc* ↔ C3H					
7 Right	23,200	2.52	?[b]	2.28	?
Left	30,200	3.28	?	2.96	(3)
11 Right	10,400	1.13	?	1.02	(1)
Left	10,200	1.11	?	1.00	(1)
13 Right	72,100	7.84	(8)	7.06	(7)
Left	60,300	6.55	?	5.90	(6)
C57BL-*Lc* ↔ AKC3					
19 Right	80,700	8.77	(9)	7.90	(8)
Left	102,000	11.09	(11)	10.18	(10)
137 Right	62,900	6.84	(7)	6.19	(6)
Left	71,200	7.74	?	7.01	(7)
C3H/HeJ controls	83,900	9.12	(9)	8.22	(8)
	82,400	9.96	(9)	8.07	(8)
(AKC3) F_4 control	114,000	12.39	?	11.18	(11)

[a]Most data from Herrup *et al.* (1984b).

[b]The presence of a question mark indicates that the calculated number of
quanta is not within 3%, our counting error, of an integral multiple.

confirms previous reports and reemphasizes the extensive mixing that
occurs among the descendants of any one progenitor.

It is important to remember, in this discussion, the way in which
the chimeric nervous system comes to be. The action of the *Lc* gene is
not observed until relatively late in cerebellar development (Caddy
and Biscoe, 1979). In the mutant, the first significant Purkinje cell
death does not begin until the end of the first postnatal week and is
largely finished by the end of the second or third postnatal month.
Preliminary observations of young postnatal lurcher chimeras leave
no reason to suspect that cell death occurs on a drastically different
time course in the chimera. Thus, the *Lc*/+ Purkinje cells in both the
mutant and the chimera are present for all but the final stages of the
maturation of the cerebellar cortex. Mutant and wild-type cells are
presumably indistinguishable during progenitor selection, cell division at the ventricular zone, and migration to the cerebellar cortical
plate. The mutant Purkinje cells even begin the early phases of den-

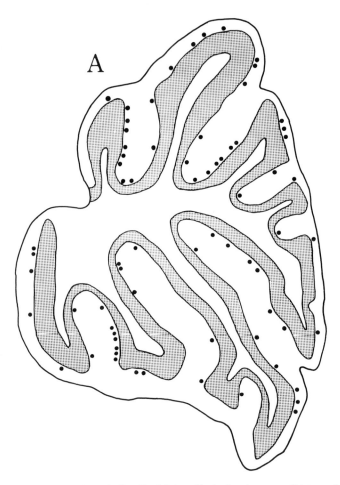

FIG. 4. Distribution of cerebellar Purkinje cells in lurcher ↔ wild-type chimeras. (A) The location of Purkinje cells in the cerebellum of a lurcher ↔ wild-type chimera in which a single Purkinje cell clone was present (chimera 11; see Table II). In this representation of a midline sagittal section, the location of each of the 70 surviving wild-type Purkinje cells has been indicated with a black dot. For reference, a comparable section from a wild-type animal would have approximately 600 cells. This picture complements the impression gained from the graphs in Fig. 4B emphasizing that, even in the sagittal plane, the descendants of a single clone do not aggregate with respect to genotype. (B) The number of Purkinje cells per section, in three lurcher ↔ wild-type chimeras, graphed as a function of the distance of each section from the midline. The area under each curve is proportional to the total number of cells in the entire half cerebellum. For each animal, the remaining (wild-type) Purkinje cells are distributed throughout the mediolateral extent of the cerebellum rather than concentrating in any one region thus emphasizing the absence of aggregation of cells of like genotype within this structure. (Both figures reprinted from Wetts and Herrup, 1982c.)

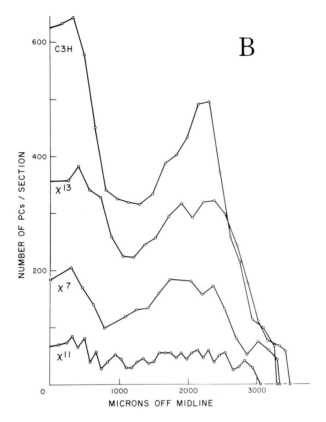

FIG. 4B.

dritic maturation (Caddy and Biscoe, 1979). This timing means that, as stated above, the lurcher gene is serving only as a cell marker–it allows the accurate counting of the Purkinje cells from the nonlurcher embryo by destroying all the Purkinje cells from the mutant embryo. The gene is not influencing the early developmental events under discussion here.

The Purkinje cell population in lurcher chimeras illustrates one way in which a clonal organization can be present numerically without being visible spatially. Yet, in many ways, it is surprising that the numerical quanta seen in the Purkinje cell population exist in the form that they do. This issue is discussed in greater detail below. For now, suffice it to say that the model proposed would seem not to allow for any *epigenetic* adjustments to cell number. These adjustments are well known to occur throughout the CNS during normal development

(Oppenheim, 1981; Cowan, 1973). To understand what a numerical clonal organization would look like in the presence of such naturally occurring cell death, it will be instructive to return to our cherry pie analogy.

The filling that we have prepared in our hypothetical developmental mixing bowl contains not only a specific number of dark cherries (the number of cans of dark cherries times the number of cherries in a can), it also contains a specific ratio of dark to light cherries. This ratio is equal to the ratio of dark to light cans used to make the filling. If we put all of the filling into the crust, then we preserve both the number of dark cherries as well as the ratio of dark to light. But notice that, once the filling is well mixed, it does not matter if you bake a 10-in. pie, an 8-in. pie, or a tart (you can even spill some filling on the floor); the ratio of dark to light cherries in the pastry is preserved. If you keep track of this ratio every time you prepare a pie, you will come to find that the fraction of dark cherries in the pie is always an integral multiple of $\frac{1}{7}$ (i.e., the inverse of the number of cans the recipe calls for). These *fractional* quanta can be thought of in the same way as the numerical quanta described above in the Purkinje cell population. The CNS population that illustrates this concept is the motor neurons of the facial nerve nucleus.

The motor neurons of the seventh cranial nerve (the facial nerve) undergo a period of naturally occurring cell death beginning before birth in the mouse (E17) and ending a few days later (Ashwell and Watson, 1983). The result is that, as in other primary motor and sensory structures, a significant fraction of the population succumbs to what is, presumably, a numerical matching process. In the adult, less than half of the originally generated cells survive. Herrup *et al.* (1984a) studied the clonal organization of this nucleus in several chimeras. There is no mutation known in the mouse (or any other mammal to my knowledge) that specifically destroys facial neurons in a manner analogous to the destruction of Purkinje cells by the lurcher gene. The enzyme polymorphism at the structural locus for the enzyme β-glucuronidase has been used as a cell marker instead.

Counts of facial neurons were performed and each cell was scored for β-glucuronidase genotype. The counts, shown in Table III, reveal two significant features. First, there is variability in the total number of facial motor neurons both from chimera to chimera and from left to right in a single animal. This is in distinction to counts of facial neurons from nonchimeric inbred mice. Second, there is no evidence of any quantal increase in the number of cells from one animal to the next. This is true whether one examines total counts or analyzes each genotype separately. However, if one examines the *fraction* of the total

TABLE III

CLONAL RELATIONSHIPS AMONG THE NEURONS OF THE FACIAL NERVE NUCLEUS[a]

Animal	Cell number		Percentage of total (quanta)	
	Gus^b/Gus^b	Gus^h/Gus^h	Gus^b/Gus^b	Gus^h/Gus^h
Chimera 7				
Left	530	2600	.168 (2.02)	.832 (9.98)
Chimera 19				
Right	1500	1030	.734 (8.81)	.266 (3.19)
Left	2100	750	.594 (7.13)	.406 (4.87)
Chimera 23				
Left	1760	380	.839 (10.07)	.161 (1.93)
Chimera 46				
Right	2330	480	.798 (9.58)	.202 (2.42)
Left	1950	500	.830 (9.96)	.170 (2.04)
Chimera 47				
Right	2100	200	.913 (10.96)	.087 (1.04)
Left	2030	710	.740 (8.88)	.260 (3.12)
Chimera 119				
Right	1770	560	.828 (9.94)	.172 (2.06)
Left	1790	380	.760 (9.12)	.240 (2.88)

[a]Data from Herrup et al. (1984a).

cells represented by one genotype or the other, these fractions do show a quantal relationship—they increase in jumps of 0.083. These jumps, as illustrated by the cherry pie analogy, are the pattern expected if there were an organization in which 12 (1/0.083) clones contribute equally to a nucleus that can vary in absolute size. The conclusion reached from this is that the neurons of the facial nucleus descend from 12 progenitor cells that were specified to be the sole source of these cells some time during early development.

The discovery that two distinct populations of CNS neurons retain, in their numerical distributions, evidence of the cell-lineage relationships that went into forming them is significant. Recently, we have begun to pursue a few of the more intriguing observations that are suggested by these data. One of the observations that has puzzled us, almost from the beginning, concerns the Purkinje cell clones. The lurcher mutation that we use as a cell marker in these experiments is maintained in our lab on a C57BL/6 (B6) background. The first set of chimeras we examined was made between B6-Lc/+ embryos and, as wild type, either C3H/HeJ or AKR/J or a hybrid line created by crossing these two inbred lines. In all of the animals examined, the number of Purkinje cells in a clone was the same: 10,200. This observation is all the more remarkable because of a characteristic of chimeric mice that is little appreciated, namely, the contribution that each of the original embryos makes to the adult mouse is not always 50% plus or minus a small standard deviation. Instead, one observes a flat distribution of all possible ratios of the two genotypes, including both non-chimeric classes. The reasons for this have been discussed elsewhere (McLaren, 1976; Falconer and Avery, 1978; Mullen and Whitten, 1971); suffice it to say here that this pattern reflects events in early mammalian development during which selections occur that commit certain subsets of the cells of the conceptus as a whole to be the sole source of the cells of the embryo proper. This variance is reflected in our own data (Table II); the contribution of the wild-type embryo to the Purkinje cell population varies from rather small (e.g., chimera 11) to fairly substantial (e.g., chimera 13). Consider for a moment only the first three animals in Table II, all of which are B6-Lc/+ ↔ C3H in composition. In each of these animals, the mix of extrinsic "influences" during development must have varied from mostly B6-like to mostly C3H-like. If any of the extrinsic influences were responsible for controlling the final adult number of Purkinje cells, one would expect that the final size of the observed Purkinje cell clone (i.e., 10,200 cells) would vary as was described for the clones in the facial nucleus.

One possible explanation for the constant observed clone size is that these extrinsic "influences" are the same in all strains of *M. musculus* and that all Purkinje cell clones, regardless of genetic background, are 10,200 cells in size. To test this we first counted the number of Purkinje cells in half cerebella of wild-type mice of B6 genetic background. All our counts were within 3% of the value of 92,000 Purkinje cells. This value is consistent with a clonal organization of 9 clones of 10,200 Purkinje cells each. To confirm this we next made a series of chimeras, identical to those shown in Table II, except that, instead of using C3H as the wild-type strain, we used B6. The results of our counts of 12 half cerebella from these lurcher ↔ B6 chimeras are shown in Table IV. As before, the counts occur in quanta, indicating a clonal organization to the Purkinje cell population. In this series of animals, however, the quanta are not 10,200 but 9200, Purkinje cells in size. Thus, the differences observed between the numbers of Purkinje cells in the B6 and C3H inbred strain of mouse are due to both different numbers of Purkinje cell clones (10 vs 8) as well as different numbers of cells in a single clone (9200 vs 10,200).

These findings take on added importance, not only in view of the substantial variation in the ratio of lurcher (B6) to wild-type (B6 or C3H) cells in the animals examined, but also in view of the large

TABLE IV

CLONAL RELATIONSHIPS OF C57BL/6 PURKINJE CELLS AS REVEALED BY
PURKINJE CELL COUNTS IN LURCHER ↔ C57BL/6 CHIMERIC MICE

Animal	Purkinje cells remaining	Assumed quantal size	
		9200	10,200
105 Right	62,900	6.82 (7)	6.17 (6)
Left	47,100	5.12 (5)	4.62 ?[a]
106 Right	90,100	9.70 (10)	8.83 (9)
Left	72,600	7.89 (8)	7.12 (7)
107 Right	90,000	9.86 (10)	8.82 (9)
Left	90,700	9.78 (10)	8.89 (9)
110 Right	18,700	2.03 (2)	1.83 ?
Left	18,000	1.96 (2)	1.77 ?
111 Right	21,800	2.37 ?	2.14 ?
Left	28,100	3.06 (3)	2.76 ?
112 Right	30,100	3.27 ?	2.95 (3)
Left	37,200	4.04 (4)	3.64 ?
C57BL/6 Controls	93,900	10.2 (10)	9.21 (9)
	93,200	10.1 (10)	9.14 (9)
	90,800	9.87 (10)	8.90 (9)
	90,200	9.80 (10)	8.84 (9)

[a]The presence of a question mark indicates that the calculated number of quanta is not within 3%, our counting error, of an integral multiple.

degree to which the cells of the two genotypes intermingle during development (see above). The implication is that, despite exposure to substantial developmental "influences" from a strain in which Purkinje cell clones are 9200 cells in size, Purkinje cell lineages of C3H genotype produce clones of cells that are characteristic of their wild-type strain, namely, 10,200 cells. This leads to the hypothesis that the size of a clone is an intrinsic property of that clone and, hence, an intrinsic property of the progenitor that gives rise to it.

VI. Implications for the Roles of Cell Lineage in CNS Development

The discussion so far has incorporated the concept of progenitor cells without being terribly specific about what is meant by this. The nature of the data we have considered is retrospective; the authors who study cell lineage in the mammal must usually sift through the ashes of adult mosaic brains and try to deduce from the patterns they find something about the developmental events that created them. Because of this retrospective approach we are restricted in our ability to know the properties of any proposed progenitor cells. The numerical studies which I have described are no exception to this. We can say with some confidence that the group of cells we are calling progenitors are destined to be the sole source of the cells of the adult population we study. We do not know the converse, that is, whether or not these same progenitors also give rise to other cells in the CNS, either neurons or glia.[2] Despite these restrictions, many exciting conclusions can be reached from the analysis of the data. These observations offer a fresh perspective on the development of the mammalian CNS and the roles that cell lineage plays in guiding this process.

A. TIMING

We can use some simple arithmetic calculations to estimate the time in development when the progenitor cells existed as single cells. For the Purkinje cell progenitors, for example, to have a single cell divide into a clone of 9200 to 10,200 cells requires 13.2–13.3 cell divisions. This assumes, of course, that all divisions are symmetric. Any divisions of the stem cell type would mean that more divisions would be needed. The work of Miale and Sidman (1961) as well as Inouye and Murakami (1980) demonstrates that the last S-phases of the mouse Purkinje cells occur during a 3-day period from embryonic day 11.5 (E11.5) to E13.5 with most Purkinje cells dividing for the last time on day 12.5. Rates of cell division have been estimated in the mouse neural tube (Atlas and Bond, 1965; Kauffman, 1968; Wilson, 1974 and 1981) and, on average, they are seen to increase during development from approximately 1 division every 7 hours at E8 to 1 division every 10 hours at E12. If we use 8.5 hours as an average time of division (which contains several obvious assumptions), 13.3 cell divisions would require 113 hours, or 4.7 days, to complete. Subtracting this figure from the time of the last cell division, we can estimate that the first cell division occured around E7.8. I have tried to indicate the large

---

[2]Because of the significant variation in genotype ratios among left and right homologues of one population, as well as among ipsilateral, anatomical near neighbors (see data in Herrup *et al.*, 1984a,b), we can be certain that our data do not simply chronicle the fact that the whole embryo descends from a small number of cells. Neither are our results consistent with the hypothesis that large fractions of the CNS descend from the group of cells we are discussing.

number of uncertainties in this estimate; its usefulness is restricted to alerting us to the fact that the events that result in the selection of the Purkinje cell progenitors must be occurring at very early times during neural development. As reference, the neural groove first closes to form the neural tube at the midtrunk region at E8.5 (Snell and Stevens, 1966).

A similar calculation can be performed for the cells of the facial nucleus. This exercise leads to the estimate that the facial neuron progenitors begin their divisions at similarly early times in development. The close concordance of these calculations is gratifying but, again, the many uncertainties in the calculation restrict the emphasis one can place on the precise result. What we can take away from these data is that important events that proscribe aspects of the fates of at least two populations of CNS neurons are occurring extremely early in development, many cell divisions before the neuroblasts leave the ventricular zone to migrate to their final location and differentiate into their adult form. This conclusion brings to mind the descriptions of the early determinative events in neural development, and there are indeed many parallels between the two sets of experimental conclusions. These parallels include timing and the independence of the left and right cell groups as well as the apparent autonomy of the fate proscribed. Here again, however, caution is advised. The events we are describing may well be a part of the early determinations studied by experimental embryology. Until direct proof is available, however, the analogies should remain just that—analogies.

B. MIXING

The projection patterns of most neuronal cell groups are highly organized spatially and presumably dependent on this organization for proper function. Rakic and colleagues (Rakic, 1971a,b; Sidman and Rakic, 1973) have hypothesized that specialized ependymal cells known as radial glia provide a simple mechanism by which the two-dimensional relationships among the cells of the ventricular zone can be reproduced with great precision on the developing cortical plate (Rakic, 1978). This highly ordered mechanism of moving cells from the germinal zone to their adult location coupled with the presumed functional requirement for spatial order stands in contrast to the dispersion of neurons of the two lineages observed in chimeric mice (Figs. 2–4). The contrast suggests, in fact, that cell lineage relationships must have very little to do with "to whom a cell talks" (see Section II).

The extent to which lineage mixing is observed in the chimeras may seem remarkable, but Bodenstein (1986) has constructed a computer model based solely on the kinetics of cellular proliferation that will generate a well-mixed two-dimensional population of cells beginning with a small number of ordered progenitors. In addition, we know that this mixing, however it is generated, cannot be completely haphazard. Neural plate/tube transplantation experiments show that the early CNS is a functional mosaic, with cell types committed to specific morphological fates and, hence, to specific regions of the adult brain (see discussion in Cooke, 1980). Any mixing must respect these commitments. In the interplay between these two competing mechanisms (ordered radial growth vs. cell-division related mixing), cell lineage relationships may yet be found to play a role. Recent experiments indicate that certain aspects of the position of a cell within a population may be related to its lineage. In reconstructions of large areas of the Purkinje cell layer of Crus II and paramedian lobule in the cerebellar hemispheres of $Gus^h/Gus^h \leftrightarrow Gus^b/Gus^b$ chimeras, Herrup and Bower (1986) observed a patchy organization in the deployment of the two genotypes of Purkinje cells. The scale of this patchiness was the same as that of the "mosaic" formed by the physiological maps of the somatosensory projections to these regions, and one large projection area from the mouse lower lip (to the paramedian lobule) corresponds consistently with a large "lineage patch" consisting of Purkinje cells of distinct genotype ratio to the population as a whole. Thus, while precision neuronotopic maps such as the highly ordered map of retina on tectum probably depend to a great extent on fiber:fiber interactions, the coarser level of connections, especially the boundaries of such maps in the target regions, may have components of lineage in their specification.

C. REGULATION OF CELL NUMBER

As discussed in the previous section, the absence of any spatial evidence of significant cell lineage relationships in the adult nervous system argues against a role for such relationships in the regulation of cell position. By this logic, the existence of numerical evidence for a clonal organization to at least two CNS populations suggests that such relationships may well be important in the regulation of cell number. This view is supported by the finding that, in the cerebellar Purkinje cell population, genetic differences in Purkinje cell number among different inbred mouse strains can be traced to differences in the number of Purkinje cell clones (the number of progenitors identified in the early nervous system) or the size of each clone (the number of Purkinje cells in one clone) or both. Since the size of a clone appears to be an intrinsic property of the lineage, the progenitor cell that gives rise to the lineage must be intrinsically programmed to produce a specific *number* of cells in addition to a specific *type* (or types) of cell. This concept reaffirms the claim made in the first part of the chapter that the number of cells present in a given cell population is as important a trait of that population as what the cells look like. It also provides, for the first time in a mammal, a hint that, in addition to extrinsic mechanisms of cell number control such as target-related cell death, there are also intrinsic control mechanisms. The lineage of a neuron may well be used to convey information about the number of like neurons that will come to be formed during ontogeny.

Together, these two mechanisms (one cellularly intrinsic, one cellularly extrinsic) act in harmony to produce a nervous system that is both suited, in overall shape and size, for the particular species in which it is to function as well as suited, in its fine details, to the particular individual in which it comes to exist. Intrinsic mechanisms create a "rough out" of the adult CNS, following which extrinsic, epigenetic forces sculpt out the fine contours to bring the various cell groups in the rough cut nervous system into balance with each other and with the other organ systems of the body. Thus, in many ways, for the developing CNS, life is just a bowl of cherries.

ACKNOWLEDGMENTS

The data from my laboratory, reported here, were gathered with support from the following grants: NS18381, NS20591, and a basic research grant (1-763) from the March of Dimes Birth defects Foundation. I would like to express my gratitude to John Gagne and Rebecca Jones for their patient help in the preparation of this manuscript.

REFERENCES

Ashwell, K. W., and Watson, C. R. R. (1983). *J. Embryol. Exp. Morphol.* **77**, 117–141.
Atlas, M., and Bond, V. P. (1965). *J. Cell Biol.* **26**, 19–24.
Berry, M., and Bradley, P. (1976). *Brain Res.* **116**, 361–387.
Berry, M., McConnell, P., and Sievers, P. (1980). *Curr. Top. Dev. Biol.* **15**, 27–40.
Black, I. B., and Patterson, P. H. (1980). *Curr. Top. Dev. Biol.* **15**, 27–40.
Bodenstein, L. (1986). *Cell Diff.* **19**, 19–34.
Boterenbrood, E. C. (1970). *J. Embryol. Exp. Morphol.* **23**, 751–759.
Bradley, P., and Berry, M. (1978). *Brain Res.* **142**, 135–141.
Caddy, K. W. T., and Biscoe, T. J. (1979). *Philos. Trans. R. Soc. London* **287**, 167–201.
Caviness, V. S., and Rakic, P. (1978). *Annu. Rev. Neurosci.* **1**, 297–329.
Chapman, V. M., Whitten, W. K., and Ruddle, F. H. (1971). *Dev. Biol.* **26**, 153–158.
Condamine, H., Custer, R. P., and Mintz, B. (1971). *Proc. Natl. Acad. Sci. U.S.A.* **68**, 2032–2036.
Cooke, J. (1980). *Curr. Top. Dev. Biol.* **15**, 373–408.
Cowan, W. M. (1973). *In* "Development and Aging in the Nervous System" (M. Rockstein, ed.), pp. 19–41. Academic Press, New York.
Crepel, F., Delhaye-Bouchaud, N., Dupont, J. L., and Sotelo, C. (1980). *Neuroscience* **5**, 333–348.

Crick, F. H. C., and Lawrence, P. A. (1975). *Science* **189**, 340–347.

DeLorenzo, R. J., and Ruddle, F. H. (1969). *Biochem. Genet.* **3**, 151–162.

Deol, M. S., and Whitten, W. K. (1972). *Nature (London) New Biol.* **238**, 159–160.

Dewey, M. J., Gervais, A. G., and Mintz, B. (1976). *Dev. Biol.* **50**, 68–81.

Falconer, D. S., and Avery, P. J. (1978). *J. Embryol. Exp. Morph.* **43**, 195–219.

Felton, J., Meisler, M., and Paigen, K. (1974). *J. Biol. Chem.* **249**, 3267–3272.

Garcia-Bellido, A. (1972). *In* "Results and Problems in Cell Differentiation" (H. Ursprung and R. Nothiger, eds.), pp. 59–91. Springer-Verlag, New York.

Gardner, R. L. (1982). *J. Embryol. Exp. Morphol.* **68**, 175–198.

Gardner, R. L., and Johnson, M. H. (1973). *Nature (London) New Biol.* **246**, 86–89.

Gardner, R. L., and Lyon, M. F. (1971). *Nature (London)* **231**, 385–386.

Gardner, R. L., and Papaioannou, V. E. (1975). *In* "Early Development of Mammals (BSDR Symposium 2)" (M. Balls and A. E. Wild, eds.), pp. 107–132. Cambridge Univ. Press, London.

Gardner, R. L., and Rossant, J. (1979). *J. Embryol. Exp. Morphol.* **30**, 561–572.

Gartler, S. M., and Riggs, A. D. (1983). *Annu. Rev. Genet.* **17**, 155–190.

Gearhart, J., and Oster-Granite, M. L. (1978). *In* "Genetics and Mosaics in Mammals" (L. B. Russell, ed.), pp. 111–124. Plenum, New York.

Goldman-Rakic, P. S. (1980). *Prog. Brain Res.* **53**, 3–19.

Goldowitz, D., and Mullen, R. J. (1982a). *Dev. Biol.* **89**, 261–267.

Goldowitz, D., and Mullen, R. J. (1982a). *J. Neurosci.* **2**, 1474–1485.

Green, M. C., Schultz, L. D., and Nedzi, L. A. (1975). *Transplantation* **20**, 172–175.

Herrup, K. (1983). *Dev. Brain Res.* **11**, 267–274.

Herrup, K., and Bower, J. (1986). *Neurosci. Abst.* **12**, 769.

Herrup, K., and Mullen, R. J. (1979). *J. Cell Sci.* **40**, 21–32.

Herrup, K., and Silver, J. (1985). *In* "The Retina: A Model for Cell Biology Studies" (R. Adler and D. Farber, eds.). Academic Press, New York, pp. 245–274.

Herrup, K., Diglio, T., and Letsou, A. (1984a). *Dev. Biol.* **103**, 329–336.

Herrup, K., Wetts, R., and Diglio, T. J. (1984b). *J. Neurogenet.* **1**, 275–288.

Horvitz, H. R., Ellis, H. M., and Sternberg, P. W. (1982). *Neurosci. Comment.* **1**, 56–65.

Hotta, Y., and Benzer, S. (1971). *Nature (London)* **240**, 527–535.

Inouye, M., and Murakami, U. (1980). *J. Comp. Neurol.* **194**, 499–504.

Jensen, K. F., and Killackey, H. P. (1984). *Proc. Natl. Acad. Sci. U.S.A.* **81**, 964–968.

Jonakait, G. M., Bohn, M. C., Markey, K., Goldstein, M., and Black, I. B. (1981). *Dev. Biol.* **88**, 288–296.

Kankel, D. R., and Hall, J. C. (1976). *Dev. Biol.* **48**, 1–24.

Katz, M. J., and Grenader, U. (1982). *J. Theor. Biol.* **98**, 501–517.

Kauffman, S. L. (1968). *Exp. Cell Res.* **49**, 420–424.

Landmesser, L. (1984). *Trends Neurosci.* **7**, 336–339.

LaVail, N. M., and Mullen, R. J. (1976). *Exp. Eye Res.* **23**, 227–245.

LeDouarin, N. (1980). "Curr. Top. Dev. Biol. **15**, 31–85.

McLaren, A. (1976). "Mammalian Chimeras." Cambridge Univ. Press, London.

Miale, I. L., and Sidman, R. L. (1961). *Exp. Neurol.* **4**, 277–296.

Mintz, B. (1962). *Am. Zool.* **2**, 432.

Mintz, B. (1965). *Science* **148**, 1232–1233.

Mintz, B. (1974). *Annu. Rev. Genet.* **8**, 411–470.

Mintz, B., and Sanyal, S. (1970). *Genetics* **64**, 43–44.

Morrow, A. G., Greenspan, E. M., and Carroll, D. M. (1949). *J. Natl. Cancer Inst.* **10**, 657–661.

Mullen, R. J. (1977a). *Nature (London)* **270**, 245–247.

Mullen, R. J. (1977b). *Soc. Neurosci. Symp.* pp. 47–65.

Mullen, R. J. (1985). Site of gene action in rodents with hereditary retinal degenerations. *In* "Heredity and Visual Development" (J. B. Sheffield and S. R. Hilfer, eds.), pp. 35–42. Springer-Verlag, New York.

Mullen, R. J., and Whitten, W. K. (1971). *J. Exp. Zool.* **178**, 165–176.

Mystkowska, E. T., and Tarkowski, A. K. (1968). *J. Embryol. Exp. Morphol.* **20**, 33–52.

Nesbitt, M. N. (1971). *Dev. Biol.* **26**, 252–263.

Oliver, C., and Essner, E. (1973). *J. Histochem. Cytochem.* **21**, 218–228.

Oppenheim, R. W. (1981). Neuronal death and some related regressive phenomena during neurogenesis: a selective historical review and progress report. *In* "Studies in Developmental Neurobiology: Essays in honor of Viktor Hamburger" (W. M. Cowan, ed.), pp. 74–133. Oxford Univ. Press, New York.

Oster-Granite, M. L., and Gearhart, J. (1981). *Dev. Biol.* **85**, 199–208.

Perry, V. H., and Linden, R. (1982). *Nature (London)* **297**, 683–685.

Phillips, R. J. S. (1960). *J. Genet.* **57**, 35–42.

Ponder, B. A. J., Wilkinson, M. M., and Wood, W. (1983). *J. Embryol. Exp. Morphol.* **76**, 83–93.

Rakic, P. (1971a). *Brain Res.* **33**, 471–476.

Rakic, P. (1971b). *Brain Res.* **33**, 471–476.

Rakic, P. (1972). *J. Comp. Neurol.* **146**, 335–364.

Rakic, P. (1978). *Postgrad. Med. J.* **54**, 25–40.

Roffler-Tarlov, S., and Herrup, K. (1981). *Brain Res.* **215**, 49–59.

Rossant, J. (1987). *Curr. Top. Dev. Biol.*, in press.

Rossant, J., and Chapman, V. M. (1983). *J. Embryol. Exp. Morphol.* **73**, 193–205.

Rossant, J., Vijh, M., Siracussa, L. D., and Chapman, V. M. (1983). *J. Embryol. Exp. Morphol.* **73**, 179–191.

Sanyal, S., and Zeilmaker, G. (1976). *J. Embryol. Exp. Morphol.* **36**, 425–538.

Sanyal, S., and Zeilmaker, G. H. (1977). *Nature (London)* **265**, 731–733.

Shatz, C. J. (1983). *J. Neurosci,* **3**, 482–499.

Sidman, R. L., and Rakic, P. (1973). *Brain Res.* **62**, 1–35.

Silver, J., and Ogawa, M. Y. (1983). *Science* **220**, 1067–1069.

Siracussa, L. D., Chapman, V. M., Bennett, K. L., Hastine, N. D.. Pietras, D. F., and Rossant, J. (1983). *J. Embryol. Exp. Morphol.* **73**, 163–178.

Slack, J. M. W. (1983). "From Egg to Embryo: Determinative Events in Early Development" Cambridge Univ. Press, London.

Snell, G. D., and Stevens, L. C. (1966). *In* "Biology of the Laboratory Mouse" (E. L. Green, ed.), 2nd Ed., pp. 205–246. Dover, New York.

Sotelo, C. (1978). *Brain Res.* **48**, 149–170.

Sulston, J., Schierenberg, E., White, J., and Thompson, N. (1983). *Dev. Biol.* **100**, 64 f.

Sutton, W. D., and McCallum, M. (1972). *J. Mol. Biol.* **71**, 633–656.

Swisher, D. A., and Wilson, D. B. (1977). *J. Comp. Neurol.* **173**, 205–218.

Tarkowski, A. K. (1961). *Nature (London)* **230**, 333–334.

West, J. D. (1976). *J. Embryol. Exp. Morphol.* **35**, 433–461.

West, J. D. (1978). Analysis of clonal growth using chimeras and mosaics. *In* "Development in Mammals" (M. H. Johnson, ed.), Vol. 3, pp. 413–460. Elsevier, New York.

Wetts, R., and Herrup, K. (1982a). *Brain Res.* **250**, 358–362.

Wetts, R., and Herrup, K. (1982b). *J. Embryol. Exp. Morphol.* **68**, 87–98.

Wetts, R., and Herrup, K. (1982c). *J. Neurosci.* **2**, 1494–1498.

Wilson, D. B. (1974). *Brain Res.* **69**, 41–48.

Wilson, D. B. (1981). *Dev. Brain Res.* **2**, 420–424.

CHAPTER 4

THE INSECT NERVOUS SYSTEM AS A MODEL SYSTEM FOR THE STUDY OF NEURONAL DEATH

James W. Truman

DEPARTMENT OF ZOOLOGY
UNIVERSITY OF WASHINGTON
SEATTLE, WASHINGTON 98195

I. Introduction

Cell death is an important force in determining the final form of the developing nervous system. This fact was first underscored by the studies of Hamburger and colleagues on the development of the motor system of the chick (see Hamburger and Oppenheim, 1982) but has since proven to be a universal phenomenon in terms of both the regions of the nervous system that show neuronal death and the range of animals that use this mechanism in the construction of their nervous system.

During recent years there has been increasing interest in the role of cell death in the development of the nervous system of invertebrates (see Truman, 1984, for a review). Features such as the presence of unique, identifiable neurons that can be studied in successive individuals coupled with the detailed knowledge of neuronal lineages available for some species have allowed neuronal death to be studied in a cellular context that has not been possible with vertebrate material. Good examples are seen in the development of the embryonic nervous system of both nematodes (White *et al.*, 1976) and locusts (Goodman and Bate, 1981; Loer *et al.*, 1983). In both cases the mature nervous system shows marked regional differences in the number and types of neurons present even though each region comes from essentially identical arrays of precursor cells. Each set of precursors produces similar or identical lineages of daughter cells, and cell death, in a region-specific manner, then carves out the mature array of neurons from this collection of immature cells. Although the lineages of some of the cells that die are well understood in locusts and nematodes, the factors responsible for triggering neuronal degeneration in the embryos of these animals are not known.

99

CURRENT TOPICS IN
DEVELOPMENTAL BIOLOGY, VOL. 21

Neuronal death also occurs during postembryonic stages of some insects. In species that have incomplete metamorphosis, such as the locust, the newly hatched insect has essentially the full complement of central neurons (interneurons and motoneurons) that it will have as an adult. The only neurons that are known to be added postembryonically are some classes of sensory interneurons that are produced as complex sensory structures such as the compound eyes that grow during larval life (e.g., Anderson, 1978). Not only is postembryonic cell birth rare in these insects, but so, too, is neuronal death in larval or adult stages. In contrast to insects with incomplete metamorphosis, those having complete metamorphosis, such as moths and flies, show extensive neuronal birth and death during postembryonic stages. In these insects, metamorphosis causes the loss of many larval neurons as well as the addition to new adult-specific cells. These changes are controlled by a number of hormones, which makes it possible to manipulate the fates of particular cells (e.g., Truman and Schwartz, 1984).

This chapter deals with neuronal death in insects, although examples from other invertebrates are presented in order to illustrate or support particular points. Since most of the insect work has been done on the regulation of neuronal degeneration during postembryonic stages of the moth, *Manduca sexta,* the major proportion of the review is devoted to this insect.

II. Background to the Regulation of Postembryonic Growth and Metamorphosis in Insects

A consideration of the cellular changes that come about during metamorphosis requires a brief description of the events, and their endocrine control, that occur during this time (see Riddiford and Truman, 1978, for a more complete review). Following hatching, *M. sexta* goes through a succession of larval molts to produce five sequential larval stages. During each molt a new cuticular exoskeleton is formed and the old one then shed, the shedding process being called ecdysis. Each molt is triggered by the appearance of the steroid hormones, ecdysteroids, the most active of which is 20-hydroxyecdysone (20-HE). The simultaneous presence of the sesquiterpenoid hormone, juvenile hormone, insures that the larva will molt into another version of what it was before, i.e., into another larval stage. At the end of larval life, juvenile hormone disappears and metamorphosis ensues. In the absence of juvenile hormone, 20-HE causes the larva to undergo the first metamorphic molt, transforming it into the pupal stage. Depending on environmental conditions the pupa then either enters a state of developmental arrest (diapause) or releases ecdysteroids (again in the absence of juvenile hormone) to start the molt to the adult stage. This final transition from the pupal to the adult form requires about 18 days. At the end of each molt, the declining titer of ecdysteroids brings about the release of the peptide hormone, eclosion hormone, that triggers the ecdysial behaviors used for the shedding of the old cuticle.

III. Major Systems for the Study of Neuronal Death in Insects

A. DEGENERATION DURING EMBRYONIC DEVELOPMENT IN LOCUSTS

The segmental ganglia of locusts each arise from a set of 61 neuroblasts and 7 midline precursor cells (Bate, 1976; Bate and Grunewald, 1981). At the end of embryonic development, each thoracic ganglion contains about 3000 neurons whereas each abdominal ganglion has about 500. This difference in the number of neurons in the ganglia from the two regions reflects, primarily, differences in cell death (Goodman and Bate, 1981). For example, the unpaired midline neuroblasts in the third thoracic segment (T3) and in the first abdominal segment (A1) generate 90–100 progeny. Subsequently, about half of the A1 progeny die whereas virtually all of the T3 cells survive.

The most detailed examination of the fate of particular neurons during locust embryogenesis concerns the two progeny of midline precursor cell 3 (MP3) which divides once to produce two neurons (Loer *et al.*, 1983). Both cells typically survive in thoracic ganglia but one or both usually die in abdominal segments. The manner by which the segment-specific survival of the MP3 progeny is determined is not known, but a mixture of genetic and epigenetic factors appears to be involved.

B. CELL DEATH DURING THE DEVELOPMENT OF THE OPTIC LOBES

The early studies by Nordlander and Edwards (1968) on the metamorphosis of the optic lobes of the monarch butterfly showed that neuronal death was a normal feature of the metamorphosis of this structure. Subsequent studies on several insects (Anderson, 1978; Mouze, 1974, 1978) and on the crustacean *Daphnia magna* (Macagno, 1979) have shown that the generation of new neurons in the first optic ganglion, the lamina, is not affected by the presence or absence of ingrowing receptor axons. Yet the subsequent survival of these new cells requires synaptic contacts from these axons. Indeed, removal of the eye results in the loss of essentially all of the laminar neurons, whereas the implantation of supernumerary eyes causes enhanced survival of these cells.

C. Death during Metamorphosis in *Manduca*

Except for the studies on the optic lobes, essentially all of the other studies of postembryonic neuronal death have concentrated on the segmental ganglia of *Manduca*. Neuronal death has at least three functions in shaping the CNS of this insect during metamorphosis.

1. Death of Unneeded Larval Neurons

The most striking type of cell death is the death of functioning neurons. In the case of the unfused abdominal ganglia, approximately 50% of the larval neurons do not survive metamorphosis. These cells die in two discrete waves, one following ecdysis to the pupal stage and the other after the ecdysis of the adult. At both times, both interneurons and motoneurons die. By following the fates of identified motoneurons (Truman, 1983; Weeks and Truman, 1986), it is clear that the spatial pattern of death is precisely determined with specific neurons invariably undergoing degeneration. The larval motoneurons that die after pupal ecdysis are cells whose target muscles degenerate during the larval–pupal transition. It should be noted, though, that not all cells that lose target muscles at this time degenerate. A substantial proportion of them become associated with differentiating adult muscles and survive to perform new functions in the adult (e.g., Levine and Truman, 1982).

The abdominal intersegmental muscles and their motoneurons survive the first wave of cell death at pupal ecdysis. These larval cells control the behaviors shown by the pupa and developing adult while other larval neurons are being reorganized and new cells are being incorporated into the adult nervous system. After adult ecdysis, these persisting larval muscles then die along with their motoneurons and approximately half of the interneurons in the segmental abdominal ganglia (Truman, 1983).

2. Sex-Specific Cell Death in the CNS

A second role for cell death is in generating sex-specific differences in the CNS. The terminal segments of the adult abdomen bear the genitalia and are strongly sexually dimorphic. Likewise, in the adult terminal ganglion, the motoneuron populations that supply the genital muscles are also dimorphic. These dimorphic sets of motoneurons arise from a pool of functional neurons that are identical in the larval stages of the two sexes (Fig. 1). From this monomorphic set of larval motoneurons, a few survive in both sexes while others survive only in males and still others only in females. Thus, the sexual differences in the

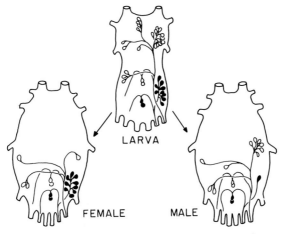

FIG. 1. Origin of sexually dimorphic arrays of motoneurons in the terminal ganglion of *Manduca sexta*. Cell body position and axon path are indicated for motoneurons that exit the ganglia through the eighth dorsal nerve (open cell bodies) and the eighth ventral nerve (filled cell bodies). Motoneuron populations are identical in the larval stages of both sexes. The adult configuration comes about through sex-specific patterns of cell death. Data from Giebultowicz and Truman (1984).

adult motor systems are made by taking a common set of functional motoneurons in the larva and using cell death to carve out the arrays characteristic of each sex (Giebultowicz and Truman, 1984).

Although this sex-specific cell death is similar to that described above in that it involves the death of functional larval neurons, it differs as to its timing (Giebultowicz and Truman, 1984). The larval neurons that die after pupal ecdysis in both sexes do so prior to the initiation of the adult molt. By contrast, many of the neurons that show sex-specific degeneration, especially those that are lost in males, wait until adult differentiation is well underway before dying. The reason for this delay is not known, but we have speculated that some signal from he differentiating periphery, perhaps associated with the forming genitalia, may be involved in determining which neurons live and which die.

3. Degeneration of Immature Neurons during Metamorphosis

The third category of postmetamorphic neuronal death is seen in clusters of immature neurons that are born during larval life (R. Booker and J. W. Truman, unpublished). Scattered neuroblasts reside in characteristic locations in the larval ganglia and generate new cells through much of larval life. After these cells are born, they apparently then arrest development at a relatively undifferentiated state and wait for metamorphosis. As the larvae grow, the number of immature neurons in the cluster associated with each neuroblast increases steadily. At the outset of metamorphosis, the cells in each cluster abruptly begin to change with some enlarging and differentiating into functional neurons while others degenerate. The factors triggering death and the manner by which particular cells in the cluster are selected for differentiation or death are not known.

IV. Temporal Patterns of Cell Death

Studies on both vertebrate and invertebrate systems show that neurons can undergo programmed death only during rather precise times in their developmental histories. This time is not the same though for all neurons which degenerate. In the nematode, *Caenorhabditis elegans,* the cells that die do so before they extend their axon (Robertson and Thomson, 1982). At the other extreme are fully mature neurons with functional synaptic contacts that die at metamorphosis in *Manduca* (Stocker *et al.,* 1978). The locust is intermediate between these two examples in that some of the motoneurons that die in the embryo grow axons that extend to the periphery but no connections are apparently made before they degenerate (Whittington *et al.,* 1982).

A striking feature of the timing of cell death in *Manduca* is its fine temporal resolution. The number of neurons in the fourth abdominal ganglion (A4) drops from about 650 before adult ecdysis to approximately 350 cells by 2 days later (Taylor and Truman, 1974; Truman, 1983). The first cells to die are interneurons that begin degeneration a few hours before adult ecdysis. Deaths among the interneurons reach peak frequency at about 18 hours after ecdysis and are essentially complete by 30 hours. Within the population of motoneurons, deaths begin at 8 hours after ecdysis and continue through the next 24 hours.

The stereotyped location of many invertebrate neurons, as well as the ability to precisely stage the animals relative to developmental or behavioral markers, allows a detailed description of the progression of death in individual cells (e.g., Stocker *et al.,* 1978; Robertson and Thomson, 1982). In *Manduca,* Stocker *et al.* (1978) followed the death of three pairs of motoneurons that reside in the anterodorsomedial region of ganglion A4. These neurons, the D-IV group, innervate ventral intersegmental muscles which persist through metamorphosis from the larval stage but then die after adult emergence. At the time of adult ecdysis, the ultrastructure of the somata of these cells was identical to that of motoneurons that were destined to live through the life of the adult. The first changes in the D-IV neurons occurred about

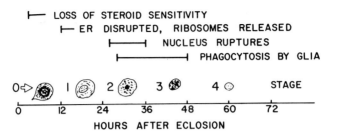

FIG. 2. Schedule of events that occur during the degeneration of the D-IV motoneurons. The timing is based on when 50% of the neurons show a particular characteristic. The drawings represent the appearance of the neuronal cell body at the various stages of degeneration. 0, Healthy cell with distinct nucleus and dark-staining cytoplasm; 1, nucleus irregular in outline and cytoplasm pale; 2, nucleus collapsed and condensed; 3, cytoplasm condensed and dark staining; 4, remains of cell shrunken and pale staining.

12 hours after ecdysis with the dissolution of the long endoplasmic reticulum channels into smaller vesicles and the release of the ribosomes. By about 16 hours a number of the other cytoplasmic organelles, such as the mitochondria, showed some degree of swelling. At about 24 hours the nuclear membrane abruptly ruptured and, by a few hours later, bits of neuronal cytoplasm were seen in the surrounding glial cells. Prior to this time the glia showed no sign of being actively involved in the death of the neurons. With succeeding time the remains of the cell shrank and eventually appeared as a tightly wrapped collection of membranes. The degeneration of the cells appeared to be a very orderly process having the characteristics of a programmed cell death rather than a necrotic response (Wyllie, 1981). Correlates of the ultrastructural changes could also be seen in material prepared for light microscopy and were the basis of a scoring system used to follow the progression of degeneration in identified cells (Fig. 2).

Although other motoneurons went through the same stages of degeneration as did the D-IV cells, they did not all begin the process at the same time (Truman, 1983). Figure 3 shows the time course of degeneration for 4 sets of identified motoneurons. Each cell began to die at its own characteristic time. The earliest cell was MN-11 at 8 hours, followed by the D-IV cells, MN-2, and MN-12 at about 36 hours. In contrast to the motoneurons, the death of the interneurons could not be followed on an individual basis. The stereotyped pattern of regions containing dying interneurons at various times suggested that the time of death in these cells was likewise individually determined. Therefore, cell death

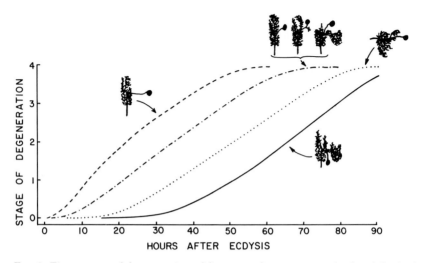

F<small>IG</small>. 3. Time course of degeneration of four sets of motoneurons in the abdominal ganglion of *M. sexta* after the ecdysis of the adult moth. The lines show the progression of degeneration for MN-11 (dashed); the D-IV neurons (dot–dash); MN-2 (dotted); and MN-12 (solid). The scoring system is as given in Fig. 2. The drawings show the central dendritic arbors of the particular cells. Data from Truman (1983).

in this insect is not a stochastic process but occurs according to a temporal program with each cell apparently having its own characteristic time to die. Although various treatments can interrupt the sequence of degeneration (Truman, 1983; Truman and Schwartz, 1984), we have found no way of rearranging the order in which the cells die.

V. Regulation of Neuronal Degeneration

Studies on the development of the compound eyes and the optic lobes in arthropods present convincing evidence that the survival of neurons in the lamina of the brain is dependent on the presence of receptor axons at a critical stage in their development (e.g., Macagno, 1979). In the ventral nerve cord there is no evidence that the formation of appropriate connections is required for neuronal survival. In fact, in locusts particular identified motoneurons survived when their limb-bud target was removed prior to the arrival of their axons in the periphery (Whittington *et al.*, 1982). The possibility remains, however, that these cells could have innervated inappropriate peripheral targets, which, in turn, could have provided factors important for motoneuron survival.

In the segmental ganglia, survival or degeneration of cells appears not to be a population phenomenon but rather is the characteristic fate of particular cells. In considering why particular cells die, it is necessary to divide the problem into two parts. The first relates to the factors that determine *which* cell will die, the second to the proximate trigger that then initiates the degeneration process in these prescribed cells. The first question deals with a fundamental problem in developmental biology, namely, how do cells acquire individualized fates? The study by Loer *et al.* (1983) on the fate of the two progeny of MP3 during locust embryogenesis examined different species of locusts as well as single broods and clones in *Schistocerca*. Their results suggested that the segmental fates of these cells depend on a combination of genetic and epigenetic factors, but the nature of these factors is unknown.

The second part of the question, the proximate trigger for degeneration, has not been determined for embryonic systems in insects but is at least partially known for the postembryonic death of larval neurons in *Manduca* (Truman and Schwartz, 1984). Insight into the factors that trigger neuronal death was provided by the simple procedure of isolating the abdomens of developing moths just before adult ecdysis. These detached abdomens lived for a number of days and showed the normal stereotyped pattern of cell death. Since the abdomens did not perform the ecdysis behavior, it is clear that cell death is not caused by the performance of this behavior even though it is normally correlated in time with it.

When abdomens were isolated 2–3 days prior to the scheduled time of ecdysis, the normal sequence of death still occurred but all of the cells died earlier than expected. This result suggested that late in adult development the cells were sustained by some influences from the anterior end of the insect; apparently abdominal isolation interrupted these influences and, hence, brought about early death. Two experiments indicated that a blood-borne factor was involved (Truman and Schwartz, 1984): transection of the ventral nerve cord just below the thorax did not result in early cell death, and ganglia implanted into the abdominal cavity of intact animals showed neuronal degeneration in concert with the host's nervous system.

This circulating factor proved to be the ecdysteroids. These steroids are produced by glands in the anterior end of the insect and occur at a high titer early in the adult molt but then gradually decline to low levels just before ecdysis (Schwartz and Truman, 1983). In both isolated abdomens and intact animals injected with the main circulating active ecdysteroid, 20-HE, the onset of neuronal death was delayed (Truman and Schwartz, 1984). The extent of the delay was proportional to the amount of 20-HE injected, and continuous infusion of low levels of 20-HE blocked cell death for the duration of the infusion (up to 6 days). Preliminary data on the larval neurons that die after pupal ecdysis suggest that these cells are also sensitive to the withdrawal of ecdysteroids but they respond to the drop that occurs after the pupal molt (Weeks and Truman, 1986).

Although the death of particular motoneurons can be blocked by ecdysteroid treatment, the death of their target muscles is also prevented by the same treatment (Schwarz and Truman, 1983). Therefore, it was not clear whether both neurons and muscle died as a direct response to the same endocrine cue or whether the death of one set of cells triggered the death of the other. By treating groups of insects with a standard dosage of 20-HE at various times, it was possible to define a *commitment point* before which steroid application delayed cell death but after which the cells died on schedule despite the treatment. The commitment point (as defined when 50% of the cells can no longer be delayed) for the ventral intersegmental muscles occurred at about 14 hours before ecdysis (Schwartz and Truman, 1983) whereas that for the D-IV motoneurons, which innervate these muscles, was 3 hours after ecdysis (Fig. 4; Truman and Schwartz, 1984). Application of exogenous 20-HE between the two times resulted in the death of the muscles but the survival of the motoneurons that supply them. Consequently, the motoneurons do not die simply because their muscles begin to degenerate; rather it is likely that both cells are targets for the steroid.

Not only did muscles and motoneurons have different commitment points, but within the CNS individual neurons differed in the times that they became committed to die (Truman and Schwartz, 1984). This is best seen in the differential response of the D-IV motoneurons and cell MN-11 (an early dying motoneuron) to treatment with 20-HE at various times (Fig. 5). Early treatments retarded the onset of degeneration of both sets of cells. A few hours later the same treatment did not save MN-11, but the D-IV cells were still delayed (Fig. 5, left). With still later 20-HE injections, both sets of cells died on schedule. Indeed, by judiciously timed treatments with 20-HE it was possible to interrupt the temporal sequence of death at essentially any point.

The temporal sequence of commitment points is the same as the sequence of cell death (e.g., Fig. 5). One possible interpretation is that the commitment point represents the time at which the ecdysteroid titer declines below the threshold needed to maintain a particular cell. If cells have different thresholds, then a gradual decline in hormone levels would bring about a sequence of cell deaths. This possibility was

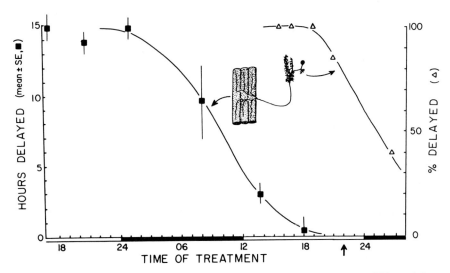

Fig. 4. The ability of injections of the steroid, 20-hydroxyecdysone (20-HE), to delay the degeneration of the D-IV neurons (△) and the ventral intersegmental muscles (■) that they innervate. Animals were injected at the times indicated and then examined 24 hours after adult ecdysis to determine the state of the muscles and motoneurons. Motoneuron delay was calculated as the percentage of animals in each treatment group that showed delayed neuronal death. The muscle delay is expressed in hours and is based on the dry weight of the muscles at the time of examination. The commitment point is taken as the 50% response time for the various tissues. Arrows represent the time of injection. The open and filled bars represent the light–dark cycle to which the insects were exposed.

explored by culturing ganglia taken from *Manduca* on the last day of adult development (Bennett and Truman, 1985). Ganglia cultured in the absence of exogenous hormones showed degeneration of both interneurons and motoneurons. The spatial pattern of degenerating interneurons seen *in vitro* was similar to that seen *in vivo* (Truman, 1983). Among the motoneurons, death was confined exclusively to those cells which normally die *in vivo*. The neurons in the cultured ganglia played out the same fates as they showed in the intact animal except that the progress of degeneration *in vitro* was slower than that seen *in vivo*.

Cultured ganglia also retained their ability to respond to ecdysteroids. When cultured with 100 ng 20-HE/ml, conditions similar to those seen *in vivo* late in adult development (Schwartz and Truman, 1983), neuronal death was almost completely prevented, indicating that 20-HE acts directly on the CNS to regulate cell number.

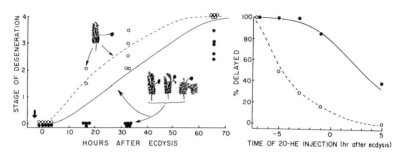

FIG. 5. The ability of 20-HE injections to delay the death of identified motoneurons. (○, - - - - -) MN-11; (●, ——) the D-IV cells. Left, Response of the neurons to injection with the steroid at 3 hours before ecdysis. The lines represent the time course of degeneration in untreated animals, the circles are scores from individual ganglia from treated individuals. Treatment at that time delays the death of the D-IV cells but not MN-11. Right, The ability of a standard dosage of 20-HE (25 μg/animal) to delay the death of the two groups of motoneurons as a function of the time of injection. The commitment points for each cell are taken as the time that the dosage caused a 50% delay.

The *in vitro* studies exclude one possible explanation for the nature of the commitment points. The explanted ganglia experienced an abrupt "step down" in the concentrations of ecdysteroid to which they were exposed. Therefore, all of the cells should have crossed their thresholds at the same time and, hence, should have all died together. However, in the cultured ganglia the normal sequence of cell death was maintained. Thus, if extrinsic factors are responsible for the sequence of cell death, they must act prior to the time that the ganglia were explanted, which was a number of hours before the commitment points of the cells.

Although endocrine cues play a major role in triggering neuronal death, other factors may also be involved. Under natural conditions *Manduca* pupate underground and the newly emerged moth must dig to the surface before it can inflate its new wings. When insects were forced to dig through the soil for extended periods of time after emergence, the degeneration of the D-IV motoneurons was delayed by about 8 hours (Truman, 1983), but the time course of death for the early cells such as MN-11 was unaffected. Thus, behavioral factors can selectively interrupt the degeneration program but they cannot do so indefinitely. Interestingly, the cells whose deaths were delayed were those that supply the muscles used for digging and wing expansion behaviors.

The ability of certain behaviors to delay the death of particular cells is of interest in terms of the commitment points of the cells involved. Most of the interneurons and MN-11 become committed to die prior to ecdysis while the insect is still encased in the old pupal skin. Therefore, these cells cannot be influenced by the conditions that the insect faces after it emerges. By contrast, the neurons involved in digging (the D-IV cells) and the wing inflation behavior (MN-2) have fates that are not irrevocably fixed until some time after ecdysis. Consequently, external stimuli can still modify their fates even after the insect emerges.

At this point it is not possible to explain the sequential death of neurons in *Manduca* in terms of a single factor. The most reasonable assumption is that at least two factors, the steroid decline and some unknown event(s) interact to determine when particular cells die. The fact that ganglia *in vitro* showed the normal ordering of cell death even after being subjected to an abrupt drop in ecdysteroid concentrations suggests that the second factor is primarily responsible for determining the sequence of the deaths. The nature of this second factor is unknown. It could reflect intrinsic differences between cells, such as in the number of steroid receptors. Such a variation could result in different amounts of hormone being present in the cells at the time of disappearance of the external ecdysteroids. If the residual hormone was then gradually released from its receptors, the time to drop below the critical concentration would be inversely proportional to the number of receptors. In contrast, the ordering of cell death might also arise from interactions between cells. Although the death of early dying cells does not trigger the death of the later cells (Fig. 5), it may nevertheless be a necessary prerequisite for their degeneration. The fact that the insect can modify the timing of the death of selected cells by behavioral means lends some support to the importance of interactive factors (Truman, 1983). One possible set of interactions are schematically presented in Fig. 6. The cells that die are divided into early, intermediate and late classes. The interactions between these neurons are envisioned as trophic interactions which prevent the cells from activating their degeneration program. Ecdysteroids also provide trophic support that inhibits degeneration. Both trophic inputs must be withdrawn before the cells are committed to die. With the drop in ecdysteroids, one source of trophic support is eliminated and the cells are maintained only by their trophic interactions. The early cells which have few or no inputs begin to die, withdrawing their support from the intermediate cells who then start to die, etc. Up until the time that a cell is actually committed, however, the replacement of ecdysteroids can rescue it. The scheme presented in Fig. 6 should be taken as a working model. Further experimentation *in vitro* involving

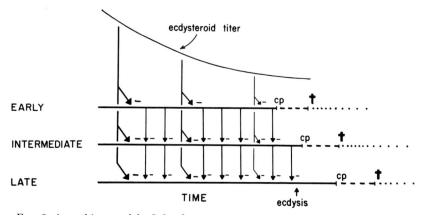

Fig. 6. A working model of the factors interacting to control the time course of degeneration of moth neurons after ecdysis. The cells are divided into groups that die at early, intermediate, and late times. The horizontal lines after each group represent the progression of the cells from healthy (solid) to past their commitment point (cp, dashed) to dying (dotted). The degeneration program in each cell is proposed to be inhibited in its expression (arrows and minus signs) by both high ecdysteroid levels and trophic interactions from other cells. The drop in the ecdysteroid titer withdraws one source of trophic support from all of the cells. The pattern of trophic interaction between the cells then determines the sequence of death within the groups. See text for further explanation.

both isolated ganglia and dissociated cell culture is required to test this model.

VI. Critical Periods for Cell Death

As described above, there is typically a characteristic time in the life of a cell when it is most susceptible to the cues which initiate degeneration. The best example is in the chick spinal cord in which the spinal motoneurons, during a critical period of their development, are dependent on peripheral targets in order to survive. After this period is past, however, they are resistant to degeneration if they subsequently lose their target. Likewise, in the optic lobes of *Daphnia*, the laminar interneurons require interactions with receptor axons when the cells are quite young, but in older individuals lesions in the retina do not result in interneuron death (Macagno, 1979).

In *Manduca* the D-IV motoneurons die in the adult stage in response to the disappearance of the ecdysteroids. However, in the preceding pupal stage these same cells may experience weeks or months of ecdysteroid-free conditions if the pupa enters diapause and yet they remain alive and healthy (Truman, unpublished). The appearance of ecdysteroids at the outset of the adult molt changes these cells so that they require ecdysteroids for their continued survival. The factors involved in rendering these cells steroid dependent are unknown.

THE MECHANISM FOR NEURONAL DEGENERATION

Most of the biochemical studies of cell death in invertebrates have used nonneural tissues (see Lockshin, 1981) for a review). Within the nervous system, the most intriguing data have been provided by genetic studies on the nematode *C. elegans*. A number of cell death (*ced*) mutants have been isolated in this worm. The most interesting of these is the *ced-3* mutant which blocks cell death. In *ced-3* individuals, cells that would normally die soon after they are born survive and differentiate into normal-appearing neurons. Thus, the activation of a particular genetic locus appears essential if the cell is to undergo its programmed degeneration.

The studies on the death of the MP3 progeny in locusts (Loer *et al.*, 1983) suggested a genetic component in determining which cells will die in this insect. In the case of the moths, the ability of certain motoneurons to die is a species-specific characteristic, with certain species, such as *Manduca,* showing cell death whereas the homologous neurons survive in the related giant silkmoths irrespective of the fact that their target muscles degenerate (Truman, unpublished).

In *Manduca* the data on MN-11 and the D-IV cells show that cells become committed to die about 10–12 hours before the first signs of death (Truman and Schwartz, 1984). This time span is sufficient for the transcription and translation of new genetic information. Also, since most, if not all, of the known actions of ecdysone come about through actions on the genome (Riddiford, 1980), this mode of action would not be unreasonable. A preliminary test of this hypothesis was made by injecting insects with large dosages of actinomycin D, an inhibitor of RNA synthesis (E. Martel and J. W. Truman, unpublished). This antibiotic preserved the cells when given through about 6 hours after the commitment point. After this time the cells subsequently died even though the inhibitor was present. These preliminary experiments are consistent with an involvement of RNA synthesis in the activation of the cell death program.

VII. Implications

The ability to follow the degeneration process in identified motoneurons has allowed a precise description of the events associated with cell death. In *Manduca* this is a highly ordered process with cells dying according to a precise temporal sequence. The removal of the trophic support provided by a steroid hormone is an important factor in caus-

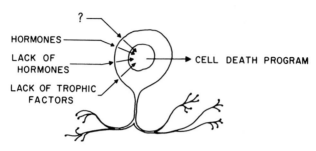

Fɪɢ. 7. A schematic representation indicating that many external factors may impinge on a cell to cause it to activate its latent degeneration program. This program likely involves the "reading out" of new genetic information.

ing death, but an analysis of the temporal pattern of degeneration also suggests that trophic interactions between cells may also be important (Fig. 6).

The possibility that the activation of new genes is involved in programmed cell death is a recurrent theme in many invertebrate systems. The studies on *C. elegans* clearly show that the activation of new genetic information is necessary if death is to occur. The RNA inhibitor studies in *Manduca* are also consistent with such a hypothesis for neuronal death in insects. In the moth *Antheraea polyphemus* the degeneration of the abdominal intersegmental muscles after ecdysis is triggered by a peptide hormone (Schwartz and Truman, 1983), but RNA and protein synthesis inhibitors can subsequently block degeneration of this tissue in a time-dependent manner (Lockshin, 1969).

In the case of muscle death in *A. polyphemus,* eclosion hormone acts as a death signal to initiate the degeneration of the cells (Schwartz and Truman, 1982). By contrast, in *Manduca* death comes about by the withdrawal of trophic support provided by a steroid. The nature of the signals that trigger death in *C. elegans* is unknown. Thus, the three systems presumably have very different ways to trigger death but they may share the common feature of having to turn on new genetic information in order for the cells to degenerate (Fig. 7). This relationship suggests that one should consider the mechanism of the degeneration separate from the factors that trigger this developmental program. Even in the case where cells are competing for postsynaptic targets such as the optic lobes or the vertebrate spinal cord, it may well be that the lack of putative trophic factors does not simply "starve" cells (Patterson and Purves, 1982) but rather turns on the latent genetic program that results in the cell's orderly demise.

In summary, in the context of development it is evident that each cell contains in it the genetic information to bring about its own destruction. Whether a cell has only one such program or a number of them is not known. Cells apparently do not have access to these programs at all times in their development but only at particular critical periods. It is the task of the developmental biologist to determine the factors that regulate when the cell can have access to this program and how external signals such as the presence or absence of hormones or tropic factors act to turn on these programs.

ACKNOWLEDGMENTS

I thank Lynn M. Riddiford for a critical reading of the manuscript. Unpublished work was supported by grants from NSF, NIH, and the McKnight Foundation.

REFERENCES

Anderson, H. (1978). *J. Embryol. Exp. Morphol.* **45**, 55.
Bate, C. M. (1976). *J. Embryol. Exp. Morphol.* **35**, 107.
Bate, C. M., and Grunewald, E. B. (1981). *J. Embryol. Exp. Morphol.* **61**, 317.
Bennett, K. L., and Truman, J. W. (1986). *Science* **229**, 58.
Giebultowicz, J. M., and Truman, J. W. (1984). *J. Comp. Neurol.,* **226**, 87.
Goodman, C. S., and Bate, M. (1981). *Trends NeuroSci.* **4**, 163.
Hamburger, V., and Oppenheim, R. W. (1982). *Neurosci. Comment.* **1**, 39.
Horvitz, H. R., Ellis, H. M., and Sternberg, P. W. (1982). *Neurosci. Comment.* **1**, 56.
Levine, R. B., and Truman, J. W. (1982). *Nature (London)* **299**, 250.
Lockshin, R. A. (1969). *J. Insect Physiol.* **15**, 1505.
Lockshin, R. A. (1981). *In* "Cell Death in Biology and Pathology" (I. D. Bowen and R. A. Lockshin, eds.), p. 79. Chapman & Hall, London.
Loer, C. M., Steeves, J. D., and Goodman, C. S. (1983). *J. Embryol. Exp. Morphol.* **78**, 169.
Macagno, E. R. (1979). *Dev. Biol.* **73**, 206.
Mouze, M. (1974). *J. Embryol. Exp. Morphol.* **31**, 377.
Mouze, M. (1978). *Wilhelm Roux Arch.* **184**, 325.
Nordlander, R. H., and Edwards, J. S. (1968). *Nature (London)* **218**, 780.
Patterson, P., and Purves, D. (1982). "Readings in Developmental Neurobiology." Cold Spring Harbor Laboratory, Cold Spring Harbor, New York.
Riddiford, L. M. (1980). *Annu. Rev. Physiol.* **42**, 511.
Riddiford, L. M., and Truman, J. W. (1978). *In* "Insect Biochemistry" (M. Rockstein, ed.), p. 307. Academic Press, New York.
Robertson, A. M., and Thomson, J. N. (1982). *J. Embryol. Exp. Morphol.* **67**, 89.
Schwartz, L. M., and Truman, J. W. (1982). *Science* **215**, 1420.
Schwartz, L. M., and Truman, J. W. (1983). *Dev. Biol.* **99**, 103.
Stocker, R. F., Edwards, J. S., and Truman, J. W. (1978). *Cell Tissue Res.* **191**, 317.
Taylor, H. M., and Truman, J. W. (1974). *J. Comp. Physiol.* **90**, 367.
Truman, J. W. (1983). *J. Comp. Neurol.* **216**, 445.
Truman, J. W. (1984). *Annu. Rev. Neurosci.* **7**, 171.
Truman, J. W., and Schwartz, L. M. (1984). *J. Neurosci.* **4**, 274.
Truman, J. W., Levine, R. B., and Weeks, J. C. (1984). *Br. Soc. Dev. Biol. Symp. 8th,* in press.
Weeks, J. C., and Truman, J. W. (1986). *J. Neurobiol.* **17**, 249.

White, J., Southgate, E., Thomson, N., and Brenner, S. (1976). *Philos. Trans. R. Soc. London Ser. B* **275,** 327.

Whittington, P., Bate, M., Seifert, E., Ridge, K., and Goodman, C. S. (1982). *Science* **215,** 973.

Wyllie, A. H. (1981). *In* "Cell Death in Biology and Pathology" (I. D. Bowen and R. A. Lockshin, eds.), p. 9. Chapman & Hall, London.

CHAPTER 5

BRAIN-SPECIFIC GENES: STRATEGIES AND ISSUES

*Robert J. Milner, Floyd E. Bloom, and J. Gregor Sutcliffe**

DIVISION OF PRECLINICAL NEUROSCIENCE AND ENDOCRINOLOGY
DEPARTMENT OF BASIC AND CLINICAL RESEARCH AND
*DEPARTMENT OF MOLECULAR BIOLOGY
RESEARCH INSTITUTE OF SCRIPPS CLINIC
LA JOLLA, CALIFORNIA 92037

I. Introduction

The physiological properties of any tissue are largely determined by the set of protein molecules made by the cells of that tissue. Functions that are unique to a particular tissue or organ therefore depend largely on the expression of a particular set of proteins, many of which are unique to that organ or tissue. Because many of the properties and functions of the brain are unique, we may expect its cells—neurons and glia—to express a wide variety of proteins that are expressed specifically or predominantly in neural tissues. These would include proteins involved in neurotransmitter metabolism (neuropeptide precursors, enzymes for neurotransmitter synthesis and degradation, and transmitter uptake systems), neurotransmitter receptors and ion channels, components of the cellular specializations unique to neurons (axons, dendrites, and synapses), unique glia structures (such as myelin), proteins that regulate neural connectivity during development, and molecules that may mediate higher mental processes (such as memory and learning). Some of these may be expressed by all neurons or glia in the brain. In addition, neurons exhibit an especially wide variety of forms and functions, and such phenotypic differences probably result from a heterogeneity of protein expression. The different types of glial cells—astrocytes, oligodendrocytes, and Schwann cells—express cell type-specific proteins and may also exhibit cellular heterogeneity. To dissect this complexity and approach an understanding of brain function, it is essential to identify and exploit the phenotypic differences between brain cell types. We believe that one key to a more detailed understanding of the mammalian nervous system requires the isolation and characterization of the protein molecules that are uniquely or predominantly expressed by its cells.

117

The unique properties of neurons and their heterogeneity of form and function are generated during the process of neuronal development. Through this process, primitive precursor cells acquire their adult phenotypes by the expression of a particular set of genes in each cell type. Furthermore, each phase of development—cell division, migration, morphogenesis, synaptogenesis, or cell death—imposes a different functional demand on the developing cells that must be reflected in a changing pattern of gene expression in response to spatial and temporal cues. To understand fully the structure and function of the brain, therefore, it is important to study not only the phenotypic differences between neurons in the adult brain, at the end point of development, but also to analyze how and when these differences arose during the formation of the brain. While proteins that are shared with other tissues must contribute to neural function, the unique properties of the brain are likely to be derived particularly from proteins which are expressed in the brain but not in other tissues. Such molecules are most appropriate for the study of neuronal differentiation.

Current knowledge of brain proteins is extremely limited, although recombinant DNA techniques have allowed more rapid progress in the past five years. For example, several molecules originally identified at the protein chemical level have been characterized further by molecular cloning of their mRNAs: tyrosine hydroxylase (Grima *et al.*, 1985), myelin basic protein (Roach *et al.*, 1983; Zeller *et al.*, 1984), glial fibrillary acidic protein (Lewis *et al.*, 1984), neurofilament protein (Lewis and Cowan, 1985), S-100 protein (Kuwano *et al.*, 1984), neuron-specific enolase (Sakimura *et al.*, 1985), and the protein precursors for many neuropeptides (reviewed by Douglass *et al.*, 1984). Another group of brain proteins has been identified by immunological approaches (reviewed by Valentino *et al.*, 1985), using either polyclonal or monoclonal antibodies, but very few of these have been identified functionally or characterized further. The characterization of functionally defined neural molecules, such as neurotransmitter receptors, has proceeded slowly due to their low abundance and the difficulty of maintaining functional integrity during extraction and purification. Even for neuropeptides, which are currently the most intensely studied neural molecules, the list is by no means complete, and major new neuropeptide systems, such as the PYY and NPY peptides (Tatemoto and Mutt, 1980; Tatemoto, 1982) and CGRP (Rosenfeld *et al.*, 1984) are still being found. Clearly, the full repertoire of the protein library of the brain has only just begun to be uncovered.

We have approached the characterization of brain proteins with a strategy (Milner and Sutcliffe, 1983; Sutcliffe *et al.*, 1983a) based on *recombinant DNA technology* (which allows us to define the problem at the level of mRNA rather than protein), *nucleotide sequence analysis* (which provides accurate and extensive structural information), and *antibodies against synthetic peptides* (which enable us to characterize proteins biochemically and immunocytochemically). These studies have resulted in descriptions of the brain-specific protein 1B236, a potential peptide precursor (Sutcliffe *et al.*, 1983a,b; Malfroy *et al.*, 1985; Bloom *et al.*, 1985; Lenoir *et al.*, 1986), rat brain proteolipid protein (Milner *et al.*, 1985), and the identifier (ID) sequence, a repetitive genetic element that may regulate brain gene expression (Sutcliffe *et al.*, 1982, 1984a,b; Milner *et al.*, 1984; Brown *et al.*, 1986; McKinnon *et al.*, 1985). The purpose of this chapter is to summarize those studies and discuss our current knowledge of gene expression in the mammalian brain.

II. Gene Expression in the Brain

A. RNA COMPLEXITY STUDIES

The complexity of an RNA population essentially provides a measure of the information content of the RNA in a particular tissue or cell and can be extrapolated, with some assumptions, to give an estimate of the number of genes from which the RNA was transcribed. Complexity is usually expressed in nucleotides and can be considered to represent the total length of unique sequence in the RNA population. Two methods are usually used to measure the complexities of RNA populations: (1) saturation hybridization, which involves determining the fraction of unique sequence genomic DNA that can form double stranded hybrids with the RNA, and (2) kinetic analysis of the hybridization of the RNA population with its complementary DNA (cDNA) (Kaplan and Finch, 1982). While saturation hybridization allows detection of mRNAs of lower abundance, kinetic analysis can be used to measure the degree of overlap between different RNA populations and to classify a RNA population into abundance classes.

For most tissues, including brain, individual mRNA species are present at a wide range of concentrations. These are conventionally, although rather arbitrarily, classified, usually by kinetic analysis, into three abundance classes:

1. *High abundance,* consisting of a few (1–10) mRNAs present at high concentration (thousands of copies per cell). These mRNAs encode the major products of the particular cell type, for example, ovalbumin mRNA represents 50% of the total mRNA population in hen oviduct (Axel *et al.*, 1976).

2. *Middle abundance,* consisting of 500–1000 mRNAs, each present at moderate concentrations (hundreds of copies per cell). These mRNAs also encode major cell products and together with the high-abundance mRNAs usually account for most of the mass of the mRNA population while making only a small contribution to its complexity.

3. *Low abundance* (also called rare copy or complex class), consisting of a large number of different mRNA species (10,000+), each present at a very low concentration (1–20 copies per cell). Most of the complexity of an mRNA population is present in this abundance class, although together these RNAs may account for only a fraction of its mass. While mRNAs present at such low concentrations might appear intuitively to be functionally irrelevant, Galau *et al.* (1977) have calculated that several mRNAs that encode proteins known to function in the liver must be present at less than 10 copies per cell. In addition, similar low concentrations have been measured directly for the mRNAs encoding transferrin (McKnight *et al.,* 1980) and metallothionin (Hager and Palmiter, 1981).

Using both saturation hybridization and kinetic approaches, a number of studies over the past decade have demonstrated that the rodent brain expresses a large number of genes, severalfold more than any other tissue. While this statement is generally true, the actual data show quite wide variation (Table I). For example, the reported complexities of the cytoplasmic or polysomal poly(A)$^+$ RNA population (usually considered equivalent to mRNA) of rat or mouse brain vary over a 9-fold range from 26×10^6 to 230×10^6 nucleotides. The higher estimates probably reflect an improved technical ability to detect mRNAs of increasingly lower abundance over the relatively lower sensitivities of earlier studies; the largest complexity estimates come from the more recent measurements. These studies indicate that a considerable fraction of the rodent genome is expressed in the brain, for example, as much as 42% of the total nonrepeated single copy genomic sequences may be expressed as brain nuclear RNA (Bantle and Hahn, 1976). Despite their wide variation, the complexities of brain mRNA populations are consistently 2- to 3-fold higher within a single study than for similar RNA populations from nonneural tissues such as liver or kidney (Chikaraishi, 1979; Chaudhari and Hahn, 1983). Most of this excess brain RNA complexity is contained within an extremely large class of low-abundance mRNAs that have been calculated to be present on average at less than one copy per cell. Given the cellular heterogeneity of the brain, these low-abundance mRNAs are likely to be present at higher abundance in particular

TABLE 1

COMPLEXITIES OF RNA POPULATIONS FROM RODENT TISSUES[a]

Species (method)[b]	Brain Cytoplasmic mRNA Total	Brain Cytoplasmic mRNA A+	Brain Cytoplasmic mRNA A−	Brain Nuclear RNA Total	Brain Nuclear RNA A+	Brain Total A+	Liver Cytoplasmic RNA	Liver Nuclear RNA	Liver Total A+	Kidney Cytoplasmic RNA	Kidney Nuclear RNA	Reference
Mouse (s)				300				100			100	Hahn and Laird (1971)
Mouse (s)				280				58			65	Brown and Church (1972)
Mouse (k)		45					21					Young et al. (1976)
Mouse (k)		26					23			22		Hastie and Bishop (1976)
Mouse (s)	140			780	500	480						Bantle and Hanh (1976)
Mouse (k)	110											Hahn et al. (1978)
Rat (s)						650			360			Kaplan et al. (1978)
Rat (s)	132			661								Grouse et al. (1978)
Rat (s)				590				410			230	Chikaraishi et al. (1978)
Rat (s)	360	170	170				86			58		Chikaraishi (1979)
Mouse (s)	255	123	117									Van Ness et al. (1979)
Mouse (k)							58					Van Ness and Hahn (1980)
Mouse (s)							65					Van Ness and Hahn (1980)
Rat (s)		230				650						Colman et al. (1980)
Rat (s)	290											Beckmann et al. (1981)
Mouse (s)	230	120		560	400		60			65		Chaudhari and Hahn (1983)

[a]The complexities of the RNA populations are given in millions of nucleotides.
[b]Method of analysis: s, saturation hybridization; k, kinetic analysis.

subpopulations of brain cells. The complexity values (Table I) can be used to calculate the number of genes active in the rodent brain, providing estimates of between 14,000 and 128,000 genes, using 1800 nucleotides as an average mRNA length. In comparison, similar calculations indicate that 12,000–48,000 genes may be active in liver and 12,000–36,000 genes in kidney. The finding, discussed below, that brain-specific mRNAs may be considerably longer on average reduces the estimates for brain.

In several of these studies attempts were made to measure the overlap between the RNA populations of brain and other tissues. Those data suggest that a significant fraction of these RNAs are expressed exclusively or predominantly in the brain. For example, Chaudhari and Hahn (1983) have demonstrated that only 35% of the low-abundance cytoplasmic mRNA sequences expressed in mouse brain are also present in mouse kidney or liver, suggesting that 65% of the brain complex-class mRNA is expressed specifically in brain. Earlier studies indicated that only 10% (Hastie and Bishop, 1976) or as much as 45% (Young et al., 1976) of brain mRNA complexity was unique to brain.

The complexity of nuclear RNA in brain is 3- to 4-fold greater than the complexity of cytoplasmic or polysomal RNA (Table I). At least part of the excess complexity may be accounted for by the presence of large intron regions in the primary hnRNA transcripts found in nuclear or "total cell" RNA preparations (Van Ness and Hahn, 1980). It is also possible that some genes may be transcribed in brain to produce hnRNA but these transcripts are not processed further to mature cytoplasmic mRNA. The complexity of brain nuclear RNA, however, is also consistently higher than that of nonneural tissues (Table I).

Within the brain, there appears to be relatively little difference between the complexities of RNA populations prepared from different dissected regions. The complexities of total RNA or hnRNA from cerebellum, hypothalamus, hippocampus, and regions of neocortex are quite similar, both to each other and to the complexity of whole-brain hnRNA (Kaplan et al., 1978; Beckmann et al., 1981). Some differences have been observed, however, between the complexities of cytoplasmic mRNA preparations from different brain regions. For example, the complexities of anterior and parietal cortex cytoplasmic poly(A)$^+$ RNA are slightly lower than those of whole brain or occipital cortex, and the complexity of cerebellar poly(A)$^+$ RNA is consistently lower than other brain regions (Beckmann et al., 1981; Bernstein et al., 1983). In any case, the methods used to measure RNA complexity have a resolution of ±5–10% (Kaplan and Finch, 1982); the regional complexity measurements would therefore not exclude regional differences in the expression of some 2000–3000 genes encoding mRNAs with an average length of 5000 nucleotides. The complexity of rat cerebellar RNA also appears to decrease during postnatal development (Bernstein *et al.*, 1983), in contrast to the general increase observed in the complexity of brain nuclear poly(A)$^+$ RNA during postnatal development in the rat (Kaplan and Finch, 1982) and the increase observed in the complexity of the complex-class mRNAs between embryonic day 17 and birth in the mouse (Chaudhari and Hahn, 1983).

No discussion of brain RNA complexity would be complete without mention of poly(A)$^-$ RNA. With some known exceptions, such as histone mRNAs (Adesnik and Darnell, 1972), most eukaryotic mRNAs are conventionally found to have poly(A) tails at their 3′ ends. Although nonpolyadenylated mRNAs can be found in many cell types, these have also been shown to be a subpopulation of the polyadenylated mRNAs (Kaufmann *et al.*, 1977). In contrast, rodent brain appears to express a population of poly(A)$^-$ RNA that is equivalent in complexity to the poly(A)$^+$ RNA population but shares no sequences in common (Chikaraishi, 1979; Van Ness *et al.*, 1979). Most poly(A)$^-$ RNAs appear in the brain only postnatally, and the complexity of this population increases until about 45 days after birth (Chaudhari and Hahn, 1983). The function of these RNAs is currently not understood; although both cDNA and genomic clones corresponding to poly(A)$^-$ RNAs have been generated (Hahn *et al.*, 1983; Brilliant *et al.*, 1984), the possible protein coding potential of the RNAs corresponding to these clones has not been characterized.

B. A CLONAL ANALYSIS OF BRAIN GENE EXPRESSION

The complexity studies demonstrate that the mammalian brain expresses a large number of distinct mRNAs and that a large fraction of these are expressed exclusively or predominantly in the brain. These measurements provide information about the bulk properties of the RNA population. To examine the properties of individual brain mRNAs, and thus to generate a complementary and more detailed description of gene expression in the brain, we used recombinant DNA techniques to isolate copies (cDNAs) of individual brain mRNAs in a form that is easy to manipulate and characterize. Although characterizing cDNA clones of every brain mRNA is at present impractical, we examined a substantial (albeit incomplete) cross section of these mRNAs selected at random. The results of these studies (Milner and Sutcliffe, 1983) agree with and extend the conclusions of the complexity measurements and provide a starting point for investigating the properties of individual molecules that mediate brain-specific functions.

We generated a cDNA library from cytoplasmic poly(A)$^+$ prepared from the brains of adult male Sprague–Dawley rats (Milner and Sutcliffe, 1983). The library was cloned in the plasmid vector pBR322, using standard techniques, and the only selective step was to enrich the collection for cDNA inserts larger than 500 base pairs (bp) in length. Individual clones were then taken at random, radioactively labeled and used to probe for the corresponding mRNA(s) on RNA (Northern) blots that carried samples of cytoplasmic poly(A)$^+$ RNA prepared from adult rat brain, liver, and kidney. The majority of clones hybridized to one or a few discrete RNA species. These experiments provided three important pieces of information about the mRNAs corresponding to each clone tested: (1) the size of the mRNA from its mobility relative to known standards, (2) the abundance of the mRNA from the intensity of the hybridization signal, and (3) the tissue distribution of the mRNA.

The tissue distribution patterns of the mRNAs could be separated into three general classes; examples of each are shown in Fig. 1. *Class I mRNAs* are present in brain, liver, and kidney RNA populations to approximately the same extent; these may encode "housekeeping proteins" that are expressed by all cells at similar levels. The prototype Class I mRNA, corresponding to cDNA clone p1B15 (Fig. 1), has been shown to maintain its constant level of expression in a wide variety of rat tissues (Sutcliffe *et al.*, 1984a), across different regions of rat brain (Sutcliffe *et al.*, 1983a), and during embryonic and postnatal development of rat brain (Lenoir *et al.*, 1985). It has been used extensively to detect degradation in RNA preparations and to normalize their concentrations.

Class II mRNAs are also present in different tissues but with a marked variation in their abundances. These mRNAs must encode proteins that function in many tissues but are required at quite different concentrations in different cell types. For example, this category would probably include the mRNAs for the several isotypes of tubulin, which are expressed in all or most cells in the body but at quite different abundances in different tissues (Havercroft and Cleveland, 1984).

Class III mRNAs are expressed in brain but are not detectable on Northern blots in liver or kidney RNA preparations. We have operationally defined these mRNAs as brain specific; their encoded proteins are likely to mediate functions that are also brain specific. Examples of three Class III clones are shown in Fig. 1. The proteins encoded by p1B236 have been studied extensively and are described in Section

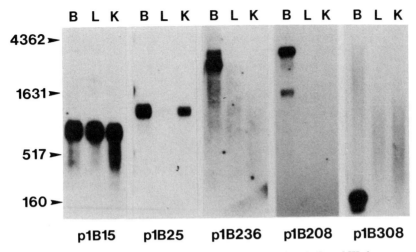

F<small>IG</small>. 1. Typical Northern blot hybridization patterns of Class I, II, and III clones to rat brain (B), liver (L), and kidney (K) cytoplasmic poly(A)⁺ RNA. p1B15, A typical Class I clone; p1B25, a Class II clone; p1B236, p1B208 and p1B308, Class III clones. The positions and lengths of DNA size standards are indicated at the left.

III,A. Clone p1B208 encodes rat brain myelin proteolipid protein and is described in Section III,B. Clone p1B308 is one of five clones from our collection that contain the ID sequence and shows the characteristic hybridization to the small brain-specific RNA BC1; the properties of BC1 and ID sequences are described in Section IV.

An additional set of clones, defined as *Class IV*, failed to detect any target on the Northern blots but did contain rat cDNA as shown by hybridization to rat genomic DNA; these presumably correspond to mRNAs present in brain at abundance levels too low (less than 0.01%) to be detected on Northern blots and may largely represent mRNAs expressed in only some brain cells. These clones correspond to the low-abundance mRNAs observed in the complexity studies discussed above.

A survey of some 191 cDNA clones by this procedure allowed us to describe the general properties of the brain mRNAs in each class (Table II). Two important conclusions emerged from the analysis of this data. First, Class III mRNAs are on average larger than Class II mRNAs, which, in turn, are on average larger than Class I mRNAs. The difference between the sizes of Class I and III mRNAs is greater than twofold, for both size and number averages. Furthermore, within Class III mRNAs, less abundant mRNAs tend to be longer than more abundant mRNAs, a phenomenon consistent with that observed for

TABLE II

CLASSIFICATION OF RAT BRAIN cDNA CLONES

Class	Number of clones	Total clones (%)	Number of mRNAs	Total RNA (%)	Size[a] (average)	Number[a] (average)
I	29	18	33	11.95	1780	1250
II	41	26	49	9.55	2350	1870
III	47	30	48	6.51	3660	2640
IV	41	26	—	—	—	—
Total	158		130	28.01	2690	1790

[a]The size average of the mRNAs was computed arithmetically; the number average is the sum of the product of the size and abundance divided by the sum of the abundance for the RNAs in each class. Taken from Milner and Sutcliffe (1983).

rare mRNAs in mouse L cells (Meyuhas and Perry, 1979). We observed that the rarest detectable brain-specific mRNAs average at least 5000 nucleotides in length. Since these mRNAs make up the bulk of brain mRNA complexity, this value is more appropriate for the calculation of gene numbers from complexity measurements than the average length of the total population, usually estimated as about 1800 nucleotides, close to our observed value for the number average size of brain mRNAs (Table II). Assuming a complexity of 1.2×10^8 nucleotides for the brain poly(A)$^+$ mRNA population (Chaudhari and Hahn, 1983) and an average length of 5000 nucleotides, then approximately 30,000 genes are expressed in the rat brain (Milner and Sutcliffe, 1983).

The second important conclusion from these studies is that a significant fraction of brain poly(A)$^+$ mRNA is expressed specifically in the brain, as defined operationally by the Northern blots; 30% of the cDNA clones tested hybridized to RNA species present in brain but not in liver or kidney. This is consistent with the degree of tissue specificity for brain mRNA determined by complexity studies (Chaudhari and Hahn, 1983).

C. "BRAIN SPECIFICITY"

It is obviously impractical to attempt characterization of all the proteins encoded by these clones. Therefore, some criteria for selection must be used. Our initial strategy was to make the simplest possible selection and elect to study brain-specific mRNAs, based on the assumption that their encoded proteins would probably mediate brain-specific functions, possibly including functions that are not yet predictable from current data. Our sampling of brain mRNAs indicated that it is rather easy to isolate clones corresponding to such mRNAs, and these provide the starting point for the characterization of their encoded proteins, using the approaches described below.

While we have operationally defined clones corresponding to mRNAs that are expressed in brain but not liver or kidney as brain specific, it is worth discussing what is meant by the terms "brain specific" or "tissue specific." For example, it is possible that liver or kidney cells express some or all of the mRNAs that we have defined as brain specific at levels too low (less than 10 copies per cell) to be detected by our Northern blot assay. What is the significance of such transcripts? On one hand, we know, as has already been discussed, that mRNAs present at 1–10 copies per cell can encode proteins that are physiologically important for those cells (Galau *et al.*, 1977). On the other hand, the mRNA for globin has been detected at low levels in many different tissues, including brain, where it is thought unlikely to have any function (Humphries *et al.*, 1976). These examples are part of a larger debate concerning the level at which control of gene expression is exerted. While it is most commonly assumed that genes are regulated by their rate of transcription, there are many other possible points of control between this initial process and the translation of the mature mRNA in the cytoplasm (Darnell, 1982). To put the arguments in their extreme forms, do cells transcribe only those genes that they need or do they transcribe a much larger pool of genes but select a subpopulation of the primary transcripts for processing and export to the cytoplasm? Evidence exists for both sides of this issue. Complexity studies, for example, show that the majority of the RNA sequences expressed in liver nuclei is also expressed in brain nuclei (Chikaraishi *et al.*, 1978) and a genomic DNA clone has been described that hybridizes to a nuclear RNA present in brain, liver, and kidney but to a polysomal RNA present only in brain (Brilliant *et al.*, 1984). On the contrary, a number of cDNA clones of "liver specific" mRNAs detect corresponding hnRNAs in liver but not in brain nuclei (Derman *et al.*, 1981). Furthermore, there are significant differences between the nuclear RNA complexities of different tissues (Chikaraishi *et al.*, 1978), indicating at least some degree of tissue-specific selectivity of transcription. In the long run the two sides of the argument may not be mutually incompatible.

From a practical point of view, however, these discussions are largely moot. It is, after all, those mRNAs that are translated in the cytoplasm that determine cellular phenotype and the existence of a large population of brain-specific cytoplasmic mRNAs has been clearly demonstrated (Chaudhari and Hahn, 1983). Furthermore, the fraction of mRNAs defined as Class III in our survey by Northern blotting is quantitatively in agreement with mRNA complexity studies. If these are not each absolutely brain specific, they are at least greatly enriched in brain over nonneuronal tissues, in most cases by more than one or two orders of magnitude. Their encoded proteins are also likely to be greatly enriched in brain and consequently of greater functional importance to brain rather than liver or kidney. Therefore, while the phrase "brain specific" may possibly be meaningless in an absolute sense, it is an extremely useful empirical concept.

The operational definition of brain-specific mRNAs, based on their expression in brain but not in liver or kidney, does not, of course, exclude the possibility that these mRNAs may be expressed in other nonneural tissues. Many "brain-specific" mRNAs may also be expressed in neurons in the peripheral nervous system and hence may be found in tissues such as gut. Other genes may be expressed in brain and certain other nonneural tissues. For example, cells of the immune system are known to express several different receptors for neuropeptides, such as β-endorphin (Hazum et al., 1979) and VIP (Ottaway and Greenberg, 1984), that have properties similar to those expressed on neurons. Conversely, proteins, such as Thy-1 (Campbell et al., 1981) and OX-2 (Clark et al., 1985), which were originally defined in the immune system, have also been found to be expressed on cells in the nervous system. This type of limited tissue distribution, particularly between the nervous and immune systems, may yield clues as to the function of the shared molecules and to functions in common between these tissues. (For a recent discussion of possible interactions between the nervous and immune systems see Goetzl, 1985.) The use of liver and kidney as nonneural tissues in our screen, apart from the ease of preparing mRNA from these tissues, probably does not exclude many of these mRNAs with interesting and potentially telling tissue distributions. However, the two Class III mRNAs that we have studied in detail, encoding the protein 1B236 and myelin proteolipid protein, are expressed only in the nervous system. Ultimately, we expect to find examples of Class III mRNAs that are shared with occasional nonneuronal tissues. An estimate of the relative frequency of such RNA species will be interesting in its own right.

III. Brain-Specific Gene Products

Several clones of brain-specific mRNAs have been selected as described above and analyzed by nucleotide sequencing. Within such sequences one can identify regions likely to encode proteins and, using the genetic code, derive the partial sequence of the putative brain-specific protein. To characterize these hypothetical proteins, we made antibodies against synthetic peptides mimicking selected regions of the deduced amino acid sequences. This section will summarize our studies on two proteins identified and characterized by this approach: the brain-specific protein 1B236 and rat-brain proteolipid protein (PLP), the major component of central nervous system mylein.

A. THE BRAIN-SPECIFIC PROTEIN 1B236

The clone p1B236 hybridizes to an mRNA of 2500 nucleotides that was present in brain with an abundance of approximately 0.01% but was undetectable in liver or kidney (Fig. 1) (Sutcliffe *et al.*, 1983a). A second mRNA, 3000 nucleotides in length, is also detectable in rat brain at lower abundance. Both mRNAs show a heterogeneous but parallel distribution in RNA preparations from different regions of rat brain, with highest concentrations in thalamus, midbrain, and pons–medulla (Sutcliffe *et al.*, 1983a). The nucleotide sequence of p1B236 provided the 318-amino acid carboxy-terminal sequence of the corresponding putative protein (1B236), which had no homologies to any previously defined sequence as determined by computer search. The original p1B236 cDNA clone lacks approximately 1000 nucleotides of the 5′ region of the 2500 nucleotide 1B236 mRNA; the relationships of the 1B236 mRNA, the cDNA clone p1B236, and the putative open-reading frame for the protein 1B236 are shown in Fig. 2A. More recently, we have generated additional, longer clones of this mRNA and determined their nucleotide sequences (Lai *et al.*, 1987a). These studies indicate that the complete open-reading frame for the 1B236 protein is 626 amino acids in length. As discussed below, this sequence contains a putative signal peptide, several possible sites for N-glycosylation and proteolytic cleavage, and a hydrophobic sequence that may form a transmembrane domain.

To detect the putative 1B236 protein and to characterize its properties, rabbit antibodies were generated against synthetic peptides P5, P6, and P7 (Fig. 2). Because the 1B236 protein has not, as far as is known, been characterized previously and we have no prior knowledge of its biochemical properties, it is essential to be very critical of any species identified by these antisera. To maximize our confidence of detecting the 1B236 protein products, we synthesized several peptides, each derived from nonoverlapping regions of the putative protein sequence. Protein species detected by antibodies against more than one of these nonoverlapping peptides are then most likely to correspond to

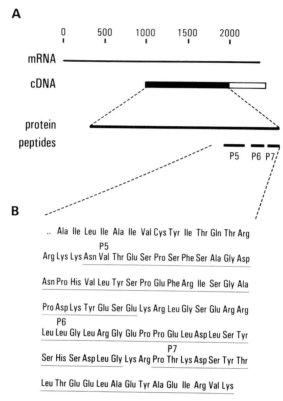

Fig. 2. Structure of the 1B236 protein. (A) The 1B236 mRNA, the p1B236 cDNA clone, and the 1B236 protein are schematically represented. The mRNA is approximately 2500 nucleotides in length; the cDNA clone p1B236 corresponds to the 3' 1500 nucleotides. The open-reading frame in the cDNA sequence is shown by the filled bar; the 3' noncoding region is shown by the open bar. Also shown are the positions of peptides P5, P6, and P7. (B) The carboxy-terminal sequence of the 1B236 protein is shown, with peptides P5, P6, and P7 underlined. Taken from Malfroy *et al.* (1985).

the putative sequence from which the peptides were derived. In addition, the presence of several pairs of basic amino acids (Arg-Arg-Lys-Lys, Lys-Arg, Arg-Arg, Lys-Arg) in the carboxy-terminal region (Fig. 2B) suggested that the 1B236 protein might be proteolytically processed to generate a number of different peptides with potential bioactivity (Sutcliffe *et al.*, 1983a). Similar sequences, particularly the dipeptide Lys-Arg, have been found previously in the sequences of neuropeptide or peptide hormone precursors and have been shown to be the sites of proteolytic processing to produce bioactive peptides (Loh *et al.*, 1984; Douglass *et al.*, 1984). We therefore chose peptides P5, P6, and P7 (Fig. 2), derived from nonoverlapping regions of sequence and the most likely cleavage products, for synthesis and antibody generation, so that these antibodies could also be used to detect potential cleavage products.

1. The 1B236 Polypeptide Exists in Brain and Has Multiple Molecular Forms

Several immunoreactive molecular forms can be detected in rat brain extracts using the antipeptide antibodies to the putative 1B236 protein in radioimmunoassay (RIA) or immunoblotting ("Western blotting"). The most abundant form is of high molecular weight (~100,000) and is detected as a diffuse band on Western blots by antibodies against each of the three peptides P5, P6, and P7, thus satisfying our criterion for detection (Fig. 3). In each case the detection of this band is blocked by preincubation of the antiserum with the appropriate synthetic peptide. The 100 kDa species is present, in varying concentrations, in all areas of the central nervous system but is not detectable in any nonneural tissue tested. Subcellular fractionation experiments indicate that this material is largely (>90%) particulate and is found in the P2 fraction of a rat brain homogenate (Malfroy et al., 1985). This form of the 1B236 protein can be solubilized, however, using detergents such as NP-40 or Triton X-100, and assayed in solution by radioimmunoassay (RIA). In the presence of the nonionic detergent NP-40, brain extracts contain large amounts of high-molecular-weight material reactive with antibodies to P5, P6, and P7 (Malfroy et al., 1985). The elution patterns of this material on a Sephadex G-75 column were very similar for each peptide RIA and are consistent with the presence of a protein species containing all three peptide sequences. The coelution of the 1B236 protein detected by RIA and by Western blotting (Fig. 4) suggests that both techniques detect the same high-molecular-weight form of the 1B236 polypeptide.

The predominant high-molecular-weight form of the 1B236 protein is glycosylated as suggested by the presence of six sites within the carboxy-terminal 318-amino acid sequence with the sequence Asn-X-Ser/Thr, the consensus sequence for asparagine-linked glycosylation (Aubert et al., 1976). Our recent determination of the full-length sequence of 1B236 reveals three further putative glycosylation sites. When brain extracts are treated with the enzyme endoglycosidase F, which removes all N-linked oligosaccharides (Elder and Alexander, 1982), and assayed for 1B236 protein by Western blotting, there is a decrease in the apparent molecular weight of the detected bands from 100,000 to 75,000, suggesting that the balance, approximately 25,000

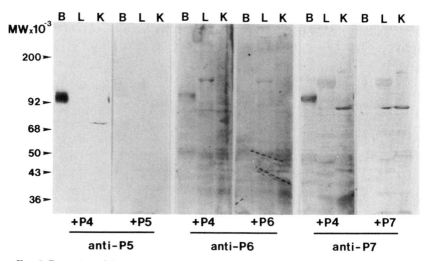

F‍ɪɢ. 3. Detection of the 1B236 protein by immunoblotting. Extracts of brain (B), liver (L), and kidney (K) were separated by electrophoresis on 8% polyacrylamide gels, electroblotted onto nitrocellulose (Towbin *et al.*, 1979), and incubated with antipeptide antibodies that were preabsorbed with synthetic peptides as indicated (P4 is a control peptide with a sequence corresponding to a more N-terminal region of 1B236). Bound antibody was visualized with goat antirabbit immunoglobulin conjugated to horseradish peroxidase and developed using diaminobenzidine and hydrogen peroxide. The positions of standard proteins and their molecular weights are indicated at the left.

Da, is due to carbohydrate (Sutcliffe *et al.*, 1983b). The apparent size of the "naked" polypeptide chain, 75,000 Da, demonstrated by this experiment is consistent with the 626-amino acid length of the complete open-reading frame determined by nucleotide sequencing of full-length cDNA copies of 1B236 mRNA (Lai *et al.*, 1987a). We have also shown that NP-40 solubilized 1B236 binds to concanavalin A or lentil lectin affinity columns and can be eluted with the sugar α-methylmannoside.

The particulate nature of the 1B236 protein and its detergent requirement for solubilization suggest that this form of the polypeptide may be membrane associated, possibly as an integral membrane protein. The sequence of 1B236 contains a highly hydrophobic region, 21 amino acids in length, which satisfies current criteria for a transmembrane domain (Eisenberg, 1984). The presence of a 14-residue sequence resembling a signal peptide at the amino terminal of the complete 1B236 sequence is consistent with its membrane location (although this would also be consistent with a possible role for the 1B236 protein as a precursor for a secreted product).

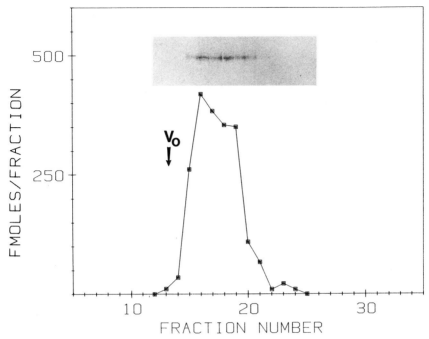

Fɪɢ. 4. Radioimmunoassay and Western blot analysis of high-molecular-weight 1B236. An extract of brain in Tris/NaCl/NP-40 was fractionated on Sephadex G-75 and fractions assayed for 1B236 by RIA for P5. In parallel, aliquots of each fraction were assayed for 1B236 by Western blotting (Towbin *et al.*, 1979) using anti-P5 antibodies; the photograph of the blot showing the 100 kDa immunoreactive band is aligned so that each gel slot corresponds to its appropriate fraction. Taken from Malfroy *et al.* (1985).

In addition, there may be a second high-molecular-weight form of the 1B236 protein that is freely soluble. Approximately 5–10% of the total 1B236 immunoreactivity in a brain homogenate is not particulate. Small, but reproducible, amounts of high-molecular-weight material reactive with RIAs for P5, P6, and P7 can be detected in brain extracts made in the absence of detergent (Malfroy *et al.*, 1985).

Fragments of the 1B236 protein can also be detected in rat brain extracts in a low-molecular-weight form of peptide size. Because peptides are often sensitive to proteolysis when extracted under neutral conditions without suitable precautions, for these studies we used extraction conditions designed to reduce or eliminate endogenous proteolysis—extraction in hot acid or rapid heating of the brain by microwave irradiation followed by extraction under neutral conditions.

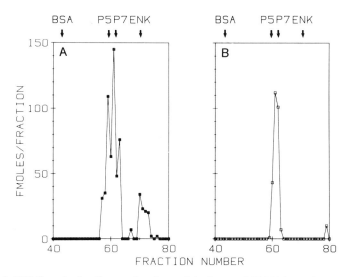

Fig. 5. HPLC analysis of low-molecular-weight forms of 1B236 in rat brain extracts. A whole rat brain was subjected to microwave irradiation immediately after dissection, extracted, and an aliquot injected on two HPLC protein analysis columns mounted in series. The eluted fractions were assayed by RIA for P5 (A) and P7 (B). The positions of standard proteins and peptides are indicated at the top. Taken from Malfroy *et al.* (1985).

When whole rat brains were extracted under either of these conditions, there was a large increase in the amount of low-molecular-weight material reactive with antibodies to P5 and P7. On an HPLC protein analysis column, this material was resolved into peaks of P5 and P7 immunoreactive, each eluting at the same position as the corresponding synthetic peptide (Fig. 5).

Thus, the 1B236 mRNA gives rise to protein molecules of two large forms as well as some smaller peptide fragments. The multiplicity of 1B236 immunoreactive forms suggest that this molecule undergoes extensive posttranslational modification, including proteolytic processing, to generate potentially bioactive peptides. Isolation of these processed peptides, determination of their exact relationship to the high-molecular-weight forms of 1B236, and the demonstration that these peptides are physiologically relevant will be necessary for further understanding of this system.

2. Immunocytochemical Analysis of 1B236

In immunocytochemical experiments, antisera against each of the three nonoverlapping peptides detected an extensive but almost identi-

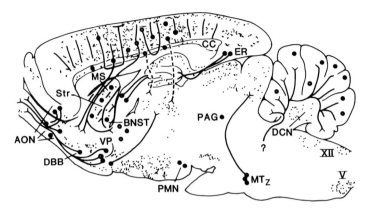

FIG. 6. Neuroanatomical distribution of 1B236; schematic overview of neuropil and perikaryonal immunoreactivity patterns. Cell body-rich areas are symbolized by circles, neuropil-rich regions by dots, presumptive fiber trajectories by solid lines, and putative pathways by dotted lines. AON, Anterior olfactory nucleus; BNST, bed nucleus of the stria terminalis; CC, corpus callosum; DBB, diagonal band of Broca; DCN, deep cerebellar nuclei; ER, entorhinal cortex; MS, medial septal nucleus; MT_z, medial trapezoid nucleus; PAG, periaqueductal grey; PMN, premamillary nucleus; Str, striatum; V, spinal trigeminal nucleus; VP, ventral pallidal area; XII, hypoglossal nucleus. Taken from Bloom et al. (1985).

cal pattern of neuronal fibers distributed throughout the adult rat CNS (Sutcliffe et al., 1983a; Bloom et al., 1985). The immunoreactivity was most pronounced in the olfactory bulb, in specific hypothalamic and preoptic nuclei, in the neostriatum, in limbic and neocortical regions, particularly in somatosensory cortex, in particular thalamic and cranial nerve nuclei, in cerebellum, and in spinal cord (Fig. 6). The reactivity was blocked in each case by preincubation of the antibodies with the appropriate peptide, again satisfying our criterion for identification of a putative protein with antipeptide antibodies. The fiber staining exhibited patterns that could be interpreted as the staining of terminal and preterminal axons, fibers of passage, and dendrites. In electron microscopy studies, at least part of the immunoreactivity in cortex was found associated with synaptic vesicles (Bloom et al., 1985).

Very few neuronal cell bodies were stained in the untreated adult rat; immunoreactive cells could be visualized, however, after pretreatment of rats with colchicine (Bloom et al., 1985), which inhibits axonal transport causing newly synthesized proteins to accumulate in cell bodies. The intensity of fiber staining was also reduced somewhat after colchicine treatment, suggesting that at least part of the 1B236 protein or peptides undergo axonal transport. Neurons with conventional morphologies were visualized in the olfactory nuclei, amygdala, and piriform cortex. A second neuronal type, termed "hairy cells," exhibited multiple dendritic-like elements and were prominent in the striatum, diencephalon, pons, and brainstem. Both cell types were found in thalamic, midbrain, and piriform structures. In general, neuronal perikarya were stained with antibodies against P5, P6, and P7. However, minor exceptions to the general pattern were occasionally observed; for example, magnocellular neurons in the paraventricular and supraoptic nuclei were stained with antisera against P7 but not P5 or P6, and cells in the medial trapezoid nucleus were stained with anti-P5 but not anti-P6 or -P7. Similarly, material apparently surrounding neurons of the deep cerebellar nuclei was detected only with antibodies against P5. In each case the reaction was specifically blocked by preincubation with the appropriate peptide. While these reactions of individual antisera could simply represent cross-reactions with antigens unrelated to 1B236, it is also possible that these staining patterns reflect differential expression of the 1B236 system at the transcriptional or posttranslational levels.

The overall immunocytochemical distribution of the cells and fibers detected by antibodies to 1B236 is unlike the distribution pattern of any other known neuronal marker, including neurotransmitters. In particular, immunoreactive cells and their processes were most pronounced in the olfactory system, in the cortical and thalamic elements of the somatosensory system, in the limbic structures, amygdala, hippocampus and basal ganglia, and to a somewhat lesser extent in the auditory and extrapyramidal motor systems. Within these systems, 1B236 is found in only a fraction of the total cells or fibers. In cases where the underlying connectivity patterns are known, the sites of 1B236 expression suggest some functional order. For example, in the olfactory system, 1B236 is absent from the cells and pathways mediating the main throughput of information from the nose to the brain—the efferent fibers from the nasal mucosa and axons and dendrites of the mitral and granule cells—but 1B236 is expressed at three sites that may modulate that flow of information: the periglomerular cells, short axons cells in the inner plexiform layer, and centrifugal afferent fibers arising from secondary and tertiary olfactory nuclei. These and similar data suggest that 1B236 is present in particular neuronal elements within functionally related central systems, and that these may have evolved from related progenitor lines.

3. Expression of 1B236 during Brain Development

The selectivity of 1B236 expression in specific neurons suggested that it would be of particular interest to determine the time course of

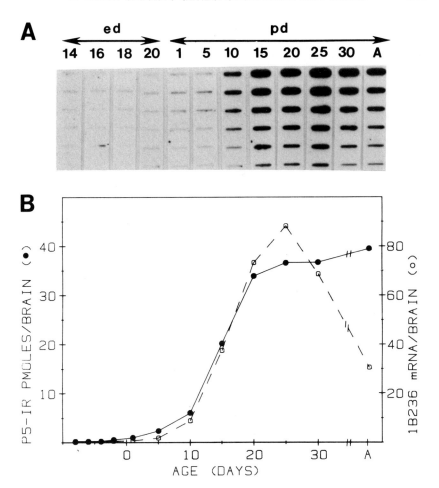

FIG. 7. Expression of 1B236 mRNA and protein during brain development. (A) "Slot-blot" analysis of 1B236 mRNA. Aliquots (2 μg) of poly(A)+ RNA, extracted from animals of the indicated embryonic (ed) and postnatal (pd) days and from adult (A) were serially diluted, applied to nitrocellulose, and hybridized with ^{32}P-labeled p1B236. The autoradiograph of the blot is shown. (B) Quantitation of 1B236 mRNA and protein. The autoradiographic intensities of the slot-blot shown above were determined by densitometric scanning and are expressed in arbitrary units of autoradiographic intensity. The data shown represent the values for one set of RNA dilutions; similar results were obtained for each set of dilutions. The amounts of 1B236 immunoreactive material in NP-40 extracts of brains of different ages were determined by RIA for 1B236 peptide P5.

1B236 gene expression during rat brain development. Our approach to the study of brain-specific molecules such as 1B236 enables us to measure 1B236 mRNAs by hybridization with clone p1B236, to measure molecular forms of the 1B236 protein using the antipeptide antibodies, and to obtain precise localization of the protein by immunocytochemistry.

To determine the time course of expression of 1B236 during brain development, we studied brain protein and mRNA extracts from whole brains and from dissected brain regions of rats between the fourteenth embryonic day (ED) to 30 days postnatal (PD) and adult (Lenoir et al., 1986). 1B236 mRNA is first detectable at 5 days after birth (PD5) and increases to a maximum concentration at PD25 (Fig. 7). In extracts of dissected brain regions, 1B236 mRNA is first detectable at PD5 in hindbrain and cerebellum, at PD9 in midbrain, but not until PD13 in telencephalon. The 2500- and 3200-nucleotide 1B236 mRNAs are both expressed with the same time course, but there appears to be a slight shift in their ratio during development. The appearance of the predominant membrane-associated 1B236 protein follows a very similar time course to that of its mRNA in both whole brain and dissected brain regions, suggesting that the expression of the protein is regulated largely by transcription of its mRNA. The appearance of detectable P5-like peptide also follows a similar time course (Malfroy, Lenoir, and Milner, unpublished observations).

The pattern of 1B236 expression was confirmed by immunocytochemical localization of 1B236 protein; immunoreactive material is detected first in spinal cord at PD3–PD5 and then appears in progressively more rostral brain regions in increasingly older animals, occurring last in cerebral cortex (Lenoir et al., 1986). Structures in several brain regions. however, that do not contain 1B236 immunoreactivity in the adult, such as optic nerve and somatic efferent cranial nerve nuclei, do show transient expression of 1B236 during postnatal development. The postnatal expression of 1B236 indicates that this protein is a marker for the terminal differentiation of particular neurons and suggests that the 1B236 protein probably mediates functions specific to the adult nervous system.

4. Structure of the 1B236 Gene

The rat genome contains a single copy of a gene encoding 1B236; in the mouse the corresponding single gene is located on chromosome 7, very close to the qv locus (C. Blatt, unpublished observations). The rat gene is greater than 15 kb in length and includes 12 introns (Lai et al., 1987b). One exon, 45 nucleotides in length, is not present in the original and some additional 1B236 cDNA clones but is found in other cDNA clones of 1B236 mRNAs, suggesting that 1B236 transcripts undergo differential exon splicing to generate at least two different major mRNA forms. The mRNAs that include this 45-nucleotide exon contain the coding capacity for a protein with an alternate carboxy terminus (Lai *et al.*, 1987b).

5. What is the Function of 1B236?

Until the development of nucleic acid-based techniques, the classic goal of the protein chemist was to determine the complete amino acid sequence of the protein under study, often requiring heroic labors and mountains of material. These sequences were used to interpret existing functional information, for example, to map the active sites of enzymes. In contrast, with the widespread application of molecular cloning and nucleotide sequencing techniques, it is now possible to obtain the complete amino acid sequence of a protein without knowing any other properties of the molecule. This is the case for the protein products of many viruses, whose genomic sequences are known, and is also the case for the no longer putative protein 1B236, whose amino acid sequence is unrelated to any other known amino acid sequence. With antibodies against synthetic peptide fragments of the 1B236 protein sequence, however, we have been able to establish a number of the properties of the 1B236 protein—its molecular forms *in vivo*, their relative distribution in different brain regions and nonneural tissues, its subcellular location, and its expression during brain development—and determine its detailed neuroanatomical distribution. This information has given clues to the function of 1B236 and suggested experiments to test these hypotheses.

One such clue, for example, was offered by the presence of several pairs of basic amino acids in the carboxy-terminal region of the 1B236 sequence and led to the hypothesis that this protein might represent a precursor for a novel family of bioactive peptides. We have demonstrated that peptides can be detected in extracts of rat brain with properties similar to those expected. The immunocytochemical localization of 1B236 in discrete neuronal populations, its association with synaptic vesicles, and the observations that colchicine treatment causes an increase in detectable 1B236 immunoreactivity in cell bodies and a decrease in terminal regions are all consistent with the view that 1B236 may represent a neurotransmitter precursor. In addition, synthetic P5 and P6 peptides have been shown to influence spontaneous or elicited activity of neurons in hippocampus and neocortex (S. J. Henriksen, personal communication), two sites where 1B236 is localized, and synthetic P5 produces changes in spontaneous locomotor activity (G. F. Koob, personal communication). On the contrary, there is little precedent for a membrane glycoprotein being a peptide precursor. However, the soluble high-molecular-weight form may be the precursor of the P5- and P7-like peptides, and the membrane-bound form may then have other functions. It is quite clear that the 1B236 story is by no means complete.

The studies of 1B236 have provided a paradigm for the characterization of other brain-specific proteins. This discussion illustrates the fact that, by integrating structural, biochemical, and neuroanatomical data, it is possible to make reasonable and experimentally testable hypotheses concerning the function of the novel brain-specific proteins uncovered by this approach. Many of these hypotheses are rapidly testable with the reagents that are a product of the approach.

B. RAT BRAIN MYELIN PROTEOLIPID PROTEIN

Our molecular approach to the study of the brain is intended primarily to detect novel brain-specific proteins that may reveal new functions or concepts of brain organization or extend existing ones. The mRNAs identified by this strategy, however, should include those encoding known brain-specific proteins. Although the class of known brain-specific proteins is rather limited, there are clearly cellular structures in the nervous system that are unique to neural tissues and are, therefore, likely to be built at least in part from brain-specific components. The most predominant such structure is myelin, accounting for 20–25% of the dry weight of rat brain (Norton and Cammer, 1984). The protein fraction of brain myelin (some 30% of the total dry weight) is composed largely of only two proteins: proteolipid protein (PLP) and myelin basic protein (MBP), each of which must constitute a significant fraction (several percent) of total brain protein (Norton and Cammer, 1984; Lees and Brostoff, 1984). It was therefore not surprising that the cDNA clone (p1B208) that hybridized to the two most abundant brain-specific mRNAs seen in our survey was later found to encode rat brain PLP (Milner et al., 1985).

The clone p1B208 hybridized to two highly abundant, brain-specific mRNAs 3200 and 1600 nucleotides in length (Milner and Sutcliffe, 1983). Both mRNAs increase in abundance in an approximate gradient from forebrain to hindbrain and are also expressed in C6 glioma cells but not in neuroblastoma or PC12 (pheochromocytoma) cells. Using p1B208 as a probe, we isolated additional clones from a cDNA library favoring full-length cDNA inserts and determined the complete nu-

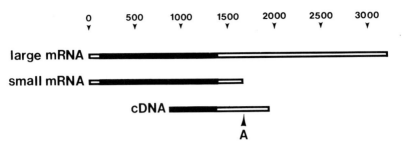

Fig. 8. Relationship between the large and small PLP rat mRNAs and the clone p1B208. Identical sequences in the three molecules are aligned vertically. Filled bars indicate protein coding regions; open bars indicate noncoding regions. The position of the putative poly(A) addition site in the sequence of p1B208 is indicated by the arrow.

cleotide sequences of both mRNAs (Milner *et al.*, 1985). The sequences of the two mRNAs begin with the same 5′ nucleotide and are identical for the entire length of the shorter mRNA. The longer mRNA then continues for another 1600 nucleotides. The region of identity between the two mRNAs contains the only open-reading frame of significant length in either sequence. The additional sequences in the longer mRNA therefore contribute solely to an extremely long 3′ noncoding region of 2066 nucleotides. The relationship between the sequences of the large and small mRNAs and the clone p1B208 are shown in Fig. 8. Because the rat appears to have only a single PLP gene, it is most likely that the two mRNA forms are generated by addition of the poly(A) tail at one of two possible polyadenylation sites in the same primary RNA transcript. All other mammalian species tested express only a 3200 nucleotide PLP mRNA, suggesting that the two mRNAs in the rat do not have profoundly different biological roles.

Both mRNAs encode identical proteins, 277 amino acids in length (Fig. 9), that can be unambiguously identified as myelin proteolipid protein (PLP), also known as lipophilin, by homology with the known sequences of rat and bovine PLP (Jolles *et al.*, 1983; Laursen *et al.*, 1984; Stoffel *et al.*, 1984). There are at least two, and possibly up to seven, amino acid differences in the complete sequences of rat and bovine PLPs (97–99% homology), indicating that the sequence of proteolipid protein has been highly conserved during evolution. The amino acid sequence translated from the nucleotide sequence is identical in length (with the exception of the amino-terminal methionine, which is presumably removed after synthesis of the polypeptide) with the mature protein, as defined by protein sequencing studies, indicating that PLP does not require a signal peptide sequence for insertion into

20
[Met]GlyLeuLeuGluCysCysAlaArgCysLeuValGlyAlaProPheAlaSerLeuValAlaThrGlyLeuCys

40
PhePheGlyValAlaLeuPheCysGlyCysGlyHisGluAlaLeuThrGlyThrGluLysLeuIleGluThrTyrPhe

60
SerLysAsnTyrGlnAspTyrGluTyrLeuIleAsnValIleHisAlaPheGlnTyrValIleTyrGlyThrAlaSer

80   100
PhePhePheLeuTyrGlyAlaLeuLeuLeuAlaGluGlyPheTyrThrThrGlySerValArgGlnIlePheGlyAsp
                                   Tyr

120
TyrLysThrThrIleCysGlyLysGlyLeuSerAlaThrValThrGlyGlyGlnLysGlyArgGlySerArgGlyGln

140
HisGlnAlaHisSerLeuGluArgValCysHisCysLeuGlyLysTrpLeuGlyHisProAspLysPheValGlyIle

160   180
ThrTyrAlaLeuThrValValTrpLeuLeuValPheAlaCysSerAlaValProValTyrIleTyrPheAsnThrTrp

200
ThrThrCysGlnSerIleAlaPheProSerLysThrSerAlaSerIleGlySerLeuCysAlaAspAlaArgMetTyr
              Ala   Thr

220
GlyValLeuProTrpAsnAlaPheProGlyLysValCysGlySerAsnLeuLeuSerIleCysLysThrAlaGluPhe

240
GlnMetThrPheHisLeuPheIleAlaAlaAlaPheValGlyAlaAlaAlaThrLeuValSerLeuLeuThrPheMetIle
                                                                   Val

260
AlaAlaThrTyrAsnPheAlaValLeuLysLeuMetGlyArgGlyThrLysPhe

Fig. 9. The amino acid sequences of rat and bovine brain proteolipid proteins. The sequence of the rat protein derived from nucleotide sequence analysis (Milner *et al.*, 1985) is shown. Residues different from the published complete sequences of bovine proteolipid protein (Laursen *et al.*, 1984; Stoffel *et al.*, 1984) are shown above the line. Residues are numbered from the amino-terminal Gly of the mature protein.

the myelin membrane. The developmental expression of the mRNAs encoding PLP follows a postnatal time course consistent with the known time course of myelinogenesis and of PLP expression in mice (Campagnoni and Hunkeler, 1980).

It is of considerable interest, and relevant to the topic of this review, that PLP is a component of myelin that is specific to the central nervous system. Despite superficial similarities in their structure and function, the protein compositions of central and peripheral myelin are quite different (Lees and Brostoff, 1984). Thus, each type of myelin contains a predominant and unique component—PLP in CNS myelin and the protein P_0 in PNS myelin—and it might be expected that these two proteins would share many features. But the recent characterization of P_0 by molecular cloning (Lemke and Axel, 1985) revealed that the sequences of the two proteins were unrelated and that their

TABLE III

COMPARISON OF THE PROPERTIES OF THE MYELIN PROTEINS
PLP AND P_0[a]

Property	PLP	P_0
Signal peptide	No	Yes
Glycosylation	No	Yes
Size of mature protein (amino acids)	276	219
Number of membrane crossing domains	3–4	1
Acylation	Yes	No
Charged carboxy terminus	No	Yes

[a]Data taken from Lees and Brostoff (1984); Lemke and Axel (1985); Milner *et al.* (1985).

other properties were different in almost every respect (Table III). Both central and peripheral myelin contain MBP, but its concentration is much higher in central myelin. Furthermore, the mouse mutant shiverer, which carries a deletion in the single gene for MBP (Roach *et al.*, 1983) and lacks any normal MBP, has highly abnormal central myelin but apparently normal peripheral myelin, indicating a more critical role for MBP in CNS myelin. It is true that the two forms of myelin are generated by different cell types—Schwann cells in the PNS and oligodendrocytes in the CNS—with different embryological origins. Nevertheless, these observations clearly indicate that quite similar neural structures in different parts of the nervous system may be the products of overlapping sets of genes displaying cell-specific expression.

IV. Control of Brain-Specific Gene Expression

Our survey of the expression of brain mRNAs in neural and non-neural tissues indicates that most of the genes expressed in the brain are regulated in either a relative sense (Class II genes) or an absolute, on/off sense (Class III genes). Fewer than 20% of brain mRNAs (Class I) surveyed showed no difference in abundance between brain, liver, and kidney (Milner and Sutcliffe, 1983). Most brain genes may therefore be regulated at the transcriptional level, and we may expect to find mechanisms of gene control that might be common to sets of brain- or neuronal-specific genes.

Clues to one possible regulatory mechanism have come from our survey of brain mRNAs. Of the 191 cDNA clones examined in our original survey, five hybridized to a small (160 nucleotide), abundant, brain-specific RNA (Sutcliffe *et al.*, 1982). In each case, the cDNA insert contained in these clones was considerably longer than the small RNA targets, suggesting that these clones were not simple copies of the small RNA. The nucleotide sequences of four of the clones showed that they shared a common 82-nucleotide sequence which was located at the 3' end of each clone and was responsible for hybridization to the small RNA target (Sutcliffe *et al.*, 1982; Milner *et al.*, 1984). The remainder of the sequences were not homologous but had some features in common, long purine and pyrimidine tracts and short repeated sequences, often encountered in intron or noncoding regions of genes. For three of the clones we have shown that these sequences correspond to single copy genomic sequences (Milner *et al.*, 1984). In contrast, the common sequence is repeated some 100,000–150,000 times in the rat genome (Milner *et al.*, 1984), and individual copies have been found in several defined genomic sequences (for complete list see Sutcliffe *et al.*, 1984b), including an intron of the rat growth hormone gene (Barta *et al.*, 1981), and in two introns of the brain-specific gene 1B236 (Milner *et al.*, 1984). We termed the common 82-nucleotide sequence an "ID" or identifier sequence because it hybridized to a small brain-specific RNA (Sutcliffe *et al.*, 1982). Based on our initial observations, we suggested that this sequence might be preferentially located in the introns of brain genes and prescribe their tissue-specific expression (Sutcliffe *et al.*, 1982).

We have subsequently shown that the small brain-specific RNA is expressed *in vivo* only by central and peripheral nervous tissue and by pituitary (Sutcliffe *et al.*, 1984a). It is present largely in brain cytoplasmic RNA and contains a poly(A) tail; we have named this species BC1 (for brain cytoplasm). A second, less abundant, brain-specific RNA, 100–110 nucleotides in length, which also hybridizes to ID, was termed BC2 (Sutcliffe *et al.*, 1984a). This RNA has a shorter poly(A) tail than BC1. The consensus ID sequence (Fig. 10), compiled from all known sequences, contains the split promotor sequences for the enzyme RNA polymerase III (pol III) which is responsible, in eukaryotes, for the transcription of small RNAs such as tRNA and 5 S RNA. The cDNA clones containing ID sequences can function *in vitro* as very efficient pol III templates, producing transcripts beginning with the 5' terminal nucleotide of the consensus ID sequence (Sutcliffe *et al.*, 1984a). In addition, isolated brain nuclei *in vitro* can generate RNA transcripts, 100–110 nucleotides in length, that contain the ID sequence and are likely to be generated by pol III (Sutcliffe *et al.*, 1984a). These observations suggest that genomic ID sequences are specifically transcribed in brain by RNA polymerase III to generate BC2 RNA.

Consensus ID   GGGGCTGGGGATTTAGCTCAGTGGTAGAGCGCTTACCTAGCAAGCACAAGGCCCTGGGTTCGGTCCCCAGCTCCGAAAAAAAAAAAAAAAAAAAA
Pol III   TGGCNNAGTGG   GGTTCGANNCC

FIG. 10. Consensus nucleotide sequence for the ID element. Where alternative bases are present in a significant fraction of sequences both possibilities are indicated. The split promotor sequences for RNA polymerase III are shown below.

The ID-containing RNAs that were copied into the original cDNA clones, however, were probably transcribed by RNA polymerase II (pol II), the enzyme that generates eukaryotic mRNA, rather than by RNA polymerase III. Brain nuclei *in vitro* generate large heterogeneous RNAs containing ID sequences in addition to the BC2-like product. The large RNAs are not produced in the presence of 0.5 μg/ml α-amanitin, suggesting that these are transcribed by RNA polymerase II (Sutcliffe *et al.*, 1984a). In contrast to brain, liver and kidney nuclei produce much lower amounts of large RNA transcripts containing ID sequences in this assay, although the total level of pol II transcription is similar for nuclei from all three tissues. This is consistent with the finding that hnRNA isolated from cortical neuronal nuclei is 5-fold enriched in ID sequences compared to hnRNA from glial or kidney nuclei (Brown *et al.*, 1986). Observations by other workers (Owens *et al.*, 1985) suggesting that there is little difference between the concentrations of ID sequences in hnRNA preparations from rodent brain, liver, or kidney may be explained by the dilution of ID sequences in cortical neuronal hnRNA by the contribution of glial hnRNA in the whole-brain RNA preparation. Despite their predominance in brain hnRNA, ID sequences are virtually absent from cytoplasmic RNA, with the exception of BC1 and BC2 (Sutcliffe *et al.*, 1982; Milner *et al.*, 1984; Owens *et al.*, 1985). Taken together these results indicate that ID sequences are contained largely in introns and are preferentially transcribed in cortical neurons by both pol II and pol III.

The transcription of ID sequences by pol II and pol III are probably regulated by different mechanisms. In the *in vitro* transcription experiments, heterogenous RNA containing ID sequences was transcribed from deproteinized liver DNA or liver nuclear extracts that had been pretreated with 0.35 *M* NaCl, a condition known to cause the removal and rearrangement of some chromosomal proteins. In neither case was there any transcription of BC2-like RNAs. These results suggest that transcription of ID sequences by pol II may be regulated by chromatin conformation or nucleosome arrangement while pol III transcription of ID may require a brain-specific transcription factor (Sutcliffe *et al.*, 1984a).

These regulatory mechanisms appear to operate during development of the rat. The small RNAs BC1 and BC2 begin to accumulate in brain cytoplasm around the time of birth and increase in concentration during the early postnatal period (Sutcliffe *et al.*, 1984b). Similarly, neuronal hnRNA transcripts show a dramatic increase in their content of ID sequences during the second week of postnatal development (Brown *et al.*, 1986). Other data from our laboratories suggest that a number of brain-specific genes, such as 1B236, may be expressed during the same postnatal time period (Lenoir *et al.*, 1986). In addition, in the first two weeks after birth, the nucleosome repeat length of neuronal chromatin decreases from 200 to 165 nucleotides, a much shorter length than is characteristic for glia and most nonneural cells (Brown, 1982; Greenwood and Brown, 1982; Kuenzle *et al.*, 1983).

Sutcliffe and colleagues (1984b) have proposed a model for the regulation of brain genes during development based on these observations. Genomic ID sequences, located in introns or adjacent to genes that are expressed during postnatal brain development, are first transcribed by pol III, under the control of a tissue-specific, trans-acting transcription factor. This activation opens up the region of chromatin surrounding each ID sequence, making adjacent promotor sequences available for interaction with pol II transcription factors and allowing transcription of the gene by pol II. Pol III transcription of ID sequences is necessary but not sufficient for subsequent pol II transcription; other, finer controls must also exist because genes such as 1B236, which are ideal candidates for control by ID sequences, are only expressed in a small fraction of all neurons in the brain. We do not know the exact mechanism of ID-dependent gene activation but recent results demonstrating that ID sequences have enhancer-like properties indicate one possible mechanism (Sutcliffe *et al.*, 1984b; McKinnon *et al.*, 1986).

Models of gene control based on moderately repetitive sequences are very attractive and have accumulated a considerable theoretical background (Britten and Davidson, 1969). Until recently, however, such models have not been experimentally accessible. We offer our models as one example of a lineage-based system for the control of tissue-specific genes during development and hope that, with its many experimental predictions, it will both stimulate research and be extended to other systems.

V. Summary and Future Prospects

The approach that we have outlined here is directed towards uncovering novel brain-specific proteins whose biochemical and neuroanatomical properties may extend existing hypotheses of neural function or reveal new ones. The direction of experiments in this approach proceeds from structure to function, in contrast to more traditional neurobiological approaches that seek the molecules that mediate defined physiological functions, as exemplified by the search for the hypothalamic releasing factors (Guillemin, 1978). While the absence of an immediate functional "handle" for the proteins we study might be seen by some as a disadvantage, the two approaches are not that far apart. In many cases, molecules that were characterized on the basis of a particular assay have often shown additional properties, suggesting unanticipated functions; for example, several of the hypothalamic releasing factors may function as neurotransmitters in regions of the brain unconnected with the pituitary (Bloom, 1985). New molecules have also been discovered as a result of studies on existing systems; thus, the putative neuropeptide CGRP was revealed in studies of the well-defined hormone calcitonin (Rosenfeld *et al.*, 1984). Establishing the function of a potential neurotransmitter requires the same set of physiological tests whatever its means of discovery (Bloom, 1985).

We hope to approach the function of novel brain-specific molecules by establishing their biochemical, cytochemical, and neuroanatomical properties in the manner that we have described for 1B236. In the studies of 1B236, we made great use of the amino acid sequence determined by translation from the nucleotide sequence of its cloned mRNA. Some features of this sequence suggested potential sites for posttranslational modifications such as proteolytic processing or glycosylation. In addition, there are a number of computerized methods for the prediction of secondary structure from primary sequences. Finer-Moore and Stroud (1984), for example, have used Fourier analysis of repeating sequence motifs to predict a fifth membrane-associated domain in the acetylcholine receptor subunits. This structure has been confirmed by experiments using antipeptide antibodies (Young *et al.*, 1985). As more protein sequences are collected and more protein crystal structures are completed, we expect that these predictive methods will become increasingly more reliable.

Even if it is not possible in some cases to proceed easily to function, the approaches described here will uncover much new information for a cross section of genes that are expressed specifically or predominantly in the nervous system. Thus, the sequences and structures of their mRNAs, and the sequences and physical properties of the encoded proteins, will be established; the anatomical locations of cells that express each protein and the subcellular locations of the proteins within those cells will be mapped; and the structure and chromosomal location of the genes themselves will be determined. In future studies it will be possible to focus more precisely on particular brain regions or cell types as more sophisticated techniques are developed, particularly subtractive cloning methods (Hedrick *et al.*, 1984), which enable one to select cDNA clones of mRNAs that are expressed in one RNA population but not another. In this way we may gain a finer resolution of cell types in the nervous system and their physiological functions, obtain insights into the mechanisms of gene regulation in the nervous system, and provide the tools for analyzing human neuropathologies.

ACKNOWLEDGMENTS

We thank our many colleagues whose work we have described here: Charles Bakhit, Elena Battenberg, Mary Ann Brow, Ian Brown, Charles Chavkin, Patria Danielson, Joel Gottesfeld, Steven Henriksen, George Koob, Mary Kiel, Cary Lai, Dominique Lenoir, Richard Lerner, Bernard Malfroy, Randy McKinnon, Klaus Nave, Wanda Reynolds, Tom Shinnick, and Ann-Ping Tsou. Our studies are supported by NIH Grants NS 20728, NS 21815, and GM 32355, NIAAA Alcohol Research Center Grant AA 06420, and McNeil Laboratories. This is publication BCR-4118 from the Research Institute of Scripps Clinic.

REFERENCES

Adesnik, M., and Darnell, J. E. (1972). *J. Mol. Biol.* **67**, 397–406.
Aubert, J.-P., Biserte, G., and Loucheux-Lefebvre, M.-H. (1976). *Arch. Biochem. Biophys.* **175**, 410–418.
Axel, R., Fiegelson, P., and Shutz, G. (1976). *Cell* **7**, 247–254.
Bantle, J. A., and Hahn, W. E. (1976). *Cell* **8**, 139–150.
Barta, A., Richards, R. I., Baxter, J. D., and Shine, J. (1981). *Proc. Natl. Acad. Sci. U.S.A.* **79**, 4942–4946.
Beckmann, S. L., Chikaraishi, D. M., Deeb, S. S., and Sueoka, N. (1981). *Biochemistry* **20**, 2684–2692.
Bernstein, S. L., Gioio, A. E., and Kaplan, B. B. (1983). *J. Neurogenet.* **1**, 71–86.
Bloom, F. E. (1985). *In* "The Pharmacological Basis of Therapeutics" (A. Goodman Gilman, L. S. Goodman, T. W. Rall, and F. Murad, eds.), 7th Ed., pp. 236–259. Macmillan, New York.
Bloom, F. E., Battenberg, E. L. F., Milner, R. J., and Sutcliffe, J. G. (1985). *J. Neurosci.* **5**, 1781–1802.
Brilliant, M. H., Sueoka, N., and Chikaraishi, D. M. (1984). *Mol. Cell. Biol.* **4**, 2187–2197.
Britten, R. J., and Davidson, E. H. (1969). *Science* **165**, 349–358.
Brown, I. R. (1982). *Biochim. Biophys. Acta* **698**, 307–309.
Brown, I. R., and Church, R. B. (1972). *Dev. Biol.* **29**, 73–84.
Brown, I. R., Danielson, P., Rush, S., and Sutcliffe, J. G. (1986). Submitted.
Campagnoni, A. T., and Hunkeler, M. J. (1980). *J. Neurobiol.* **11**, 355–365.
Campbell, D. G., Gagnon, J., Reid, K. B. M., and Williams, A. F. (1981). *Biochem. J.* **195**, 15–30.
Chaudhari, N., and Hahn, W. E. (1983). *Science* **220**, 924–928.
Chikaraishi, D. M. (1979). *Biochemistry* **18**, 3249–3256.
Chikaraishi, D. M., Deeb, S. S., and Sueoka, N. (1978). *Cell* **13**, 111–120.

Clark, M. J., Gagnon, J., Williams, A. F., and Barclay, A. N. (1985). *EMBO J.* **4,** 113–118.
Colman, P. D., Kaplan, B. B., Osterburg, H. H., and Finch, C. E. (1980). *J. Neurochem.* **34,** 335–345.
Darnell, J. E. (1982). *Nature (London)* **297,** 365–371.
Derman, E., Krauter, K., Walling, L., Weinberger, C., Ray, M., and Darnell, J. E. (1981). *Cell* **23,** 731–739.
Douglass, J., Civelli, O., and Herbert, E. (1984). *Annu. Rev. Biochem.* **53,** 665–715.
Eisenberg, D. (1984). *Annu. Rev. Biochem.* **53,** 595–623.
Elder, J. H., and Alexander, S. (1982). *Proc. Natl. Acad. Sci. U.S.A.* **79,** 4540–4544.
Finer-Moore, J., and Stroud, R. M. (1984). *Proc. Natl. Acad. Sci. U.S.A.* **81,** 155–159.
Galau, G. A., Klein, W. H., Britten, R. J., and Davidson, E. H. (1977). *Arch. Biochem. Biophys.* **179,** 584–599.
Goetzl, E. J. (1985). *J. Immunol.* **135,** 739s–865s.
Greenwood, P. D., and Brown, I. R. (1982). *Neurochem. Res.* **7,** 965–976.
Grima, B., Lamouroux, A., Blanot, F., Biguet, N. F., and Mallet, J. (1985). *Proc. Natl. Acad. Sci. U.S.A.* **82,** 617–621.
Grouse, L. D., Schrier, B. K., Bennett, E. L., Rosenzweig, M. R., and Nelson, P. G. (1978). *J. Neurochem.* **30,** 191–203.
Guillemin, R. (1978). *Science* **202,** 390–402.
Hager, L. J., and Palmiter, R. D. (1981). *Nature (London)* **291,** 340–342.
Hahn, W. E., and Laird, C. D. (1971). *Science* **173,** 158–161.
Hahn, W. E., Van Ness, J., and Maxwell, I. H. (1978). *Proc. Natl. Acad. Sci. U.S.A.* **75,** 5544–5547.
Hahn, W. E., Chaudhari, N., Beck, L., Wilber, K., and Peffley, D. (1983). *Cold Spring Harbor Symp. Quant. Biol.* **48,** 465–475.
Hastie, N. D., and Bishop, J. O. (1976). *Cell* **9,** 761–774.
Havercroft, J. C., and Cleveland, D. W. (1984). *J. Cell Biol.* **99,** 1927–1935.
Hazum, E., Chang, K.-J.. and Cuatrecasas, P. (1979). *Science* **205,** 1033–1035.
Hedrick, S. M., Cohen, D. L., Nielsen, E. A., and Davis, M. M. (1984). *Nature (London)* **308,** 149–153.
Humphries, S., Windass, J., and Williamson, R. (1976). *Cell* **7,** 267–277.
Jolles, J., Nussbaum, J. L., and Jolles, J. (1983). *Biochim. Biophys. Acta* **742,** 33–38.
Kaplan. B. B., and Finch, C. E. (1982). *In* "Molecular Approaches to Neurobiology" (I. R. Brown, ed.), pp. 71–98. Academic Press, New York.
Kaplan, B. B., Schacter, B. S., Osterburg, H. H., de Vellis, J. S., and Finch, C. E. (1978). *Biochemistry* **17,** 5516–5524.
Kaufmann, Y., Milcarek, C., Berissi, H., and Penman, S. (1977). *Proc. Natl. Acad. Sci. U.S.A.* **74,** 4801–4805.
Kuenzle, C. C., Heizmann, C. W., Hubscher, U., Hobi, R., Winkler, G. C., Jaeger, A. W., and Morgenegg, G. (1983). *Cold Spring Harbor Symp. Quant. Biol.* **48,** 493–499.
Kuwano, R., Usui, H., Maeda, T., Fukui, T., Yamanari, N., Ohtsuka, E., Ikehara, M., and Takahashi, Y. (1984). *Nucleic Acids Res.* **12,** 7455–7465.
Lai, C., Brow, M., Watson, J. B., Noronha, A. B., Quarles, R., Bloom, F. E., Milner, R. J., and Sutcliffe, J. G. (1987a). Submitted.
Lai, C., Nave, K.-A., Brow, M., Sutcliffe, J. G., and Milner, R. J. (1987b). In preparation.
Laursen, R. A., Samiullah, M., and Lees, M. B. (1984). *Proc. Natl. Acad. Sci. U.S.A.* **81,** 2912–2916.
Lees, M. B., and Brostoff, S. W. (1984). *In* "Myelin" (P. Morell, ed.), pp. 197–224. Plenum, New York.
Lemke, G., and Axel, R. (1985). *Cell* **40,** 501–508.
Lenoir, D., Battenberg, E. L. F., Kiel, M., Bloom, F. E., and Milner, R. J. (1986). *J. Neurosci.* **6,** 522–530.

Lewis, S. A., and Cowan, N. J. (1985). *J. Cell Biol.* **100**, 843–850.

Lewis, S. A., Balcarek, J. M., Krek, V., Shelanski, M., and Cowan, N. J. (1984). *Proc. Natl. Acad. Sci. U.S.A.* **81**, 2743–2746.

Loh, Y. P., Brownstein, M. J., and Gainer, H. (1984). *Annu. Rev. Neurosci.* **7**, 189–222.

McKinnon, R. D., Shinnick, T. M., and Sutcliffe, J. G. (1985). *Proc. Natl. Acad. Sci. U.S.A.* **83**, 3751–3755.

McKnight. G. S., Lee, D. C., and Palmiter, R. D. (1980). *J. Biol. Chem.* **258**, 14632–14637.

Malfroy, B., Bakhit, C., Bloom, F. E., Sutcliffe, J. G., and Milner, R. J. (1985). *Proc. Natl. Acad. Sci. U.S.A.* **82**, 2009–2013.

Maxwell, I. H., Maxwell, F., and Hahn, W. E. (1980). *Nucleic Acids Res.* **8**, 5875–5894.

Meyuhas, O., and Perry, R. P. (1979). *Cell* **16**, 139–148.

Milner, R. J., and Sutcliffe, J. G. (1983). *Nucleic Acids Res.* **11**, 5497–5520.

Milner, R. J., Bloom, F. E., Lai, C., Lerner, R. A., and Sutcliffe, J. G. (1984). *Proc. Natl. Acad. Sci. U.S.A.* **81**, 713–717.

Milner, R. J., Lai, C., Nave, K.-A., Lenoir, D., Ogata, J., and Sutcliffe, J. G. (1985). *Cell* **42**, 931–939.

Norton, W. T., and Cammer, W. (1984). *In* "Myelin" (P. Morell, ed.), pp. 147–195 Plenum, New York.

Ottaway, C. A., and Greenberg, G. R. (1984). *J. Immunol.* **132**, 417–423.

Owens, G., Chaudhari, N., and Hahn, W. E. (1985). *Science* **229**. 1263–1265.

Roach, A., Boylan, K., Horvath, S., Prusiner, S., and Hood, L. E. (1983). *Cell* **34**, 799–806.

Rosenfeld, M. G., Amara, S. G., and Evans, R. M. (1984). *Science* **225**, 1315–1320.

Sakimura, K., Kushiya, E., Obinata, M., and Takahashi, Y. (1985). *Nucleic Acids Res.* **13**, 4365–4378.

Stoffel, W., Hillen, H., and Giersiefen, H. (1984). *Proc. Natl. Acad. Sci. U.S.A.* **81**, 5012–5016.

Sutcliffe, J. G., Milner, R. J., Bloom, F. E., and Lerner, R. A. (1982). *Proc. Natl. Acad. Sci. U.S.A.* **79**, 4942–4946.

Sutcliffe, J. G., Milner, R. J., Shinnick, T. M., and Bloom, F. E. (1983a). *Cell* **33**, 671–682.

Sutcliffe, J. G., Milner, R. J., and Bloom, F. E. (1983b). *Cold Spring Harbor Symp. Quant. Biol.* **48**, 477–484.

Sutcliffe, J. G., Milner, R. J., Gottesfeld, J. M., and Lerner, R. A. (1984a). *Nature (London)* **308**, 237–241.

Sutcliffe, J. G., Milner, R. J., Gottesfeld, J. M., and Reynolds, W. (1984b). *Science* **225**, 1308–1315.

Tatemoto, K. (1982). *Proc. Natl. Acad. Sci. U.S.A.* **79**, 5485–5489.

Tatemoto, K., and Mutt, V. (1980). *Nature (London)* **285**, 417–418.

Towbin, H., Staehlin, T., and Gordon, J. (1979). *Proc. Natl. Acad. Sci. U.S.A.* **76**, 4350–4354.

Valentino, K., Winter, J., and Reichardt, L. F. (1985). *Annu. Rev. Neurosci.* **8**, 199–232.

Van Ness, J., and Hahn, W. E. (1980). *Nucleic Acids Res.* **8**, 4259–4269.

Van Ness, J., Maxwell, I. H., and Hahn, W. E. (1979). *Cell* **18**, 1341–1349.

Young, B. D., Birnie, G. D., and Paul, J. (1976). *Biochemistry* **15**, 2823–2829.

Young, E. F., Ralston, E., Blake, J., Ramachandran, J., Hall, Z. W., and Stroud, R. M. (1985). *Proc. Natl. Acad. Sci. U.S.A.* **82**, 626–630.

Zeller, N. K., Hunkeller, M. J., Campagnoni, A. T., Sprague, J., and Lazzarini, R. A. (1984). *Proc. Natl. Acad. Sci. U.S.A.* **81**, 18–22.

CHAPTER 6

CHANGES IN INTERMEDIATE FILAMENT COMPOSITION DURING NEUROGENESIS

*Gudrun S. Bennett**

DEPARTMENT OF ANATOMY
SCHOOL OF MEDICINE
THE UNIVERSITY OF PENNSYLVANIA
PHILADELPHIA, PENNSYLVANIA 19104

I. Introduction

Among the many aspects of differentiation that postmitotic neuroblasts undergo, one of the most dramatic must be the development, over a period of days to weeks, of the highly asymmetric shape that is characteristic of all neurons and, in detail, is specific to individual types of neurons. While the elaboration of neurites—axon and dendrites—requires an appropriate substrate, and although the direction and extent of growth may depend as well on trophic substances, the capacity to develop neurites must depend on some intrinsic property specific to neurons, for no other cell type can assume such a shape. In particular, it is unlikely that neuronal precursor cells, the bipolar neuroepithelial cells of the embryonic neural tube, are themselves capable of assuming a neuronal shape.

Since the differentiation of specific phenotypic properties of different cell types entails the expression of the relevant genes exclusively in the appropriate cell type, the following question arises. Which genes are expressed only in neurons that give these cells the specific capacity to elaborate neurites, or, from the perspective of gene products, which macromolecules do only neurons synthesize that are required for the development of neuronal morphology?

The system of filaments and fibers that collectively constitute the intracellular cytoskeleton largely determines the shape of all cells and mediates changes in form, movement of the cell as a whole, and probably much of the traffic within the cell as well. The critical role of two members of the cytoskeleton, microfilaments and microtubules, in

* Present address: Department of Zoology, The University of Florida, Gainesville, Florida 32611.

CURRENT TOPICS IN
DEVELOPMENTAL BIOLOGY, VOL. 21

neurite growth has been appreciated for some time and has been treated in several reviews (e.g., Johnston and Wessells, 1980; Carbonetto and Muller, 1982, in this series). Similarly, the evidence suggesting an active role of microfilaments in the changes in cell shape occurring during neurulation has been analyzed (Burnside, 1973; Jacobson, 1981). The third cytoskeletal system, intermediate filaments, is less well understood in terms of function. They differ from both microfilaments and microtubules in at least one important property: they do not undergo reversible assembly and disassembly under physiological conditions but exist as essentially completely polymerized filaments with virtually no free monomer in the cell.

If the phenotypic shape, organization, and motile behavior of neurons versus other types of cells are based on differential gene expression, it is among the macromolecules comprising the cytoskeleton that it is appropriate to look first. Is there any difference in the composition of these structures in neurons versus other definitive cells or neuronal precursors? β- and γ-actin, the subunits of cytoplasmic microfilaments, differ slightly in primary structure from the several actin isoforms that comprise the related thin filaments in muscle. However, presently available evidence suggests that there is no further heterogeneity in actin among various nonmuscle cells. The same situation apparently exists with respect to myosin. The microtubule subunits, α- and β-tubulin, also have long been considered to be extremely highly conserved in primary structure. Nevertheless, persistent suggestions of structural and functional heterogeneity have recently been extended by the identification of multiple tubulin genes. It is among the intermediate filament subunits, however, that the greatest diversity is known to exist.

My colleagues and I have been concerned with the analysis of intermediate filament isoforms during development. In this review, we summarize some of our studies, and related investigations in other laboratories, pertaining to the intermediate filament isoforms of the nervous system, particularly those of neurons and their precursors. Results to date, relating the expression of different intermediate filament isoforms to known events in the generation and differentiation of different cell types, support the contention that qualitative differences in these structures may constitute one basis for phenotypic cytoskeletal properties. Although less detailed developmental information is available, it is likely that isoform diversity in tubulin will also turn out to have important functional consequences. In addition, increasing numbers of potential regulatory proteins associated with one or more of the cytoskeletal systems are being identified, and, in a few cases, it has been possible to relate the differential expression of some of these to specific aspects of cytoskeletal organization.

II. Intermediate Filament Isoforms in the Mature Nervous System

The identification of the protein subunits of the 10-nm intermediate filaments (Fig. 1) lagged considerably behind the recognition of actin and tubulin as the subunits of microfilaments and microtubules, respectively. However, beginning about 10 years ago, enormous progress has been made in their biochemical analysis. Particularly striking was the realization that, in contrast to the only subtle variation in composition of microfilaments and microtubules in all adult and embryonic cells, the composition of intermediate filaments shows obvious, substantial variation among different types of cells. A large number of subunits differing in primary sequence have been identified, each expressed in only certain cell types. Recent reviews have appeared that detail the classification of intermediate filament subunits into five groups, the cell-type distribution of each, the common as well as variable features of their amino acid sequences and assembly into filaments, and the developmentally regulated expression of different subunits during myogenesis and epidermal differentiation (Holtzer et al., 1981; Franke et al., 1981; Lazarides, 1982; Weber et al., 1983; Wang et al., 1985; Steinert and Parry, 1985).

A. POLYPEPTIDE DISTRIBUTION AMONG DEFINITIVE CELL TYPES

The intermediate filaments of neurons are particularly conspicuous, especially in large diameter axons (Fig. 1A,B), and they have long been known as neurofilaments (NF) (Peters et al., 1976). In birds and mammals they consist of three different subunits, distinguished on the basis of apparent size: NF-L (68–70 kDa), NF-M (145–160 kDa), and NF-H (180–210 kDa) (Hoffman and Lasek, 1975; Liem et al., 1978; Schlaepfer and Freeman, 1978; Bennett et al., 1981, 1982; Lasek et al., 1985). The possibility that they may have a specific role in neuronal shape is suggested by the fact that the three neurofilament polypeptides are found *only* in neurons and in no other adult cell type, while, with a single exception, no other intermediate filament polypeptide is expressed in mature neurons (Fig. 2). Immunohistochemical studies and biochemical analyses both suggest that, for the most part, the three NF polypeptides are codistributed, although the stoichiometry need not be invariant. From the results of immunohistochemistry and antibody decoration experiments at the electron microscope level

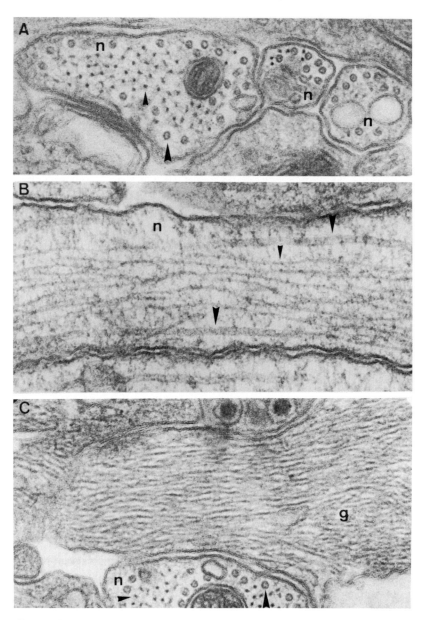

Fig. 1. Electron micrographs of neurites and a glial process in a culture of dissociated chick embryo spinal cord cells. (A) A cross section through 3 neurites (n) demonstrates the typical appearance of microtubules (large arrowheads) and neurofilaments (small arrowheads). (B) Longitudinal section through a neurite. The largest neurite in A and the one in B display the arrangement characteristically found in axons: microtubules predominantly around the periphery and abundant neurofilaments more centrally located. Thin projections linking neurofilaments and extending from microtubules can be

(Willard and Simon, 1981; Sharp *et al.*, 1982; Hirokawa *et al.*, 1984), it has been proposed that the core of neurofilaments is composed of just NF-L, with NF-M and NF-H distributed periodically along the core. In particular, NF-H appears to be associated with the side arms that project from, and link, adjacent neurofilaments. Such links and projections are characteristic of neurofilaments and are not evident among intermediate filaments composed of other isoforms. More recent amino acid sequencing data raise the possibilities, however, that both NF-M and NF-H are actually copolymerized with NF-L to form the core, and that only the long carboxy-terminal tails project laterally (Weber *et al.*, 1983).

Astrocytes also contain abundant intermediate filaments (Fig. 1C), which are composed of their unique isoform, the glial fibrillar acidic protein (GFAP) (reviewed by Eng and DeArmond, 1982). They may contain, in addition, vimentin (Fig. 2E). The latter isoform, originally identified in fibroblasts (Bennett *et al.*, 1978; Fellini *et al.*, 1978a; Franke *et al.*, 1978), is the only intermediate filament isoform that is not restricted to a histologically or embryologically defined set of cell types. Vimentin is the only isoform expressed in ependymal cells and in Schwann cells (Tapscott *et al.*, 1981a; Shaw *et al.*, 1981; Yen and Fields, 1981). In some mammals it is also present, together with 1–3 NF polypeptides, in the horizontal cells of the neural retina (Dräger, 1983; Shaw and Weber, 1983, 1984), these neurons being the only ones to contain, in the adult, an additional isoform. Oligodendrocytes appear to be devoid of any detectable intermediate filament isoforms, consistent with their lack of obvious intermediate filaments in electron micrographs (Shaw *et al.*, 1981; S. J. Tapscott, unpublished observations). Finally, desmin, the isoform specific to smooth, skeletal, and cardiac muscle, is, within the nervous system, restricted to the smooth muscle cells of blood vessels (Tapscott *et al.*, 1981a).

B. HETEROGENEITY IN INTRACELLULAR DISTRIBUTION:
   LOCALIZED POSTTRANSLATIONAL MODIFICATION

Although immunohistochemistry usually demonstrates colocalization of all three NF polypeptides, we and others have noticed that in the

---

seen. The neurofilaments are composed of 3 polypeptides, NF-L, NF-M, and NF-H. (C) Longitudinal section through a glial cell process (g) adjacent to a cross-sectioned neurite. The glial process (probably that of an astrocyte) is filled with intermediate filaments that lack projections, and they are more densely packed than are the neurofilaments in neurites. These filaments are composed of vimentin and/or GFAP. ×75,000. (Micrographs courtesy of Dr. Parker B. Antin.)

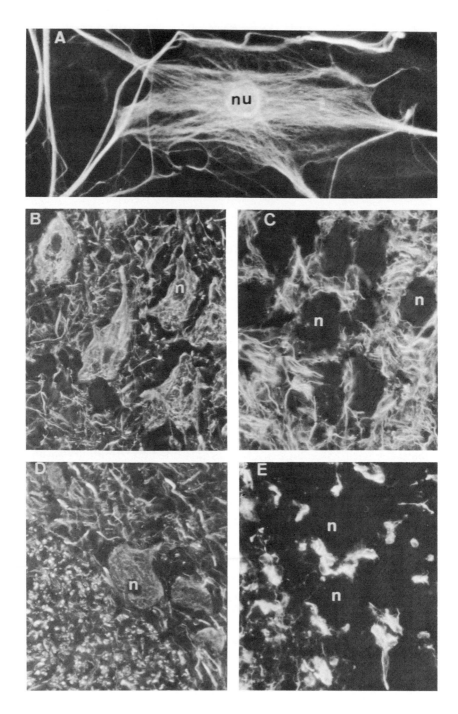

case of certain large neurons, some polyclonal and monoclonal antibodies against NF-M (Bennett *et al.*, 1984a; Dahl, 1983) or NF-H (Shaw *et al.*, 1981; Goldstein *et al.*, 1983; Hirokawa *et al.*, 1984) stain *either* the axons *or* the soma and dendrites, but not both (Fig. 2C). This apparent heterogeneity in NF polypeptide distribution, not only among different types of neurons, but within the various morphologically different parts of a single kind of neuron, was difficult to interpret until the nature of the epitopes recognized by the antibodies became clear.

It has been known for some time that all intermediate filament isoforms are phosphorylated to varying degrees. The two larger neurofilament proteins are particularly highly phosphorylated (Jones and Williams, 1982; Julien and Mushynski, 1982; Wong *et al.*, 1984). Sternberger and Sternberger (1983) reported that many monoclonal antibodies against mammalian neurofilament polypeptides, especially NF-H, were specific to *either* the phosphorylated *or* dephosphorylated polypeptides. Such antibodies also produced different immunohistochemical staining patterns; those specific to the phosphorylated protein often staining only axons, but not soma or dendrites, and those specific to the dephosphorylated protein producing the converse. In another laboratory (Carden *et al.*, 1985), nearly all monoclonal antibodies against mammalian NF-H, and 50% of those against NF-M, were directed against phosphate-dependent epitopes. We found that a polyclonal anti-chicken NF-M was specific to completely phosphorylated NF-M (Bennett and DiLullo, 1985c) and failed to bind to the soma of spinal cord motor neurons. Interestingly, this antibody detected an increasing gradient of immunoreactivity over the course of the first few hundred micrometers of the axon (Bennett *et al.*, 1984a).

Fɪɢ. 2. (A) Fluorescence micrograph showing the distribution of intermediate filaments in a cultured spinal cord neuron as visualized at the light microscope level by "staining" with anti-NF-H. Fine filaments can be resolved in this unusually flattened neuronal cell body. Neurites are intensely stained, but the confluent nonneuronal cells are negative. nu, Neuronal nucleus. (B–E) Cross sections through the ventral horn of a 19-day chick embryo spinal cord stained with anti-NF-L (B), anti-NF-M (C), anti-NF-H (D), or anti-vimentin (E). The adult pattern of expression of the different intermediate filament polypeptides has been established by this time. Note that neurites are stained by all three NF antibodies but not by vimentin. Filaments in neuronal soma (n) are stained by anti-NF-L (B) and anti-NF-H (D), but not by anti-NF-M (C) or by anti-vimentin (E). The failure of anti-NF-M to stain filaments in neuronal soma is due to the fact that NF-M is not phosphorylated in the soma, and the antibody recognizes only the phosphorylated polypeptide. Anti-vimentin stains only glia, which are not stained by any of the NF-antibodies. White matter is included in the lower left of the field shown in (D) and left in that in (E). (A) ×550; (B–E) ×440. (A, From Bennett *et al.*, 1981; B and C, from Bennett *et al.*, 1984a.)

Additional information has come from studies on the biosynthesis of NF polypeptides. Using intact cultured chick spinal cord neurons, we found that NF-M is progressively phosphorylated over an unusually long time course (24–48 hours), and only after a delay of 5–8 hours after translation (Bennett and DiLullo, 1985c). These results, together with the immunohistochemical distribution, suggested that NF-M is not phosphorylated in the neuronal soma (the site of synthesis), but only after it has entered the axon, where phosphorylation occurs during the course of transport down the axon. Similar findings have been reported with respect to both NF-M and NF-H in rat sensory neurons *in vivo* (Oblinger and Shick, 1985). Furthermore, although present information does not rule out the complete absence of NF-M and/or NF-H from dendrites, it is also possible that one or both of these polypeptides are indeed transported into dendrites but never become phosphorylated there. This difference in the content and/or degree of phosphorylation of NF-M and/or NF-H has been linked to the organizational and morphological differences between axons, especially large-diameter axons, and dendrites.

It is worth pointing out that while in mammals NF-M is considerably less highly phosphorylated than NF-H, chicken NF-M is probably as highly phosphorylated as mammalian NF-H. Both NF-M and NF-H migrate anomalously on SDS gels, exhibiting a mobility slower than that corresponding to their true molecular weight (Kaufmann *et al.*, 1984; Scott *et al.*, 1985). Upon removal of phosphate, the mobility of mammalian NF-H increases substantially (to an effective MW approximately 30,000–40,000 less than the fully phosphorylated form), and that of mammalian NF-M increases only slightly (Julien and Mushynski, 1982; Kaspi and Mushynski, 1985; Wong *et al.*, 1984; Carden *et al.*, 1985). Thus, the magnitude of the difference in apparent size between the phosphorylated and dephosphorylated polypeptides correlates with the amount of phosphate. Although the phosphate content of avian NF polypeptides has not yet been determined, the difference in apparent size that we observed between newly synthesized (nonphosphorylated) chicken NF-M and the fully phosphorylated form (Bennett and DiLullo, 1985c) is as large as that observed with mammalian NF-H, thus suggesting that chicken NF-M is as highly phosphorylated as mammalian NF-H. Although it is not known what functional significance might be associated with these and other (Shaw *et al.*, 1984: Lasek *et al.*, 1985) differences among the NF polypeptides from different animals, they should be kept in mind in the following sections where different results have been reported during development of the chicken versus mammals.

III. Initiation of Neurofilament Polypeptide Expression during Neurogenesis

Because of the unparalleled diversity of IF isoforms and their cell-type restriction in the mature nervous system as compared with other cell types, it is of particular interest to examine the expression of the neurofilament and other intermediate filament subunits during the development of the nervous system, particularly at the time of generation of postmitotic neuroblasts and the initiation of neurite outgrowth. We have investigated this in some detail during the embryonic development of the chicken, with interesting results. Although the stages of development, and regions of the nervous system, that have been examined in mammals (primarily rodents) do not completely overlap with our studies on the chick, the analyses available to date suggest similar dynamics in general but some possible differences in detail.

A. CORRELATION WITH EMERGENCE OF POSTMITOTIC NEUROBLASTS

In initial surveys of the embryonic chick spinal cord (Tapscott *et al.*, 1981a; Bignami *et al.*, 1980), NF polypeptides were undetectable in the replicating neuroepithelial cells of the neural tube. Positive immunoreactivity with the antibodies to these proteins could first be detected in thin processes in the nascent marginal zone at the time at which the earliest postmitotic neuroblasts are known to arise (Fig. 3A,B,E). This was not particularly surprising in view of the general finding that, within the CNS, expression of many neuron-specific macromolecules is initiated in the postmitotic neuroblast and does not occur in the precursor cell. Of particular interest was the result that probably *all* neuroepithelial cells contain intermediate filaments of the vimentin type (Tapscott *et al.*, 1981a). Vimentin, present well before the time of neural tube closure (see Section IV), is present throughout the bipolar neuroepithelial cells but is particularly concentrated in the broad basal endfeet (Fig. 3G). During mitosis, when the basal process detaches from the outer limiting membrane and is withdrawn, vimentin forms a cap on the basal side of the chromatin mass.

Similar results have been obtained with rodents (Schnitzer *et al.*, 1981; Bignami *et al.*, 1982; Shaw and Weber, 1983; Houle and Fedoroff, 1983; Cochard and Paulin, 1984), although vimentin could not be detected until a day or more after neural tube closure. It may be that this delayed detection is due to a lesser sensitivity of the reagents and/or detection methods used in the mammalian studies, for otherwise one must conclude that early mammalian neuroepithelial cells contain either no intermediate filaments or an as yet undescribed isoform. On the other hand, certain real differences may exist between

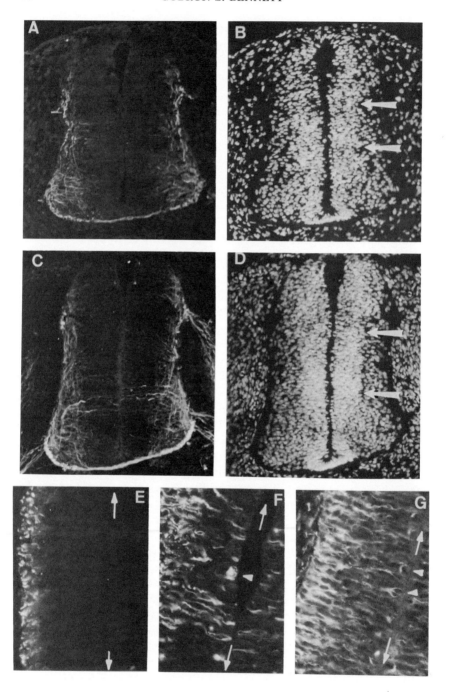

birds and mammals in the sequence of switches occurring during gastrulation and neurulation (see below).

Radial glia in the chicken (Tapscott *et al.*, 1981a) and in rodents (Bignami *et al.*, 1982; Cochard and Paulin, 1984) contain only vimentin, and during the time that these cells are prominent, anti-vimentin staining results in their clear visualization. They do not contain GFAP. The latter isoform first appears at the time that definitive astrocytes are generated (Raju *et al.*, 1981; Eng and DeArmond, 1982) and remains restricted to this subset of glia. Primates apparently constitute an exception in that radial glia and a subpopulation of ependymal cells contain GFAP (Choi and Lapham, 1978; Levitt *et al.*, 1981, 1983).

The observations that (1) all replicating neuroepithelial cells contain vimentin but not NF polypeptides, (2) mature neurons contain the NF triplet but not vimentin, and (3) NF isoforms are first readily detectable in postmitotic neuroblasts, suggested that a switch in intermediate filament isoform expression occurs with the birth of postmitotic neuroblasts. This situation was analogous to the one we had found earlier in the case of skeletal myogenesis, in which vimentin is the only isoform expressed in the replicating precursor cells and desmin synthesis is initiated with the birth of the postmitotic myoblasts (Fellini *et al.*, 1978b; Bennett *et al.*, 1979: Holtzer *et al.*, 1981; see also Gard and Lazarides, 1980). More recent analyses suggest that the initiation of desmin synthesis actually occurs during the terminal cell cycle (probably G_2) of the immediate precursor (Dlugosz *et al.*, 1983). The switch from vimentin to NF polypeptide synthesis constitutes the only instance known to date in which the synthesis of a new cytoskeletal gene product accompanies the emergence of the *neuronal* cell type. It

---

FIG. 3. Fluorescence micrographs of chick embryonic spinal cord showing the expression of intermediate filament polypeptides during the period of neurogenesis (4 days incubation). (A–D) Low power views of cross sections stained with anti-NF-L (A) or anti-NF-M (C), and counterstained with H33258 to visualize nuclei (B and D; same fields as A and C, respectively). Arrows in B and D indicate the lateral extent of the replicating cell layer. (E–G) High power views of longitudinal sections stained with anti-NF-L (E), anti-NF-M (F), or anti-vimentin (G). The neurocoel extends between the arrows; the full lateral extent of the spinal cord is included to the left (but not the right) side. Both NF polypeptides are present in postmitotic neuroblasts that have migrated out of the replicating layer to form the emerging mantle and marginal layers. Within the replicating cell layer, no NF-L is detectable (A, E). In contrast, in a short stretch of the ventral spinal cord, phosphorylated NF-M (C, F) is present in processes extending through this layer to the neurocoel and in occasional mitotic cells (F, arrowhead). At all levels, vimentin (G) is present in virtually all processes and mitotic cells. A–D, ×140; E–G, ×315. (A and B, From Bennett and DiLullo, 1985a.)

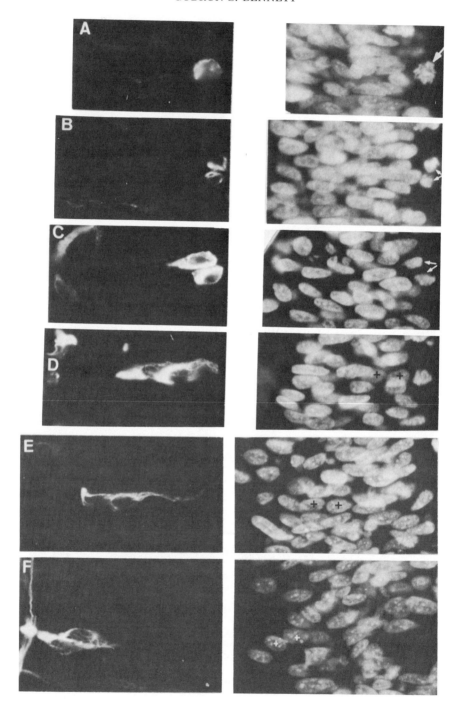

was of interest to examine more closely the temporal relation of this switch to the birth of the neuron. Using both normal embryos and embryos treated with colcemid to accumulate cells arrested in metaphase, we carried out a more detailed examination of NF-L, NF-M, and NF-H expression during the period of neurogenesis in the spinal cord, neural retina, cerebral cortex, and optic tectum. Correlation of these results with known temporal and spatial patterns of the birth of different groups of neuroblasts revealed considerable variability in the exact timing and sequence with which expression of the three NF polypeptides was initiated.

Motor neuroblasts of the brainstem and spinal cord begin expressing detectable levels of the three NF polypeptides at times very close to the terminal mitosis, but not precisely simultaneously (Tapscott et al., 1981b; Bennett and DiLullo, 1985a). One subpopulation of motor neuroblasts in the spinal cord begins expressing NF-L first, immediately after or even during the terminal mitosis, and these can be visualized by staining with anti-NF-L during their migration through the layer of replicating cells. In the case of an apparently different subpopulation of motor neuroblasts, NF-M (but not NF-L or NF-H) expression can be detected in the immediate precursor cell and continues to be the only detectable NF isoform during migration of the daughter neuroblasts across the germinal zone (Fig. 3C,D,F; Fig. 4). In both cases, expression of the remaining two NF polypeptides probably occurs approximately at the time migration is complete. In the neural retina as well (Bennett and DiLullo, 1985b), at least in the case of the retinal ganglion cells, NF-M expression appears to begin during the terminal cell cycle of the immediate precursor, with the addition of NF-L in migrating postmitotic neuroblasts and NF-H in the emerging neurites of the nascent ganglion cell axon layer.

Other neuroblasts in the spinal cord and all in the cerebral cortex and optic tectum do not express detectable levels of any neurofilament polypeptides until some time after the terminal mitosis, and none is

---

FIG. 4. Fluorescence micrographs of sections of trunk neural tube from 2.5- to 4-day chick embryos, demonstrating the expression of phosphorylated NF-M by cells in different stages of the mitotic cycle and pairs of migrating postmitotic neuroblasts at progressively increasing distances from the luminal surface. Sections were stained with anti-NF-M and counterstained with Hoechst H33258 to visualize nuclei/chromatin. Each pair of micrographs shows the same field photographed through filters selective for the rhodamine-labeled antibody (left) and H33258 (right). (A) Metaphase; (B) telophase; (C) mitosis just completed; (D–F) pairs of interphase cells (nuclei identified by +), presumably postmistotic neuroblasts. The neurocoel is at the right edge in A–E and just beyond the right edge in F. ×800. (Adapted from Bennett and DiLullo, 1985a.)

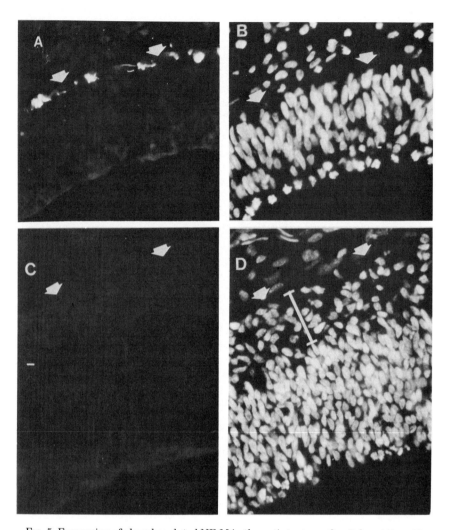

FIG. 5. Expression of phosphorylated NF-M in the optic tectum after 3 days (A) and in the cerebral vesicles after 5 days (C) of embryonic development. The position of the counterstained nuclei in both fields is presented in B and D, respectively. Arrows indicate the outer extent of the neural tube; the neurocoel is at the bottom. In the optic tectum, postmitotic neuroblasts have just begun to appear, and mantle and marginal layers are not yet clearly defined. No mitotic cells, interphase cells, or migrating postmitotic neuroblasts express phosphorylated NF-M; the few neurites that have begun to extend at the outer surface do express this polypeptide (as well as NF-L and NF-H). Thus, expression of phosphorylated NF-M does not begin until some time after the terminal mitosis but probably occurs soon after completion of migration. No qualitative change in this pattern occurs through the sixth day of development. In the cerebral vesicles, the mantle layer of postmitotic neuroblasts (indicated by the bar in D) has begun to appear, but there is no detectable NF-M (or NF-L or NF-H) associated with these cells. Most of these neurons eventually do express all 3 NF polypeptides, but their synthesis is initiated only after a relatively long delay (several days) following terminal mitosis and migration of the cell body. ×380. (From Bennett and DiLullo, 1985b.)

detectable in processes within the germinal zone. In the case of the earliest tectal neurons to emerge, neurofilament expression appears to commence with the growth of axons in the marginal zone (Fig. 5A,B); cerebral cortical neurons do not contain the neurofilament proteins until several days later (Fig. 5C,D). In both cases, the three NF isoforms appear to be expressed simultaneously (Bennett and DiLullo, 1985b).

Although the onset of NF polypeptide expression during cerebellar neurogenesis and development has not yet been thoroughly examined, we have found, in the chick, that Purkinje cells exhibit an interesting change in distribution of NF polypeptides in the course of differentiation. NF-L and NF-H are readily detectable in the soma and growing dendrite of the embryo, while the adult Purkinje cell soma and dendrites contain only traces of these isoforms (Bennett et al., 1984a).

In the embryonic rodent spinal cord and brainstem, the three NF polypeptides have been detected at times comparable to those we have found in the chick (Raju et al., 1981; Cochard and Paulin, 1984). Both NF-L-positive and NF-M-positive newborn neuroblasts were observed traversing the neuroepithelium at the earliest time of neurogenesis, with NF-H appearing by the time that neurites appear in the nascent marginal zone (Cochard and Paulin, 1984). Differences among various subpopulations of spinal cord neuroblasts were not noted, but, interestingly, anti-NF-M was reported to stain all *mitotic* cells in the neural tube and throughout the embryo in a diffuse manner (Cochard and Paulin, 1984).

There are several reports suggesting that NF-H first appears in some regions of the developing mammalian nervous system several days later than NF-L and NF-M (whole brain: Shaw and Weber, 1982; retina/optic nerve: Shaw and Weber, 1983; Willard and Simon, 1983; Pachter and Liem, 1984). This is in contrast to our observation that NF-H is detectable in chick retinal ganglion cell axons as soon as they begin to form the optic nerve fiber layer, although we have not detected NF-H in postmitotic neuroblasts migrating through the germinal zone.

In both the chicken and in rodents at most rostrocaudal levels of the neural tube, premigratory and early migratory neural crest cells contain no neurofilament immunoreactivity (Payette et al., 1984; Ziller et al., 1983; Cochard and Paulin, 1984). NF-L, NF-M, and NF-H can be detected as soon as clusters of cells have arrived at the positions of the sensory ganglia primordia and the primary sympathetic chain (Cochard and Paulin, 1984; Payette and Bennett, unpublished). Some of these NF-positive cells within the ganglia are in mitosis, suggesting that the expression of NF polypeptides begins prior to the cessation of replication. In contrast to the CNS, where expression of neuron-specific enzymes does not occur in replicating cells, several enzymes for transmitter metabolism are expressed in cells of sympathetic ganglia that are still capable of replication (see Payette *et al.*, 1984, for references). In cultures of neural crest, however, replication of the NF-positive subpopulation has not been observed (Ziller *et al.*, 1983).

The rhombencephalic region of the neural crest in the chicken appears to constitute an interesting exception to the sequence of events in the rest of the neural crest. In this region alone, a small number of NF-M-positive cells, some in mitosis, were found in the position of both the premigratory and early migrating neural crest (Payette *et al.*, 1984). These cells migrate into the mesenchyme of the third, fourth, and fifth branchial arches, and ultimately into the gut where they form the enteric nervous system. Throughout this migratory path, substantial numbers of NF-M-positive cells were observed, and it is likely that this subset of crest cells begins expression of NF-M (and probably NF-L and NF-H as well) at a very early time and continues to do so through several cell generations before the definitive neuroblast progeny appear. NF-M-positive cells are also observed within the epithelium of the epibranchial placodes. These cells presumably also contribute to the population of replicating, NF-M-positive cells within the branchial mesenchyme, subsequently giving rise to the petrosal and nodose ganglia. Whether or not this region of the neural crest in mammals exhibits the same pattern is not known.

The time at which vimentin synthesis in postmitotic neuroblasts ceases also may vary. Neurites emerging from the ventral spinal cord are already vimentin negative (Cochard and Paulin, 1984) or become so at least within a day or two (Bignami *et al.*, 1982; Bennett, unpublished). The same is true of rat retinal ganglion cell axons (Shaw and Weber, 1983). Similarly, when spinal cord neuroblasts are placed in culture during the earliest period of neurogenesis (i.e., 3-day chick embryo; 13-day rat embryo), they can be stained with anti-vimentin for the first 1–3 days but not thereafter (Bennett *et al.*, 1982; Bignami *et al.*, 1982). It is not known, however, whether the vimentin detectable for this brief period is actually synthesized or "inherited" from the precursor cells and only slowly degraded.

Neural crest-derived neurons, however, present a more complex picture. When examined *in situ,* some neurites in the sensory ganglia of mouse embryos at early stages contained vimentin, but within a few days no vimentin could be detected in neuronal structures in either sensory or sympathetic ganglia (Cochard and Paulin, 1984). We have found virtually no neuronal vimentin in sections of spinal sensory ganglia of 6-day chick embryos. In contrast, sympathetic neurons from chick embryos ranging in age from 7 to 17 days, when placed in culture continued to synthesize vimentin, as did sensory ganglion neurons from the younger, but not the older, embryos (Jacobs et al., 1982). Neurons in neural crest cultures were similarly found to express both vimentin and neurofilament polypeptides (Ziller et al., 1983). In recent experiments, we have found that sensory ganglion neurons from 6- to 8-day-old chick embryos continue to synthesize vimentin for over a month in dissociated cultures. These results suggest that some, but not all, types of neurons can retain the ability, for varying periods of time, to synthesize vimentin in culture, whereas they do not normally do so. Similar observations have been made in other cell types which do not contain vimentin *in situ* but gradually synthesize increasing amounts in culture (Franke et al., 1979). The factors that might regulate vimentin expression remain to be elucidated.

B. TRANSIENT EXPRESSION IN REPLICATING CELLS

When we examined IF isoform expression in the cranial neural tube at very early times, prior to the time at which postmitotic neuroblasts are present, we obtained some surprising results (Bennett and DiLullo, 1985b). What we found was that in certain—but not all—regions of the very early neural tube, generations prior to the appearance of postmitotic neuroblasts, *all* neuroepithelial cells contain NF-M immunoreactivity as well as vimentin (Fig. 6). These cells and/or their progeny then become NF-M negative over a period of a day or two, still before any significant number of postmitotic neuroblasts have appeared. This phenomenon is restricted to the dorsal and lateral forebrain, the entire optic vesicle, and a portion of the dorsal hindbrain. It is not observed in the ventral neural tube anywhere along its length, nor in the dorsal midbrain (future optic tectum). An example of the progression from widespread, transient NF-M expression at an early time to sustained reexpression in a subset of cells at a later time is shown by the series of events in the optic cup (Fig. 7). Also compare Figs. 6A and 5A.

The component of the early neural tube that binds anti-NF-M was shown to comigrate with adult chicken NF-M on two-dimensional immunoblots and thus is not likely to be an unrelated cross-reacting antigen. Even more surprising was the finding of transient NF-M expression in several regions of the early embryo, some of which were clearly of unrelated embryonic derivation. These sites included cardiac myoblasts, lens epithelial cells, and an unidentified group of cells in the limb bud. The significance of this *transient* expression is not clear at this time. As discussed below, it is not obviously lineage related,

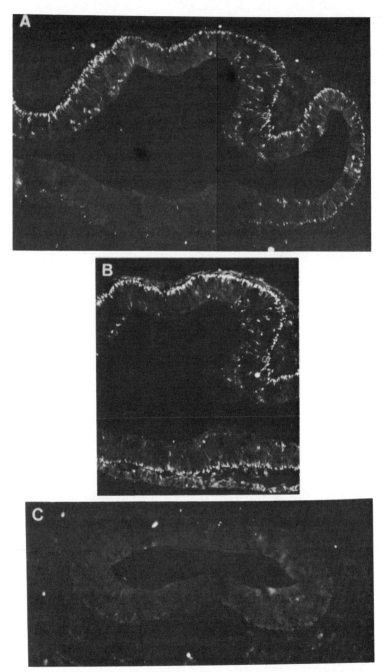

FIG. 6. Cross sections through the neural tube at the level of the forebrain and optic vesicles (A and B) and the midbrain (C) of a stage 11 (40- to 45-hour-old) embryo demonstrating the distribution of cells *transiently* expressing phosphorylated NF-M. Sections shown in A and C were stained with anti-NF-M. After photographing, the one in A was stained with anti-vimentin (B). The basal processes of the neuroepithelial cells

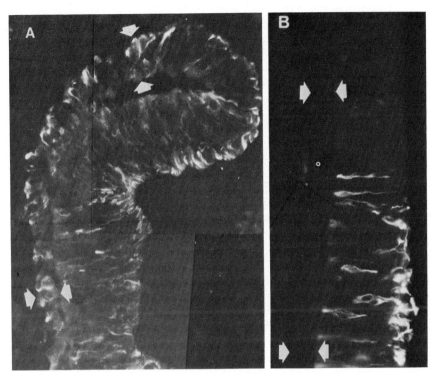

FIG. 7. Changes in NF-M expression during the development of the eye. (A) Section through the optic cup during the third day of incubation. Phosphorylated NF-M is present in both the outer wall (between arrows) and the inner wall. (B) Section through the back of the eye at the end of the fourth day of incubation. The cells of the outer wall (between arrows), which are developing into pigment epithelial cells, have become NF-M negative. The inner wall has also become NF-M negative (as in the upper half of the field) except in a short region where postmitotic neuroblasts (mostly retinal ganglion cells) have begun to emerge (lower half of the field). In this region, phosphorylated NF-M is detected in a subpopulation of mitotic cells (not present in this field), in the leading and trailing processes of postmitotic neuroblasts (often in pairs), and in the neurites of the growing optic nerve axons (right edge). This pattern is similar to that in the ventral spinal cord. ×380. (From Bennett and DiLullo, 1985b.)

---

of the entire dorsal forebrain and optic vesicles contain phosphorylated NF-M; none of those in the ventral forebrain (A) or in the midbrain region (C) expresses this NF polypeptide. All neuroepithelial cells contain vimentin, including those of the ventral forebrain (B). Positive immunoreactivity at the luminal surface is associated with cells in mitosis. Comparison of A and B shows that, within the region containing NF-M, *all* basal processes and mitotic cells that can be stained with anti-vimentin also are stained with anti-NF-M. Thus, there is no vimentin-positive and NF-M-negative subpopulation. Descendants of these NF-M-positive cells subsequently become NF-M-negative prior to the time that significant numbers of neuroblasts have appeared (compare with Figs. 5C and 7). ×380. (From Bennett and DiLullo, 1985b.)

even within the neural tube. The cranial neural tube of mammals has not been examined at comparably early stages.

C. Changes in IF Isoform Expression versus Posttranslational Events

Virtually all of the developmental studies were carried out prior to the realization that some antibodies to neurofilament polypeptides are specific to phosphate-dependent epitopes. The demonstration that our anti-chicken NF-M is specific to completely phosphorylated NF-M (Bennett and DiLullo, 1985c) opens up a new set of interpretations of our previously obtained immunohistochemical findings in early embryos. The positive immunoreactivity clearly reflects not only NF-M *expression* but its phosphorylation as well. Since these events need not be coupled, alternative possibilities exist. For example, (1) "transient" NF-M expression may be even more widespread but the phosphorylation limited to specific groups of cells, or (2) the early expression of NF-M in parts of the CNS long before the appearance of postmitotic neuroblasts may not be transient at all, but the phosphorylation may instead be transient. In this regard, it is interesting that several polyclonal and monoclonal antibodies to mammalian NF-M diffusely stained all mitotic cells in all embryonic mouse tissues (Cochard and Paulin, 1984). This distribution is unlike what we have found in the chick embryo with anti-chicken NF-M in regard to both localization within the cell and to distribution of positive cells in the embryo. However, since anti-*mammalian* NF-*M* antibodies frequently do not discriminate between the dephosphorylated and the phosphorylated polypeptide (Carden *et al.*, 1985), it is possible that the antibodies used by Cochard and Paulin recognized both forms. Thus, their results, together with ours, suggest the intriguing possibility that NF-M may indeed be transiently expressed throughout the embryo but phosphorylated only in certain cells. It is also noteworthy that, in the one-day (whole) postnatal rat brain, the level of NF-M was already at 20% of the adult level while that of NF-L was less than 10% of adult values (Shaw and Weber, 1982), suggesting that either NF-M expression had begun earlier or its rate of synthesis was initially greater than that of NF-L.

Similarly, the delayed appearance of NF-H in mammals is based, at least in part, on the immunochemical detection of NF-H with antisera whose specificity for phosphorylated or nonphosphorylated NF-H was not known at the time. At least some have been shown subsequently to be phosphate dependent (Glicksman and Willard, 1985; Shaw *et al.*, 1985). Whether it is the expression of the NF-H polypeptide or the onset of its phosphorylation that is delayed is therefore unresolved. Analysis of radioactively labeled and transported proteins

(Willard and Simon, 1983) and of the polypeptide composition of total filament preparations (Pachter and Liem, 1984) also demonstrates the absence of a component having the mobility of phosphorylated NF-H in optic nerve for the first few days after birth. However, because of the large change in mobility upon phosphorylation, the presence of non- (or partially) phosphorylated NF-H has not been definitively excluded.

It is also interesting that, in the case of motor neuroblasts, phosphorylation of NF-M initially occurs throughout the cell (Fig. 4), for the positive immunoreactivity is found around the nucleus as well as in both leading and trailing processes of the migrating neuroblast. Only later (probably soon after completion of migration) is phosphorylated NF-M restricted to the axon. This raises the additional question of whether the kinase (or associated activating mechanism) becomes restricted to the axon or an inhibitor (or inhibitory mechanism) to the soma and dendrites. Clearly, it will be of great interest to identify and characterize the relevant kinase(s) as well as to examine its expression during embryonic development and distribution in mature neurons.

IV. Intermediate Filament Isoform Divergence and the Differentiation of Neuroectoderm from Surface Ectoderm

Recently, we have turned our attention to earlier events in the embryonic development of the nervous system—to the divergence of neuroectoderm and surface ectoderm from the primitive ectoderm and the formation of neural plate and neural tube during gastrulation and neurulation. Epidermal epithelial cells, like most epithelial cells, express several of the many cytokeratin-type IF isoforms but no vimentin. No cells in the adult or embryonic CNS express any cytokeratins. The finding that avian neuroepithelial cells contain vimentin IF by the time of neural tube closure (Tapscott et al., 1981a; Bennett and Di-Lullo, 1985b) raised the questions of how early in development these cells or their progenitors begin to express vimentin and at what point a divergence in IF isoform expression between neuroectoderm vs surface ectoderm occurs.

Using a polyclonal antiserum against human prekeratin (Sun and Green, 1978) that cross reacts with at least some chicken cytokeratins, we assessed the immunohistochemical distribution of vimentin and cytokeratin in frozen sections of the chick embryo before and during gastrulation (Bennett et al., 1984b). It is clear that by the time a definitive neural plate is present (18–24 hours of incubation), and prior to the formation of neural folds, the presumptive neuroepithelial cells are vimentin positive and cytokeratin negative, and the neural plate is distinctly demarcated from the lateral ectoderm, which is

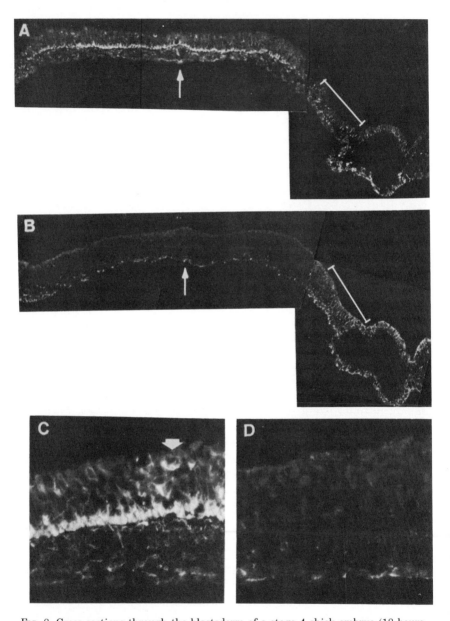

FIG. 8. Cross sections through the blastoderm of a stage 4 chick embryo (18 hours incubation, definitive primitive streak) just anterior to Hensen's node. Sections were stained with anti-vimentin (A) or with anti-prekeratin (B). The section shown in C and D was double stained with both antibodies, and the same field, located near the midline, photographed through optics selective for rhodamine anti-vimentin (C) and for fluores-

cytokeratin positive and vimentin negative (Fig. 8). The neuroepithe-lial cells are already elongated at this time, and the distribution of vimentin within them is asymmetric with a pronounced concentration in the broad basal process. Although the exact sequence of events occurring during the first 18 hours of incubation has not yet been determined precisely, it appears that the divergence of the ectoderm into presumptive neuroepithelial cells expressing only vimentin, and presumptive epidermal cells expressing only cytokeratin, may have begun by the time that gastrulation is just being initiated—i.e., prior to the definitive primitive streak stage in the chick. As gastrulation proceeds, the appearance of vimentin-positive, cytokeratin-negative cells in the presumptive neural plate region also coincides with the elongation of the cells in this region.

Our initial results suggest that earlier blastoderm cells of the chick may be homogeneous with respect to IF isoform content and express low levels of both cytokeratin and vimentin. When dissociated and placed in culture, such early blastoderm cells rapidly give rise to prog-eny that express *either* vimentin *or* cytokeratins (Biehl *et al.*, 1985). In this respect, the chick embryo would appear to differ from the mouse embryo, in which the presence of only cytokeratins could be demon-strated until gastrulation, at which time vimentin expression was de-tected in the mesodermal cells (Franke *et al.*, 1982). Development of the neural plate has not been followed in the mouse, although, as noted above (Section III,A), there does not appear to be any vimentin in the neural tube until some time after closure. On the other hand, vimentin, as well as cytokeratins, have been reported in *Xenopus* oocytes and early embryos (Godsave *et al.*, 1984). The significance of these possible differences in IF isoform switches in different animals remains to be resolved. Nevertheless, in the chick, the IF·isoform

---

cein anti-prekeratin (D). The arrows in A and B indicate the position of the nascent notochord (midline). The entire central region of the ectoderm (the neural plate) is vimentin positive and cytokeratin negative, the brightest fluorescence occurring in the basal processes of the elongated cells. The high-power view of this region, C and D, includes a mitotic cell (arrow in C) with its cap of vimentin on the basal side of the chromatin mass. In the lateral region of the embryonic ectoderm (brackets indicate the future epidermis), vimentin content diminishes while cytokeratin content increases. Beyond this area, the fields shown in A and B extend into the area opaca, which will give rise to the extraembryonic membranes. The ectoderm in this region is essentially vimen-tin negative and brightly cytokeratin positive. The mesodermal cells of the embryo proper are also vimentin positive and cytokeratin negative, the only cells containing substantial cytokeratin reactivity being the single layer of endodermal cells extending across the lower surface. The lower layer of cells in the area opaca contains both vimen-tin and cytokeratin. A and B, ×83; C and D, ×280.

complement provides one of the few, if not the only, examples known to date of the differential expression of a specific set of genes that accompany the divergence of embryonic ectoderm into presumptive neuroectodermal and surface ectodermal cells. Furthermore, the correlation of the time of elongation of presumptive neural plate cells with the switching on of vimentin expression (or switching off of cytokeratin expression) is the only clear instance of a *qualitative* change in the expression of a cytoskeletal component that accompanies this critical change in cell shape.

V. Other Cytoskeletal Components

A. Tubulin

α- and β-tubulin are essential for cell replication, and they are constitutively expressed in virtually all animal and plant cells. Their primary structures are highly conserved through evolution. Nevertheless, indications of microheterogeneity have repeatedly appeared in the literature. These include electrophoretic heterogeneity (e.g., Ginzburg *et al.*, 1983), the production of antibodies that recognize only a subset of microtubules (Thompson *et al.*, 1984) or the tubulin of only a single type of cell (Murphy *et al.*, 1986), and the demonstration of variation among some cells in drug or temperature sensitivity of microtubules to depolymerization (Brinkley and Cartwright, 1975; Brady *et al.*, 1984). Some of these observations are likely due to posttranslational modifications, such as phosphorylation and tyrosylation. However, recent DNA analyses have revealed the existence of several α- and β-tubulin genes in a given organism that are differentially expressed in various adult cells and tissues and show complex changes during embryonic development (e.g., Kalfayan and Wensink, 1982; Kemphues *et al.*, 1982; Havercroft and Cleveland, 1984). Of particular interest is the observation that one chicken β-tubulin gene (of four β-tubulin genes identified) is the overwhelmingly predominant gene expressed in brain. Analysis of cultured cells highly enriched for either neurons or nonneuronal cells suggests that this pattern of expression is characteristic of neurons rather than glia (Havercroft and Cleveland, 1984). Future work may well identify specific tubulin isoforms that subserve different functions in different cells and within different subcellular compartments. At present, details of the expression of different tubulin isoforms during neurogenesis are not available.

B. Cytoskeleton-Associated and Regulatory Proteins

It is clear that the operation of the cytoskeletal components is not simply a function of their individual properties but of their interactions with each other, with the cell membrane, and with other cytoplasmic organelles as well. Intermediate filaments and microtubules exhibit a particularly complex interrelationship, most dramatically illustrated by the aggregation of some types of intermediate filaments into densely packed cables or caps upon disruption of microtubules (e.g., Croop and Holtzer, 1975). In the case of neurons, filamentous links can often be observed between neighboring neurofilaments (e.g., Fig. 1) and between neurofilaments and microtubules (e.g., Hirokawa et al., 1984). Interactions between the two types of filaments can be demonstrated in vitro (e.g., Liem et al., 1985; Williams and Aamodt, 1985). Recent attention is being focused increasingly on identifying the macromolecules that may mediate such interactions or otherwise regulate cytoskeletal functions. The steadily multiplying list of such proteins cannot be adequately dealt with here. Furthermore, information regarding their cell-type specificity and developmental regulation is fragmentary at present. Two cases, however, that are particularly relevant to neurogenesis and differentiation, however, deserve mention.

A monoclonal antibody was described (Ciment and Weston, 1982, 1985) that recognized an epitope restricted to neurons in the mature chicken and, during embryonic development, displayed a distribution virtually identical to the one we obtained with anti-NF-M. In particular, this epitope recognized a subset of neural crest cells apparently identical to the subset found by Payette et al. (1984) to express NF polypeptides. Further studies to identify the corresponding antigen suggest that it is an intermediate filament associated protein, 73 kDa in size, that can bind to vimentin intermediate filaments or to neurofilaments (Ciment et al., 1986).

The other case concerns recent developments in the analysis of microtubule associated proteins (MAPs) in the nervous system, especially those of high molecular weight (approximately 300,000) (Bloom et al., 1985; Bernhardt et al., 1985). Previously thought to be ubiquitous, several monoclonal antibody studies have now provided evidence that different MAPs have different cell-type distributions and, furthermore, are asymmetrically distributed within neurons. The MAP-1 polypeptides appear to be expressed in most cells and are found throughout neurons. MAP-2, however, is neuron specific, highly concentrated in dendrites and soma, and virtually absent from axons. A recently identified 180 kDa MAP, MAP-3. while not specific to neurons, is highly concentrated in neurofilament-rich axons (Huber et al., 1985). Developmental studies have thus far focused on the rodent cerebellum (Bernhardt et al., 1985) where the sequences of expression and distribution of each of the MAPs, with respect to the differentiation and growth of axons and dendrites of different classes of neurons, have been examined. It is not yet known, however, which MAPs are present in the replicating neuroepithelial cells of the early neural tube, whether there is heterogeneity in MAP expression among these cells, and what changes in expression may occur as lineages diverge and different classes of neurons appear. Especially provocative, however, is the observation that MAP-2 may be present in a subpopulation of replicating cells obtained from newborn rat brain, in which it can codistribute with either microtubules or with intermediate filaments, the latter consisting of vimentin (Bloom and Vallee, 1983). External granule cells of the cerebellum apparently do not contain detectable levels of any of the high-molecular-weight MAPS, their postmitotic neuronal progeny first expressing MAP-1, -2, and -3 after migration to the internal granule cell layer. Interestingly, MAP-3 expression in these neuroblasts is transient (Bernhardt et al., 1985). It is likely that additional developmental studies will, in the near future, elucidate a dynamic situation with respect to MAP expression that will have important implications for the ability of different generations of cells in diverging lineages to assume their phenotypic morphologies.

VI. Implications of Switches in Intermediate Filament Isoform Expression

A. LINEAGE ANALYSIS

Determination of the lineage pathways in the vertebrate central nervous system has been a particularly difficult question to tackle. That is, it has not been possible to use direct observation or microinjection to follow the fate of individual replicating cells and learn at what point the lineages for neurons and glia, and for different classes of neurons, diverge. The known temporal sequence in which various postmitotic neuroblasts and glial cells emerge from an apparently homogeneous population of replicating neuroepithelial precursor cells has most often been presented as the result of a series of successive asymmetric divisions in which one daughter becomes a postmitotic neuroblast and the other continues to replicate, giving rise, in later generations, to different types of neuroblasts, and lastly to glioblasts (e.g., Jacobson, 1978; Cooke, 1980, this series). No direct evidence exists, however, to support this scheme, rather than one in which asymmetric divisions (lineage branch points) occur at an earlier time, and the morphologically homogeneous neuroepithelial cells are, in fact, heterogeneous (in terms of the type of their definitive progeny) generations prior to the ultimate birth of the definitive cell type. Indeed, Sauer (1935) and Hamburger (1948) argued that, on the basis of the rapid decline in mitotic density in the neural tube, at least in the early stages of neurogenesis, most divisions must produce two postmitotic daughters. Thus, the lineage options, at least of the immediate precursor cell, are already restricted. This is the situation that most likely exists in the erythrogenic and skeletal myogenic lineages (Dienstman and Holtzer, 1975).

The expression of a cell-type-specific "marker" in a replicating precursor cell presents an immediate temptation to conclude that this precursor cell has become "committed" to that lineage and has lost the option to give rise to progeny of a different phenotype. The consistent presence of a small number of NF-M-positive mitotic cells in the spinal cord for a brief time and in a specific location, together with the indication that both daughters continue to express this NF polypeptide and migrate across the replicating zone (Bennett and DiLullo, 1985b, Section III,A above), does indeed suggest that the immediate precursor of *this type* of (motor) neuroblast has already become restricted to the neuronal lineage and undergoes a symmetrical terminal mitosis to yield *two* postmitotic neuroblasts. The observation that the immediate precursors of most types of neurons do not express NF polypeptides, however, renders it impossible to distinguish among these possibilities: (1) commitment to the neuronal lineage is not invariably reflected in NF polypeptide expression; and (2) other types of neuroblasts are not the result of a symmetrical terminal division, in which case the immediate precursor has not been restricted to a single lineage. The early, apparently sustained, expression of NF polypeptides in a subpopulation of neural crest cells also suggests a certain degree of lineage restriction at a developmental time prior to what had previously been thought.

The presence of the astrocyte intermediate filament isoform, GFAP, in a subpopulation of the replicating ventricular cells of primate embryos at times prior to and during the generation of postmitotic neuroblasts has also been interpreted (Levitt *et al.*, 1981, 1983) as evidence for the early divergence of neuronal and glial lineages. The failure to detect GFAP in neuroepithelial or ventricular cells of other vertebrates, including rodents, again raises two possibilities: (1) the lineage relationships differ in primates from those in lower vertebrates, or (2) restriction to a glial lineage is not reflected in GFAP expression. The fact that radial glia in rodents and birds also do not express GFAP, whereas those in primates do, is consistent with the second alternative.

On the other hand, the pattern of transient expression of (phosphorylated) NF-M that we observed (summarized in Fig. 9), not only in the cranial neural tube but in other sites in the embryo, is inconsistent with a conclusion that NF-M polypeptide expression, by itself, signifies restriction to a neuronal lineage. *All* cells in certain regions of the cranial neural tube and optic vessicle express NF-M, and, thus, their progeny must include not only neurons but glia and retinal pigment epithelial cells as well. Furthermore, by the time that significant numbers of postmitotic neuroblasts are being generated, the replicating cells have virtually all become NF negative. Thus, in most cases, even the immediate precursors of neuroblasts do not express phosphorylated NF-M. The significance of the transient NF-M expression is not clear at present. However, while not obviously related to lineage or to morphological features, it does indicate regional biochemical heterogeneity in the early neural tube for which there previously has been little, if any, evidence.

The IF isoform divergence in the derivatives of the embryonic ectoderm in the chick appears to coincide not only with the divergence in cell shape but also with the time during which presumptive neuroectoderm and surface ectoderm acquire the irreversible capacity to give rise to only neuroepithelial *or* epidermal definitive cells, respectively, as suggested by classical studies. The availability of a biochemical correlate of this event provides a new means of analyzing this phenomenon. One can now ask to what extent the switch in IF isoform in these cells is dependent on replication, for example, or on specific intracellular or extracellular molecular signals.

B. Morphological Differentiation of Postmitotic Neuroblasts

If we focus, now, on the initiation of "sustained" NF polypeptide expression, it appears that this event does not occur at the same time, with respect to the terminal mitosis, in all neuroblasts (Fig. 9). With what known event(s), then, does this initiation of neurofilament expression coincide? One possibility is that it may be related to the time at which a true axon begins to differentiate, an event that is known to vary with respect to the time of terminal mitosis, migration of the perikaryon, and site relative to perikaryon and leading and trailing processes (e.g., Cajal, 1960: Lyser, 1964, 1968; Domesick and Morest, 1977a,b). In motor neurons, for example, migration commences immediately, and the leading process of the migrating neuroblast becomes the axon. In the case of other neuroblasts, such as those of the optic tectum, migration of the cell body may be delayed and the differentiation of processes more complex. The question of expression vs phosphorylation will be an important one to explore in this regard. The

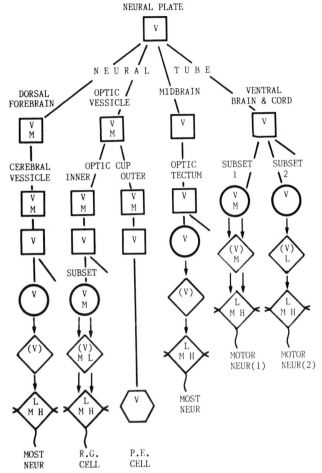

FIG. 9. Schematic summary of the progressive changes in intermediate filament (IF) expression during the establishment of the neural tube, selected regions of the primordial CNS, and the generation of some types of neurons. All replicating neuroepithelial cells in the indicated regions go through the sequences shown, but not necessarily synchronously, and the length of time spent at each step can vary. The relationship between switches in IF polypeptide expression and cell generations in cells *prior* to the terminal cell cycle is not known. A few examples are also given of likely sequences of changes in IF expression during the progression from the terminal cell cycle of the immediate precursor to the neuroblast daughter cell(s) and definitive neuron. Again, the time required to progress from one step to the next can vary greatly, and the examples given do not all occur simultaneously. Double arrows indicate a symmetrical division where our results suggest that both daughters become postmitotic neuroblasts and exhibit the same course of IF expression. In most cases (single arrow), it is not possible to determine whether or not the division is symmetrical. IF expression in glial pathways has been omitted. □, Replicating cell prior to terminal cell cycle; ○, cell in terminal mitosis; ◇, migrating postmitotic neuroblast; ⊱⊰, mature neuron; ○, definitive nonneuronal cell. V, vimentin; L, M, H, neurofilament L, M, and H, respectively; R.G. Cell, retinal ganglion cell; P.E. Cell, retinal pigment epithelial cell.

biochemical event that accompanies the initiation of an axon could very well be the phosphorylation, rather than the expression, of NF-M and/or NF-H. Certainly the availability of a *potential* biochemical correlate of this aspect of morphological differentiation will permit further experiments in which one can investigate the degree to which the two events are coupled and the nature of their regulation.

VII. Conclusions

The past few years have seen an enormous growth in basic information about the nature of intermediate filaments in different cells. Exciting new information is likely to be forthcoming as ongoing investigations in many laboratories focus on, and integrate, details of the initiation of expression, regulation, subcellular compartmentalization, significance of posttranslational modifications, and interactions among intermediate filament isoforms. The increasing awareness of possible cell-type-specific isoforms of other cytoskeletal macromolecules, especially tubulins and associated regulatory proteins, will undoubtedly be accompanied by further developmental analyses. Even with the limited number of studies presently available, there is a striking complementarity in the correlation of (phosphorylated) NF-M and/or NF-H expression with axon differentiation, on the one hand, and MAP-2 expression with dendritic differentiation on the other. It is not improbable that, within the near future, the divergence of, and progression through, lineages during embryonic development and steps in the differentiation of definitive cells will be associated with specific sequences in the switching on and off of many cytoskeleton-related genes, such that the shape, intracellular organization, and motility of cells at each step can be understood on the basis of the expression of a set of cytoskeletal proteins unique for each. It will then be possible to ask questions concerning the mechanisms and regulation of the progression—to what extent it is autonomous (intrinsically programmed) or subject to extrinsic factors.

ACKNOWLEDGMENTS

Work by the author and colleagues in Howard Holtzer's laboratory was supported by NIH Grants NS-17715, HL-15835, CA-18194, RR-05415 and by the Muscular Dystrophy Association.

REFERENCES

Bennett, G. S., and DiLullo, C. (1985a). *Dev. Biol.* **107,** 94.
Bennett, G. S., and DiLullo, C. M. (1985b). *Dev. Biol.* **107,** 107.
Bennett, G. S., and DiLullo (1985c). *J. Cell Biol.* **100,** 1799.
Bennett, G. S., Fellini, S. A., Croop, J. M., Otto, J. J., Bryan, J., and Holtzer, H. (1978). *Proc. Natl. Acad. Sci. U.S.A.* **75,** 4364.

Bennett, G. S., Fellini, S. A., Toyama, Y., and Holtzer, H. (1979). *J. Cell Biol.* **82,** 577.

Bennett, G. S., Tapscott, S. J., Kleinbart, F. A., Antin, P. B., and Holtzer, H. (1981). *Science* **212.** 567.

Bennett, G. S., Tapscott, S. J., and Holtzer, H. (1982). *In* "Changing Concepts of the Nervous System" (A. R. Morrison and P. L. Strick, eds.), p. 131. Academic Press, New York.

Bennett, G. S., Tapscott, S. J., DiLullo, C., and Holtzer, H. (1984a). *Brain Res.* **304,** 291.

Bennett, G. S., DiLullo, C., Biehl, J. B., and Holtzer, H. (1984b). *J. Cell Biol.* **99,** 317a.

Bernhardt, R., Huber, G., and Matus, A. (1985). *J. Neurosci.* **5,** 977.

Biehl, J., Holtzer, S., Bennett. G., Sun, T., and Holtzer, H. (1985). *Ann. N.Y. Acad. Sci.* **455,** 158.

Bignami, A., Dahl, D., and Seiler, M. W. (1980). *Dev. Neurosci.* **3,** 151.

Bignami, A., Raju, T., and Dahl, D. (1982). *Dev. Biol.* **91,** 286.

Bloom, G. S., and Vallee, R. B. (1983). *J. Cell Biol.* **96,** 1523.

Bloom, G. S., Luca, F. C., and Vallee, R. B. (1985). *Ann. N.Y. Acad. Sci.* **455,** 18.

Brady, S. T., Tytell, M., and Lasek, R. J. (1984). *J. Cell Biol.* **99,** 1716.

Brinkley, B. R., and Cartwright, J. (1975). *Ann. N.Y. Acad. Sci.* **253,** 428.

Burnside, B. (1973). *Am. Zool.* **13,** 989.

Cajal, S. R. (1960). "Studies on Vertebrate Neurogenesis" (L. Guth, trans.). Thomas, Springfield, Illinois.

Carbonetto, S., and Muller, K. J. (1982). *Curr. Top. Dev. Biol.* **17,** 33.

Carden, M. J., Schlaepfer, W. W., and Lee, V. M.-Y. (1985). *J. Biol. Chem.* **260,** 9805.

Choi, B. H., and Lapham, L. W. (1978). *Brain Res.* **148,** 295.

Ciment, G., and Weston, J. A. (1982). *Dev. Biol.* **93,** 355.

Ciment, G., and Weston, J. A. (1985). *Dev. Biol.* **111,** 73.

Ciment, G., Ressler, A., Letourneau, P. C., and Weston, J. A. (1986). *J. Cell. Biol.* **102.** 246.

Cochard, P., and Paulin, D. (1984). *J. Neurosci.* **4,** 2080.

Cooke, J. (1980). *Curr. Top. Dev. Biol.* **15,** 373.

Croop, J., and Holtzer, H. (1975). *J. Cell Biol.* **65,** 271.

Dahl, D. (1983). *Exp. Cell Res.* **149,** 397.

Dienstman, S. R., and Holtzer, H. (1975). *In* "Cell Cycle and Cell Differentiation" (J. Reinert and H. Holtzer, eds.), p. 1. Springer Verlag, New York.

Dlugosz, A. A., Tapscott, S. J., and Holtzer, H. (1983). *Cancer Res.* **43,** 2780.

Domesick, V. B., and Morest, D. K. (1977a). *Neuroscience* **2,** 459.

Domesick, V. B., and Morest, D. K. (1977b). *Neuroscience* **2,** 477.

Dräger, U. C. (1983). *Nature (London)* **303,** 169.

Eng, L. F., and DeArmond, S. J. (1982). *Adv. Cell. Neurobiol.* **3,** 145.

Fellini, S. A., Bennett, G. S., Toyama, Y., and Holtzer, H. (1978a). *Differentiation* **12,** 59.

Fellini, S. A., Bennett, G. S., and Holtzer, H. (1978b). *Am. J. Anat.* **153,** 451.

Franke, W. W., Schmid, E., Osborn, M., and Weber, K. (1978). *Proc. Natl. Acad. Sci. U.S.A.* **75,** 5034.

Franke, W. W., Schmid, E., Winter, S., Osborn, M., and Weber, K. (1979). *Exp. Cell Res.* **123,** 25.

Franke, W. W., Schmid, E., Schiller, D. L., Winter, S., Jarasch, E. D., Moll, R., Denk, H., Jackson, B. W., and Illmensee, K. (1981). *Cold Spring Harbor Symp. Quant. Biol.* **46,** 431.

Franke, W. W., Grund, C., Kuhn, C., Jackson, B. W., and Illmensee, K. (1982). *Differentiation* **23,** 43.

Gard, D. L., and Lazarides, E. (1980). *Cell* **19,** 263.

Ginzburg, I., Scherson, T., Rybak, S., Kimhi, Y., Neuman, D., Schwartz, M., and Littauer, U. Z. (1983). *Cold Spring Harbor Symp. Quant. Biol.* **48**, 783.
Glicksman, M. A., and Willard, M. (1985). *Ann. N.Y. Acad. Sci.* **455**, 479.
Godsave, S. F., Anderton, B. H., Heasman, J., and Wylie, C. C. (1984). *J. Embryol. Exp. Morphol.* **83**, 169.
Goldstein, M. E., Sternberger, L. A., and Sternberger, N. H. (1983). *Proc. Natl. Acad. Sci. U.S.A.* **80**, 3101.
Hamburger, V. (1948). *J. Comp. Neurol.* **88**, 221.
Havercroft, J. C., and Cleveland, D. W. (1984). *J. Cell Biol.* **99**, 1927.
Hirokawa, N., Glicksman, M. A., and Willard, M. B. (1984). *J. Cell Biol.* **98**, 1523.
Hoffman, P. N., and Lasek, R. J. (1975). *J. Cell Biol.* **66**, 351.
Holtzer, H., Bennett, G. S., Tapscott, S. J., Croop, J. M., Dlugosz, A., and Toyama, Y. (1981). *In* "International Cell Biology 1980–1981" (H. G. Schweiger, ed.), p. 293. Springer-Verlag, Berlin.
Holtzer, H., Bennett, G. S., Tapscott, S. J., Croop, J. M., and Toyama, Y. (1982). *Cold Spring Harbor Symp. Quant. Biol.* **46**, 317.
Houle, J., and Fedoroff, S. (1983). *Dev. Brain Res.* **9**, 189.
Huber, G., Alaimo-Beuret, D., and Matus, A. (1985). *J. Cell Biol.* **100**, 496.
Jacobs, M., Choo, Q. L., and Thomas, C. (1982). *J. Neurochem.* **38**, 969.
Jacobson, A. G. (1981). *In* "Morphogenesis and Pattern Formation" (T. G. Connelly, L. L. Brinkley, and B. M. Carlson, eds.), p. 233. Raven, New York.
Jacobson, M. (1978). "Developmental Neurobiology." Plenum, New York.
Johnston, R. N., and Wessells, N. K. (1980). *Curr. Top. Dev. Biol.* **16**, 165.
Jones, S. M., and Williams, R. C., Jr. (1982). *J. Biol. Chem.* **257**, 9902.
Julien, J. P., and Mushynski, W. E. (1982). *J. Biol. Chem.* **257**, 10467.
Kalfayan, L., and Wensink, P. C. (1982). *Cell* **29**, 91.
Kaspi, C., and Mushynski, W. E. (1985). *Ann. N.Y. Acad. Sci.* **455**, 794.
Kaufmann, E., Geisler, N., and Weber, K. (1984). *FEBS Lett.* **170**, 81.
Kemphues, D., Kaufman, T. C., Raff, R. A., and Raff, E. C. (1982). *Cell* **31**, 655.
Lasek, R. J., Phillips, L., Katz, M. J., and Autilio-Gambetti, L. (1985). *Ann. N.Y. Acad. Sci.* **455**, 462.
Lazarides, E. (1982). *Annu. Rev. Biochem.* **51**, 219.
Levitt, P., Cooper, M. L., and Rakic, P. (1981). *J. Neurosci.* **1**, 27.
Levitt, P., Cooper, M. L., and Rakic, P. (1983). *Dev. Biol.* **96**, 472.
Liem, R. K. H., Yen, S.-H., Salomon, G. D., and Shelanski, M. L. (1978). *J. Cell Biol.* **79**, 637.
Liem, R. K. H., Pachter, J. S., Napolitano, E. W., Chin, S. S. M., Moraru, E., and Heimann, R. (1985). *Ann. N.Y. Acad. Sci.* **455**, 492.
Lyser, K. M. (1964). *Dev. Biol.* **10**, 433.
Lyser, K. M. (1968). *Dev. Biol.* **17**, 117.
Murphy, D. B., Grasser, W. A., and Wallis, K. T. (1986). *J. Cell Biol.* **102**, 628.
Oblinger, M. M., and Schick, B. (1985). *J. Cell Biol.* **101**, 27a.
Pachter, J. S., and Liem, R. K. H. (1984). *Dev. Biol.* **103**, 200.
Payette, R. F., Bennett, G. S., and Gershon, M. D. (1984). *Dev. Biol.* **105**, 273.
Peters. A., Palay, S. L., and Webster, H., de F. (1976). "The Fine Structure of the Nervous System." Saunders, Philadelphia.
Raju, R., Bignami, A., and Dahl, D. (1981). *Dev. Biol.* **85**, 344.
Sauer, F. C. (1935). *J. Comp. Neurol.* **62**, 377.
Schlaepfer, W. W., and Freeman, L. A. (1978). *J. Cell Biol.* **78**, 653.
Schmid, E., Tapscott, S., Bennett, G. S., Croop, J., Fellini, S. A., Holtzer, H., and Franke, W. W. (1979). *Differentiation* **15**, 27.

Schnitzer, J., Franke, W. W., and Schachner, M. (1981). *J. Cell Biol.* **90**, 435.

Scott, D., Smith, K. E., O'Brien, B. J., and Angelides, K. J. (1985). *J. Biol. Chem.* **260**, 10736.

Sharp, G. A., Shaw, G., and Weber, K. (1982). *Exp. Cell Res.* **137**, 403.

Shaw, G., and Weber, K. (1982). *Nature (London)* **298**, 277.

Shaw, G., and Weber, K. (1983). *Eur. J. Cell Biol.* **30**, 219.

Shaw, G., and Weber, K. (1984). *Eur. J. Cell Biol.* **33**, 95.

Shaw, G., Osborn, M., and Weber, K. (1981). *Eur. J. Cell Biol.* **26**, 68.

Shaw, G., Debus, E., and Weber, K. (1984). *Eur. J. Cell Biol.* **34**, 130.

Shaw, G., Banker, G. A., and Weber, K. (1985). *Eur. J. Cell Biol.* **39**, 205.

Steinert, P. M., and Parry, D. A. D. (1985). *Annu. Rev. Cell Biol.* **1**, 41.

Sternberger, L. A., and Sternberger, N. H. (1983). *Proc. Natl. Acad. Sci. U.S.A.* **80**, 6126.

Sun, T.-T., and Green, H. (1978). *Cell* **14**, 469.

Tapscott, S. J., Bennett, G. S., Toyama, Y., Kleinbart, F. A., and Holtzer, H. (1981a). *Dev. Biol.* **86**, 40.

Tapscott, S. J., Bennett, G. S., and Holtzer, H. (1981b). *Nature (London)* **292**, 836.

Thompson, W. C., Asai, D. J., and Carney, D. H. (1984). *J. Cell Biol.* **98**, 1017.

Wang, E., Fischman, D., Liem, R. K. H., and Sun, T.-T. (1985). *Ann. N.Y. Acad. Sci.* **455**.

Weber, K., Shaw, G., Osborn, M., Debus, E., and Geisler, N. (1983). *Cold Spring Harbor Symp. Quant. Biol.* **48**, 717.

Willard, M., and Simon, C. (1981). *J. Cell Biol.* **89**, 198.

Willard, M., and Simon, C. (1983). *Cell* **35**, 551.

Williams, R. C., Jr., and Aamodt, E. J. (1985). *Ann. N.Y. Acad. Sci.* **455**, 509.

Wong, J., Hutchison, S. B., and Liem, R. K. H. (1984). *J. Biol. Chem.* **259**, 10867.

Yen, S.-H., and Fields, K. L. (1981). *J. Cell Biol.* **88**, 115.

Ziller, C., Dupin, E., Brazeau, P., Paulin, D., and Le Douarin, N. M. (1983). *Cell* **32**, 627.

CHAPTER 7

PLASMALEMMAL PROPERTIES OF THE SPROUTING NEURON

*Karl H. Pfenninger**

DEPARTMENT OF ANATOMY
COLUMBIA UNIVERSITY
NEW YORK, NEW YORK 10032

I. Introduction

The development of neuronal plasmalemma is particularly complex because it combines a number of features not usually associated with cellular differentiation. Once a neuron has undergone its terminal mitosis, its ultimate goal is to establish on its cell surface a series of specific functional domains characterized by the presence or absence of particular receptors, ion channels, junctional molecules, etc. (for review, see Bodian, 1967; Pfenninger, 1978). If we restrict this discussion to the (unmyelinated) axon and immediately adjacent regions, these domains include the axon hillock (separating the axon from the perikaryon), the axon shaft, and the nerve terminal with its highly differentiated presynaptic membrane. However, these domains are not evident during development. Prior to the mature stage, the postmitotic neuron is in a specialized phase of axon growth. During this phase the axon bears a highly differentiated tip, the nerve growth cone, which is capable of advancing between stationary cells, of pathfinding, and of recognizing the target cell(s) (Fig. 1) (see, e.g., Landis, 1983). Therefore, plasmalemmal differentiation of the postmitotic neuron is not simply a matter of generating cellular polarity and of massive cell surface expansion; it includes a highly specialized, transient stage during which unique membrane domains have to be established and maintained for a limited period of time.

II. The Emergence of Specific Membrane Domains

The emergence of the first growth cone from the postmitotic neuron initiates the development of this cell's extreme polarity. Within hours,

* Present address: Department of Cellular and Structural Biology, University of Colorado Health Sciences Center, Denver, Colorado 80262.

185

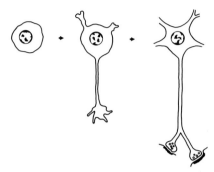

FIG. 1. Stages of neuronal differentiation. Upon its terminal mitosis, the initially rounded neuron (left) proceeds to form a neurite tipped by a nerve growth cone, a growth-specific, developmentally regulated structure (center). Contact of the growth cone with an appropriate target triggers synaptogenesis and the establishment of mature structures and cellular domains (right).

a nerve growth cone may be seen at a distance of tens or hundreds of micrometers from the perikaryon, interconnected with it by a thin, immature neurite. It is quite logical to postulate that, concomitant with the morphological polarization of the cell, there is regionalization of the plasmalemma into domains that contain specific membrane components. Indeed, attempts at demonstrating regional differences in plasmalemmal properties have been successful. Freeze–fracture studies on growing neurites *in vitro* (cf. Fig. 14) and *in vivo* have shown that, once an axon has reached a certain length, the population of plasmalemmal intramembrane particles (IMPs) is quite different at the growth cone compared to that at the perikaryon or the synaptic nerve terminal (Pfenninger and Bunge, 1974; Small and Pfenninger, 1984): Densities of IMPs (especially those of large size) in growth cone plasmalemma are much lower than in more proximal regions. Another approach that has led to similar conclusions is the use of ferritin-conjugated lectins to map the densities of specific surface oligosaccharides as a function of the distance from the perikaryon (Pfenninger and Maylié-Pfenninger, 1981a; cf. Denis-Donini *et al.*, 1978; Schwab and Landis, 1981). While some lectin receptors are more abundant in the plasmalemma of the perikaryon than in that of growth cone, others are more densely distributed at the growth cone than at the perikaryon. As in the case of the IMPs, there appears to exist a gradient between the two extreme points of the growing cell.

An early biochemical study of proximal (mainly perikaryon) vs distal (neurites plus growth cones) fractions microdissected from explant cultures of superior cervical ganglion failed to reveal differences in protein composition (Estridge and Bunge, 1978). However, a more

recent, similar attempt using sprouting explant cultures of chick retina has shown that the polypeptide pattern of the distal neurite is quite
different from that of the proximal neurite plus perikaryon (Schlosshauer and Schwarz, 1983: Schlosshauer, 1985). Furthermore, microdissected proximal (retinal) segments of the optic tract of the chick
were shown to contain the lower molecular-weight forms of the neural
cell adhesion molecule, N-CAM, whereas its larger, polysialic-acid-
containing form was found in the distal segments (Schlosshauer et al.,
1984). Other, more functional data come from studies regarding the
distribution of receptors for nerve growth factor (NGF) in peripheral
neurons. Campenot (1977) has found that NGF must be present near
the growth cone, but not the perikaryon, in order to maintain neurite
growth. Consistent with this observation is the finding that NGF stimulates phospholipid transmethylation in a preparation of distal neurites and growth cones but not in that of perikarya and proximal
neurites (Pfenninger and Johnson, 1981). Indeed, high-affinity receptors for NGF seem to be more concentrated in distal than in proximal
portions of the neuron (Carbonetto and Stach, 1982).

   Yet another parameter studied is [^{3}H]saxitoxin ([^{3}H]STX) binding,
a measure for the presence of voltage-dependent Na^+ channels. High-
affinity binding of [^{3}H]STX is considerably lower in the distal compared to the proximal neurite (Strichartz et al., 1984). The isolation, in
bulk, of growth-cone fragments, so-called growth-cone particles
(GCPs), from fetal rat brain (Pfenninger et al., 1983) has made it
possible to study various biochemical parameters of growth-cone membranes. Electron micrographs of GCPs are shown in Figs. 2 and 3. The
GCP preparation is also being used to study [^{3}H]STX receptors. Preliminary results indicate that high-affinity binding sites for [^{3}H]STX
are very sparse in GCPs, especially when compared to synaptosomes
(May and Pfenninger, 1986). This finding is consistent with electrophysiological observations on growing axons (Llinas and Sugimori,
1979; Spitzer, 1979; Grinvald and Farber, 1981; MacVicar and Llinas,
1985; Meiri et al., 1981). In summary, there are now several parameters (morphological, biochemical, and physiological) which indicate
that the growth-cone region of plasmalemma is different from more
proximal membrane areas of the growing neuron (Table I). However, it
should be pointed out that growth-cone plasmalemma itself may be
subdivided into different domains as suggested by freeze–fracture
studies (Small and Pfenninger, 1984). On the other hand, there is no
evidence for a sharp segregation of membrane domains between neurite and growth cone. Rather, a gradient of changing membrane properties seems to span the entire length of the growing neurite. The expression of cellular polarity at the membrane level raises the question of

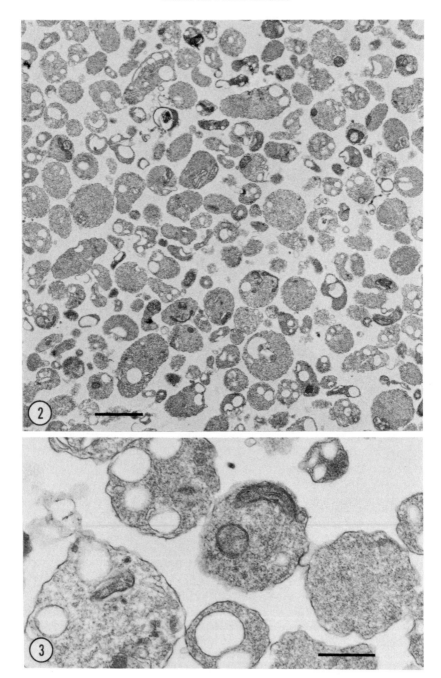

TABLE I

POLARIZATION OF MEMBRANE PROPERTIES IN THE GROWING NEURON[a]

Neuron type	Parameter	Plasmalemmal density (μm^{-2})		Reference[b]
		GC (distal)	PK (proximal)	
SCG	WGA receptors	2220	2560	1
SC	WGA receptors	880	1210	1
	RCA II receptors	850	420	1
RGC	N-CAM polyNANA	High	Low	2
DRG	NGF receptors	High	Low	3
ON	STX receptors	6.4[c]	9.7[d,e]	4
ON	IMPs (P-face)	830/185	2591[f]	5
		480[c]	1080[d]	
SC	IMPs (P-face)	78	high[g]	6
			(445)[h]	
SCG	IMPs (P-face)	Low	660[i]	6

[a]GC, Growth cone; PK, perikaryon; SCG, superior cervical ganglion (rat); SC, spinal cord anterior horn (rat); RGC, retinal ganglion cells/optic tract (chick); DRG, dorsal root ganglion (chick); ON, olfactory nerve (bullfrog); RCA I and II, WGA, agglutinins from *Ricinus communis* and wheat germ, respectively; N-CAM polyNANA, polysialylated form of neural cell adhesion molecule; NGF, nerve growth factor; STX, saxitoxin, a Na$^+$ channel ligand; IMPs (P-face), intramembrane particles as observed by freeze–fracture electron microscopy in the inner, protoplasmic membrane leaflet.

[b]1, Pfenninger and Maylié-Pfenninger (1981a); 2, Schlosshauer *et al.* (1984); 3, Carbonetto and Stach (1982); 4, Strichartz *et al.* (1984); 5, Small and Pfenninger (1984); 6, Pfenninger and Bunge (1974).

[c]Distal shaft.

[d]Proximal shaft.

[e]Comparable values for growth cones and perikarya are not available at present; however, extrapolation of values results in considerably higher receptor density for the perikaryon and lower values for the growth cone; binding studies on a growth-cone fraction isolated froom fetal rat brain (GCPs) show very low levels of STX binding (see text).

[f]Growth cones in this system exhibit domains with a relatively high IMP density (830 μm^{-2}) next to IMP-poor membrane areas.

[g]Measurements are not available for perikarya of SC neurons, but shaft values increase with time in culture and reach high levels.

[h]Average value for neurite shaft at 13 days in culture.

[i]Measurements are not available for growth cones of SCG neurons.

FIGS. 2 AND 3. Fragments sheared off nerve growth cones—growth cone particles (GCPs)—isolated by subcellular fractionation from fetal rat brain. The low-power view (Fig. 2) shows the uniformity of the fraction; the higher magnification (Fig. 3) reveals structural detail, which is consistent with that of growth cones *in vivo* or in culture. Many of the biochemical results reviewed and discussed in this chapter are based on the analysis of this GCP fraction. Bars, 2 μm (Fig. 2) and 0.5 μm (Fig. 3).

how specialized domains and intervening gradients are being established. This question will be discussed further below (Section IV).

III. Developmentally Regulated Membrane Components

The observations just reviewed state that the growth cone's plasmalemmal properties are different from those in other regions of the growing neuron but do not reveal whether some of the membrane components concentrated in the growth cone are expressed specifically during neurite growth. Such developmentally regulated components may be identified, for instance, by studying proteins synthesized and transported down the axon in the growing vs the mature neuron (cf. Willard et al., 1985). They can then be detected in the neurons' target area with or without isolation of growth-cone particles and synaptosomes, respectively. Such studies have led to the identification of so-called growth-associated proteins (GAPs), in particular, GAP-43 (e.g., Skene and Willard, 1981; Benowitz et al., 1981). GAP-43 is highly enriched in the growth-cone particle fraction and identical to one of the growth cone's major membrane proteins, a phosphorylated polypeptide of 46,000 Da (pp46) (Ellis et al., 1985a; Katz et al., 1985; Willard et al., 1985; Simkowitz et al., 1986; Meiri et al., 1986). Similar studies have shown that all three major membrane proteins of growth cones (M_r = 46, 38, 34 kDa; Fig. 4) are developmentally regulated, i.e., growth dependent (Simkowitz and Pfenninger, 1983; Simkowitz et al., 1986). However, the specificity of their association with the membrane of the growth cone rather than that of the more proximal shaft is not known at the present time. This question can be answered only by quantitative immunolocalization with appropriate antibodies.[1]

The study of carbohydrate incorporation into dividing vs sprouting (NGF-stimulated) PC12 cells has led to the identification of a "NGF-inducible large external" glycoprotein (NILE) also present in neurons of the normal peripheral (about 210 kDa) and central (about 230 kDa) nervous systems (Salton et al., 1983). However, it seems that the increased synthesis of NILE glycoprotein during growth is largely a function of surface expansion rather than a specific association of this molecule with the growth process.

Another approach to the identification of growth-specific membrane components is the immunochemical comparison of growth-cone particles with synaptosomes. This comparison has led to the identification of an antigen (5B4) which is highly enriched in growth cone mem-

---

[1] Indirect immunofluorescence with anti-GAP-43 antibody on primary cultures reveals particularly bright fluorescence in the growth-cone-containing halo, suggesting concentration GAP-43/pp46 in the growing tips (Meiri et al., 1986).

CB
GCP   LYS   GCM

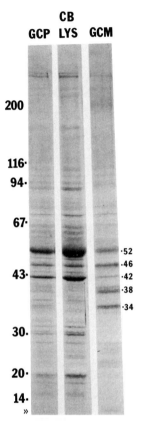

200

116·
94·

67·

·52
·46
43·
·42
·38
·34

30·

20·

14·
»

FIG. 4. SDS–polyacrylamide gel electrophoresis of proteins from whole GCPs, GCP lysate (LYS), and salt-washed GCP membranes (GCM); Coomassie Blue staining. Lanes are from a 5–15% acrylamide gradient slab gel, and the position of molecular mass markers (in kDa) is indicated on the left. Note the many bands in GCP and LYS as opposed to the simplified pattern of GCM. The bands at 42 and 52 kDa contain, or consist of, actin and tubulin, respectively. For further description, see text. (From Ellis *et al.*, 1985b.)

branes but virtually absent from synaptosomal membranes (Ellis *et al.*, 1985b; Wallis *et al.*, 1985) (Figs. 5 and 6). This antigen is a large, polysialylated glycoprotein (M_r = 185,000–255,000 which belongs to N-CAM/BSP-2/D2, a family of developmentally regulated, sometimes neuron-specific molecules (Jorgensen, 1981; Hoffman *et al.*, 1982; Rougon *et al.*, 1982; cf. also Schlosshauer, 1985). The 185- to 255-kDa 5B4 antigen is expressed exclusively in postmitotic neurons and during neurite growth. Immunofluorescence shows that the antigen is abundant in growth cones but also can be found in other domains of the

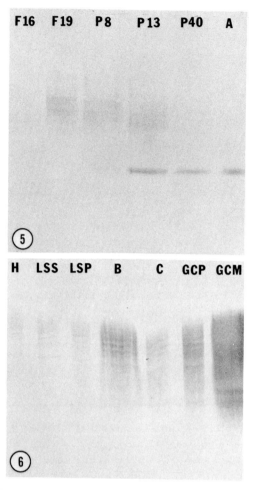

Figs. 5 and 6. Properties of the 5B4 antigen, a neuronal membrane glycoprotein, as studied by Western blot (from 5% acrylamide gels). Figure 5 shows immunoreactivity in crude membrane fractions of developing rat brain. Between approximately 16 days of gestation and 13 days postnatal, the antibody recognizes a broad band of about 185 to 255 kDa. This is the major antigen by far. During postnatal development, a second, relatively sharp band appears at 140 kDa. This minor antigen persists into adulthood, but it is essentially absent from synaptosomes. Figure 6 shows the enrichment of the fetal 185- to 255-kDa antigen in subfractions of fetal (17–18 days gestation) rat brain. H, Homogenate; LSP, low-speed pellet; LSS, low-speed supernatant (parent fraction of all subsequent fractions); GCP, growth cone particles; B and C, heavier, heterogeneous subfractions of LSS; GCM, salt-washed membranes prepared from GCPs. (From Ellis *et al.*, 1985b.)

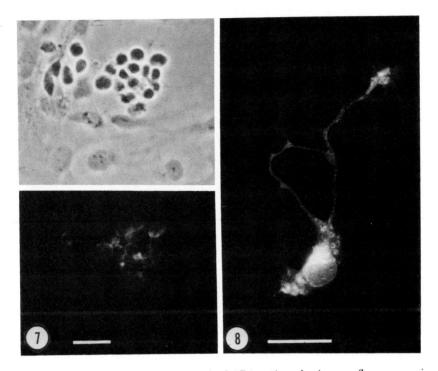

FIGS. 7 AND 8. Localization of the fetal 5B4 antigen by immunofluorescence in cultures of neurons from fetal rat cortex. It is important to note that detection of the antigen necessitates permeabilization of membranes prior to staining. Figure 7 shows a group of recently explanted neurons that have not yet produced recognizable neurites. 5B4 staining is minimal. Furthermore, nonneuronal cells in the background of this figure are not stained. In sprouting neurons (Fig. 8), bright fluorescence is evident in the growth cone and in the perikaryon at putative sites of synthesis of the antigen. The neuron shown appears to have a second growth cone projecting back onto the perikaryon; this may account for part of the intense fluorescence in this region. Bars, 20 μm. (From Wallis *et al.*, 1985.)

sprouting neuron (Figs. 7–11). Further biochemical comparisons of growth-cone particles with synaptosomes have resulted in the identification of developmentally regulated calmodulin binding proteins (Hyman and Pfenninger, 1985) and protein kinase substrates (Ellis *et al.*, 1985a; Katz *et al.*, 1985). Particularly abundant in growth-cone particles but not synaptosomes are substrates of a calcium/phospholipid-dependent protein kinase (protein kinase C) and the major substrate of a calcium/calmodulin-dependent protein kinase, pp46, the 46,000-Dalton major membrane protein mentioned above (Fig. 12). Because these proteins and associated kinase activities are growth

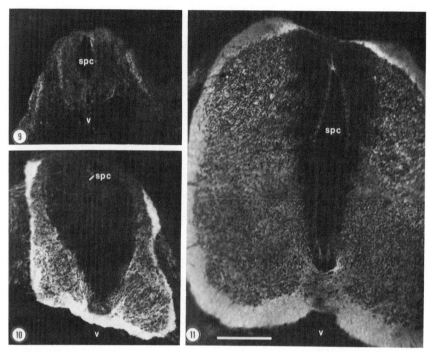

FIGS. 9–11. Localization by immunofluorescence of the (fetal) 5B4 antigen in the developing spinal cord of the rat. Cross sections of fetus of 11 (Fig. 9), 13 (Fig. 10), and 15 (Fig. 11) days of gestation are shown. spc, Spinal canal; v, ventral. Note the virtual absence of immunoreactivity at 11 days, followed by bright fluorescence in ventrolateral regions of the cord at 13 days of gestation. The fluorescent regions coincide with those containing the sprouting motor neurons and growing axons. The central and more dorsal parts of the spinal cord, which consist of dividing neuroblasts and nonneuronal cells, are not fluorescent. At 15 days of gestation, with the increase in the number of differentiated neurons and growing axons, the fluorescent area is greatly expanded toward the center, but the intensity in fluorescence is somewhat decreased. At 18 days of gestation and later, fluorescence is greatly reduced in the spinal cord. Bar, 200 μm. (From Wallis *et al.*, 1985.)

regulated and have been identified in growth-cone particles, they are likely to play a major role in growth-cone membrane function. However, we do not have quantitative data on their regional distribution in the sprouting neuron.

Yet another type of biochemical analysis has recently produced additional evidence indicating the unusual properties of growth-cone membranes. A complete analysis of lipids and protein in growth cone membranes, largely stripped of soluble contaminants at high ionic strength, has revealed a high lipid-to-protein ratio of about 3.6 to 1, similar to that of myelin. Growth-cone membrane lipid includes a sub-

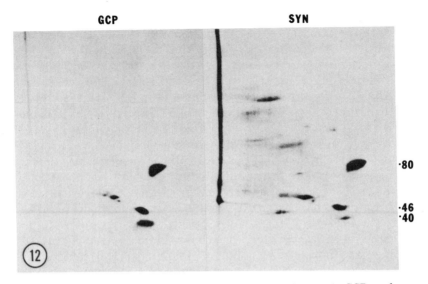

FIG. 12. Major substrates of calcium-dependent protein kinases in GCPs and synaptosomes (SYN) as resolved by two-dimensional gel electrophoresis (isoelectric focusing followed by SDS–PAGE). Note the three major substrates of GCPs at 80, 46, and 40 kDa, with isoelectric points of 4.0, 4.3, and 4.2, respectively. The tubulin complex is seen at about 52 kDa. The same phosphoproteins are observed in synaptosomes but here they are minor substrates, and much longer exposure times are required for their detection. (From Katz *et al.*, 1985.)

stantial complement of cholesterol, gangliosides, and neutral glycosphingolipids (Sbachnig-Agler, Ledeen, and Pfenninger, 1987). The presence of neutral glycosphingolipids (see also Zurn, 1982; Dodd *et al.*, 1984) is particularly unusual because these components are essentially absent from synaptosomes and, in the mature brain, are believed to occur almost exclusively in glial cells (myelin galactocerebroside). Whether specific lipid components are enriched in the growth-cone membrane relative to other plasmalemmal regions of the developing neuron is not known. Unfortunately, there are no methods available at the present for isolating, in large quantity, cellular regions other than the growth cone from developing neurons. Thus, a complete regional analysis of membrane components is currently not possible.

IV. Mechanism of Membrane Expansion

The development of neuronal polarity has to be seen in the context of the dramatic increase in cell surface experienced by the sprouting neuron. At least in theory, membrane expansion may occur in random fashion or as a localized phenomenon by the insertion of membrane components at a specific site (cf. Pfenninger, 1979b). In a vectorially

growing cell such as the sprouting neuron, insertion sites may be located at the perikaryon or the growing tip. The work of Hughes (1953)
and of Bray (1970, 1973) has suggested that there is an internal pool of
membrane available in the distal neurite and that the addition of new
membrane may occur at the growth cone. Several lines of evidence
have established more recently that this is indeed the case. Pulse–
chase experiments with a lipid precursor, [3H]glycerol, have demonstrated that most of the newly synthesized phospholipid is rapidly
transported into the distal neurite by an energy-dependent and col-
chicine-sensitive process (Fig. 13) (Pfenninger and Johnson, 1983). In
complementary autoradiographic studies, radiolabeled phospholipid
has been seen to appear first over clusters of large, clear vesicles in the
growth cone and only later over the plasma membrane (Pfenninger,
1980). These vesicles, also called "growth-cone vesicles" or "plasmalemmal precursor vesicles," are characterized by paucity of IMPs,
just like growth-cone plasmalemma (Pfenninger and Bunge, 1974;
Small and Pfenninger, 1984).[2] Additionally, it was demonstrated that
the plasmalemma covering clusters of such vesicles is the site where
lectin receptors appear during neurite growth (Figs. 14–16) (Pfenninger and Maylié-Pfenninger, 1981b; Feldman et al., 1981). Therefore, little doubt remains that the perikaryon synthesizes a specialized
type of membrane that is (presumably) sequestered from the Golgi
apparatus in the form of plasmalemmal precursor vesicles. These vesicles are then exported into the neurite periphery by rapid axoplasmic
transport for insertion, probably by an exocytosis-like mechanism (see
Fig. 17).

The principle just described is suitable for generating at and near
the growing tip a membrane domain that consists of most recently
synthesized and, possibly, specialized components. However, along the
shaft of the neurite, plasmalemmal components may mix by lateral
diffusion. Quantitative analysis of the densities of IMPs of specific size
has revealed in growing axons the presence of a series of proximodistal
gradients whose slope is dependent upon IMP size (Small and Pfen-

---

[2] Hasty and Hay (1978) and Rees and Reese (1981) have claimed that these vesicle
clusters are fixation artifacts. While fixation may alter the morphology of these structures in an unknown way, they are highly likely to represent functionally important
structures of the growth cone for the following reasons (cf. Pfenninger, 1979a): correlation with structures visible by phase-contrast microscopy in live growth cones; abundance in GCPs where plasmalemmal precursor vesicles constitute >33% of total membrane present (Pfenninger et al., 1983); lack of internalization of extracellular tracer
into the vesicles under appropriate conditions of fixation (see, however, Bunge, 1977);
and selective labeling in pulse–chase experiments with [3H]glycerol and with lectins as
summarized here.

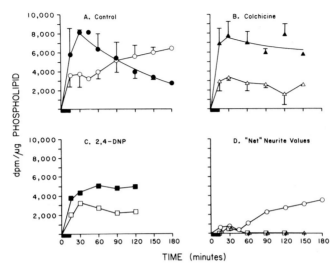

FIG. 13. Incorporation of [³H]glycerol into phospholipid in superior cervical ganglion neurons sprouting in culture. Following a 15-minute pulse with tritiated glycerol, the neurons were chased for different periods of time. Perikarya and neurites were then separated by microdissection and the radioactivity of phospholipids measured. ●, ■, Values for perikarya; ○, □, values for neurites. Note the rapid rise and decline of phospholipid counts in control perikarya vs the delayed increase observed for neurites. Colchicine, the axoplasmic transport poison, blocks the increase in radioactivity in the neurite and maintains the high perikaryal level. The metabolic blocker 2,4-dinitrophenol (2,4-DNP) reduces perikaryal incorporation somewhat and blocks the increase in neurite radioactivity completely. "Net" neurite values are obtained by subtraction of the radioactivity present in the neurite fraction at 15 minutes and not sensitive to colchicine (probably incorporated into contaminating supporting cells). The gradual increase observed in the control (○) is not evident after colchicine or 2,4-DNP treatment (△ and □, respectively). These findings indicate that newly synthesized phospholipid is rapidly transported into the growing neurite by an energy- and axoplasmic-transport-dependent process, presumably for insertion into the plasmalemma at the growth cone. (From Pfenninger and Johnson, 1983.)

ninger, 1984; Small *et al.*, 1984). These gradients resemble exponential curves (Gaussian tails) but, in fact, are much better fitted by equations that describe the diffusion of components in a system whose distal boundary is moving at a constant rate (Figs. 18, 19). The latter mathematical model suggests that, as particle-poor membrane is inserted at the growing tip concomitant with axon elongation, the components constituting IMPs diffuse at a very rapid rate (about 10^{-8} cm²/second) from perikaryal plasmalemma into the axolemma. This model predicts that (1) early on, the plasmalemma of the *emerging* growth cone is rich in IMPs because IMPs can very rapidly diffuse into it, and (2) the IMP-

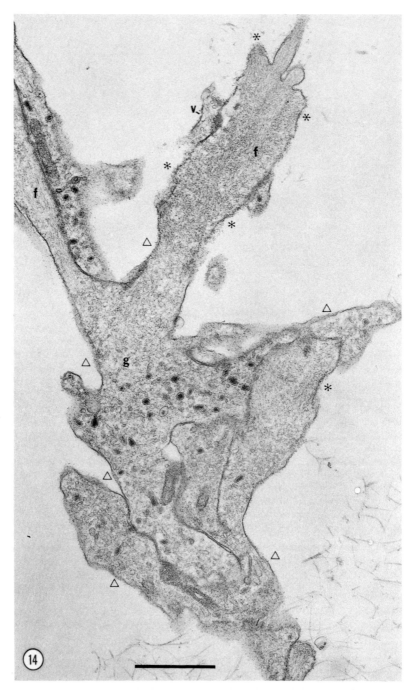

FIG. 14. Electron micrograph of a nerve growth cone in culture, from a pulse–chase experiment with wheat germ agglutinin (WGA) and its ferritin conjugate (F-WGA). Exposed surface receptors of the lectin were first saturated with (the invisible) WGA. After washing, the cultures were allowed to survive for 3 minutes. Lectin receptors

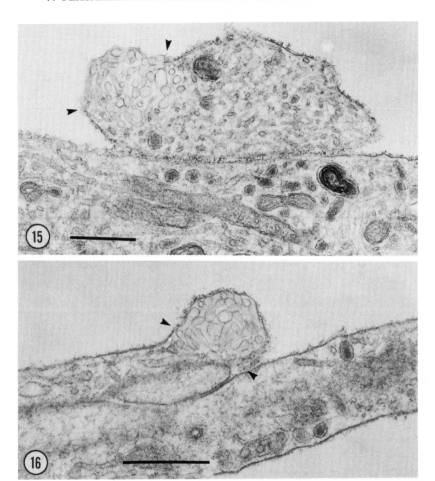

FIGS. 15 AND 16. Electron micrographs of distal regions of growing neurites from pulse–chase experiments with WGA. The cellular regions shown in Fig. 15 were pulsed with F-WGA for 5 minutes and then chased for 15 minutes. Note the absence of label from the surface of the mound-like, vesicle-filled protrusion. The neurites shown in Fig. 16 were first labeled with WGA, then chased for 15 minutes and subsequently relabeled with F-WGA (same experiment as in Fig. 14). These experiments indicate that, during neurite growth, surface components are moved away from areas overlying clusters of large, clear vesicles, and that this is due to the insertion of new lectin receptors. This suggests that the vesicles are the plasmalemmal precursor. Bar 0.5 μm. (From Pfenninger and Maylié-Pfenninger, 1981b.)

---

newly inserted into the cell surface during this chase period were then labeled with F-WGA. The asterisks mark areas, primarily on filopodia (f) and over vesicle clusters (v), where a relatively high density of the ferritin label is observed. △, Regions of low labeling density (usually proximal). Bar, 1 μm. (From Pfenninger and Maylié-Pfenninger. 1981b.)

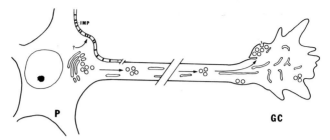

F$_{IG}$. 17. Membrane biogenesis in the growing neuron, schematic of hypotheses. Most newly synthesized membrane components, especially the lipids, are sequestered from the Golgi apparatus in the form of large, clear vesicles and then rapidly transported to the growth cone for plasmalemmal insertion by an exocytosis-like mechanism. Other components, presumably certain integral membrane proteins forming IMPs, are inserted by an unknown mechanism at the perikaryon and are then free to diffuse within the plasmalemma into the axon shaft.

poor properties characteristic of growth cones distant from the perikaryon are only gradually established, as axon elongation and axolemmal expansion outstrip lateral diffusion of IMPs. Freeze–fracture studies on neurons that are just starting to sprout and others that have been advancing for some distance have indeed borne out these points (Figs. 20 and 21). These findings suggest that the establishment of the growth cone's specialized plasmalemmal domain is the result of a *dynamic equilibrium* between the addition of new membrane at the growing tip and lateral diffusion of membrane components from the perikaryon (Fig. 17). Such a model predicts that neurites have to grow for tens of microns until plasmalemmal properties characteristic of growth cones are fully expressed at the tip. A further corollary is that the axon hillock as a domain separating axon shaft and perikaryon does not exist in the sprouting neuron.

V. Synaptogenesis: Plasmalemmal Maturation and Emergence of New Domains

During synaptogenesis, growth-cone components are gradually replaced by those of the presynaptic ending; in particular, the presynaptic membrane with its cytoplasmic and surface (i.e., junctional) specializations is established (see, e.g., Pfenninger and Rees, 1976; Pfenninger, 1979b). Although not understood in any detail, this process undoubtedly involves the insertion of new components into the membrane. It is likely, but not known, that such components are targeted specifically to the synaptic site. The molecular changes of the maturing axon tip are also evident from the comparison described

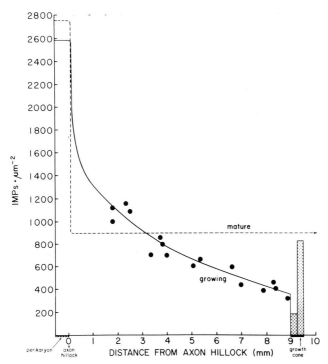

FIG. 18. Density profiles of intramembrane particles as a function of distance from the perikaryon, in growing (solid line) and mature (dashed line) olfactory axons of the bullfrog. Measurements are for IMPs in the protoplasmic leaflet (P-face) only, but the relative particle densities are the same for external-leaflet particles. In the perikaryon, IMP densities are very similar or identical in growing vs mature neurons. In the mature neuron, there is a sharp drop in IMP density at the axon hillock, followed by a constant level in the axolemma. In the growing neuron, however, IMP density decreases continually, forming a gradient. This gradient resembles a Gaussian tail but is fitted most accurately by the function describing diffusion in a moving-boundary system (cf. Small et al., 1984). At the growth cone, the plasmalemma can be subdivided into distinct domains characterized by relatively high or relatively low IMP densities.

earlier of proteins synthesized by growing vs synapsing neurons. The synthesis of growth-cone membrane components is reduced whereas that of other components is greatly increased (cf. Benowitz et al., 1981; Skene and Willard, 1981; Simkowitz et al., 1986). At this stage, rapid axon elongation has ceased, i.e., the distal boundary of the axolemmal diffusion system has become stationary, so that components of the axon's shaft (between axon hillock and synaptic ending) become evenly distributed (Fig. 18) (Small et al., 1984). However, a comparison of plasmalemmal IMPs at the perikaryon and the proximal shaft now

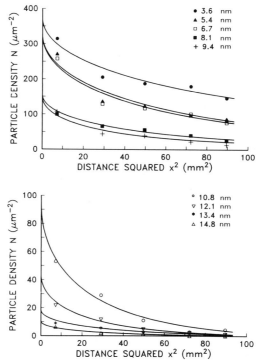

FIG. 19. The density of axolemmal IMPs in growing olfactory nerve of the bullfrog, plotted as a function of the distance from the perikaryon and as a function of size. The density distributions for the different IMP sizes have been fitted with theoretically derived gradients of components diffusing in a moving-boundary system. (From Small *et al.*, 1984.)

reveals a sharp drop in particle density, indicating that an effective barrier to diffusion has been established at the axon hillock. Although no direct evidence is available, it seems reasonable to suggest that the establishment of this barrier involves the use of submembrane proteins to anchor membrane components of the axon hillock in place. Likewise, the regional specialization of the presynaptic membrane is

FIGS. 20 AND 21. Freeze–fracture electron micrographs of a newly formed sprout (s) of a sympathetic neuron (Fig. 20) and of a more advanced growth cone (gc) of an olfactory-bulb neuron (Fig. 21), both grown in culture. Note the much higher IMP density in the new sprout compared to that in the well-developed growth cone. The arrowheads point to IMPs in the protoplasmic leaflet of the membrane. n, Neurite shaft; p, perikaryon. Bars, 1 μm. (From Small *et al.*, 1984; Pfenninger and Bunge, 1974.)

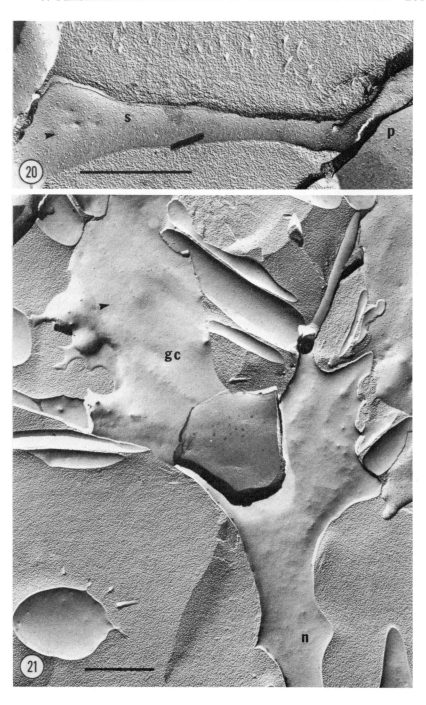

maintained by anchoring and/or cross-linking its components and, possibly, by their interaction (in the synaptic cleft) with molecules of the postsynaptic membrane.

VI. Conclusions

Various lines of evidence indicate that sprouting, i.e., the development of morphological polarity of the neuron, is accompanied by a transient regional membrane differentiation that is different from the one observed in the maturing neuron. This phenomenon appears to be the result of the insertion of a specialized type of membrane at or near the growing tip. In terms of their topography and composition, the plasmalemmal domains of the sprouting neuron, essentially that of the growth cone and distal neurite vs that of the perikaryon and proximal neurite, are quite different from the domains distinguishable in the mature neuron. Furthermore, rather than being static, i.e., the result of anchoring of membrane components, the growing neuron's membrane regions appear to be the result of a dynamic equilibrium between the increase in surface area, i.e., the addition of membrane of one particular type at the growth cone, and the proximodistal diffusion of perikaryal membrane constituents along the axon shaft. The cessation of growth and synaptogenesis appears to lead to the collapse of this dynamic equilibrium (or relaxation of the gradients involved), and new, static membrane domains are established by the differentiation of the presynaptic membrane and the establishment of the axon hillock's diffusion barrier between the perikaryon and axon shaft. It should be stressed that this model of emergence, maintenance, and alteration of membrane domains in the developing neuron may be incomplete or inaccurate because of the limited amount of data available today. We cannot exclude the possibility that mechanisms other than those mentioned may significantly contribute to the phenomena discussed. Nevertheless, the identification of a specific plasmalemmal domain at the growth cone and the increasing knowledge of its components are expected to yield major new insights into growth cone function, i.e., growth control, pathfinding, and synaptogenesis.

ACKNOWLEDGMENTS

The research of my laboratory reviewed here would not have been possible without the participation of my past and present associates. Their many contributions are gratefully acknowledged. I would also like to thank Ruth Carpino and Linda B. Friedman for their expert assistance with the preparation of this manuscript. This research is and has been supported by grants from the NIH (NS 13466 and NS 21729), the NSF (BNS-83-10248), the Stifel Paralysis Research Foundation, the National Spinal Cord Injury Association, and by an I. T. Hirschl Career Scientist Award.

REFERENCES

Benowitz, L. I., Shashoua, V. E., and Yoon, M. G. (1981). *J. Neurosci.* **1**, 300–307.

Bodian, D. (1967). *In* "The Neurosciences, A Study Program" (G. C. Quarton, T. Melnechuk, and F. O. Schmitt, eds.), pp. 6–24. Rockefeller Univ. Press, New York.

Bray, D. (1970). *Proc. Natl. Acad. Sci. U.S.A.* **65**, 905–910.

Bray, D. (1973). *J. Cell Biol.* **56**, 702–712.

Bunge, M. B. (1977). *J. Neurocytol.* **6**, 407–439.

Campenot, R. B. (1977). *Proc. Natl. Acad. Sci. U.S.A.* **74**, 4516–4519.

Carbonetto, S., and Stach, R. W. (1982). *Dev. Brain Res.* **3**, 463–473.

Denis-Donini, S., Estenoz, M., and Augusti-Tocco, G. (1978). *Cell Differ.* **7**, 193–201.

Dodd, J., Solter, D., and Jessell, T. M. (1984). *Nature (London)* **311**, 469–472.

Ellis, L., Katz, F., and Pfenninger, K. H. (1985a). *J. Neurosci.* **5**, 1393–1401.

Ellis, L., Wallis, I., Abreu, E., and Pfenninger, K. H. (1985b). *J. Cell Biol.* **101**, 1977–1989.

Estridge, M., and Bunge, R. (1978). *J. Cell Biol.* **79**, 138–155.

Feldman, E. L., Axelrod, D., Schwartz, M., Heacock, A. M., and Agranoff, B. W. (1981). *J. Neurobiol.* **12**, 591–598.

Grinvald, A., and Farber, I. (1981). *Science* **212**, 1164–1169.

Hasty, D. L., and Hay, E. D. (1978). *J. Cell Biol.* **78**, 756–768.

Hoffman, S., Sorkin, B. C., White, P. C., Brackenbury, R., Mailhammer, R., Rutishauser, U., Cunningham, B. A., and Edelman, G. M. (1982). *J. Biol. Chem.* **257**, 7720–7729.

Hughes, A. (1953). *J. Anat.* **87**, 150–162.

Hyman, C., and Pfenninger, K. H. (1985). *J. Cell Biol.* **101**, 1153–1160.

Jorgensen, O. S. (1981). *J. Neurochem.* **37**, 939–946.

Katz, F., Ellis, L., and Pfenninger, K. H. (1985). *J. Neurosci.* **5**, 1402–1411.

Landis, S. C. (1983). *Annu. Rev. Physiol.* **45**, 567–580.

Llinas, R., and Sugimori, M. (1979). *Prog. Brain Res.* **51**, 323–334.

MacVicar, B. A., and Llinas, R. R. (1985). *J. Neurosci. Res.* **13**, 323–335.

May, N., and Pfenninger, K. H. (1986). In preparation.

Meiri, H., Parnas, I., and Spira, M. (1981). *Science* **211**, 709–711.

Meiri, K., Pfenninger, K. H., and Willard, M. B. (1986). *Proc. Natl. Acad. Sci. U.S.A.* **83**, 3537–3541.

Pfenninger, K. H. (1978). *Annu. Rev. Neurosci.* **1**, 445–471.

Pfenninger, K. H. (1979a). *In* "Freeze-Fracture: Methods, Artifacts and Interpretations" (J. E. Rash and C. S. Hudson, eds.), pp. 71–80. Raven, New York.

Pfenninger, K. H. (1979b). *In* "The Neurosciences: Fourth Study Program" (F. O. Schmitt, ed.), pp. 779–795. MIT Press, Cambridge, Massachusetts.

Pfenninger, K. H. (1980). *Soc. Neurosci. Abstr.* **6**, 661.

Pfenninger, K. H., and Bunge, R. P. (1974). *J. Cell Biol.* **63**. 180–196.

Pfenninger, K. H., and Johnson, M. P. (1981). *Proc. Natl. Acad. Sci. U.S.A.* **78**, 7797–7800.

Pfenninger, K. H., and Johnson, M. P. (1983). *J. Cell Biol.* **97**. 1038–1042.

Pfenninger, K. H., and Maylié-Pfenninger, M. F. (1981a). *J. Cell Biol.* **89**, 536–546.

Pfenninger, K. H., and Maylié-Pfenninger, M. F. (1981b). *J. Cell Biol.* **89**, 547–559.

Pfenninger, K. H., and Rees, R. P. (1976). *In* "Neuronal Recognition" (S. H. Barondes, ed.), pp. 131–178 and 357–358. Plenum, New York.

Pfenninger, K. H., Ellis, L., Johnson, M. P., Friedman, L. B., and Somlo, S. (1983). *Cell* **35**, 573–584.

Rees, R. P., and Reese, T. S. (1981). *Neuroscience* **6**, 247–254.

Rougon, G., Deagostini-Bazin, H., Hirn, M., and Goridis, C. (1982). *EMBO J.* **10**, 1239–1244.

Salton, S. R. J., Richter-Landsberg. C., Greene, L. A., and Shelanski, M. L. (1983). *J. Neurosci.* **3**, 441–454.

Sbachnig-Agler, M., Ledeen, R., and Pfenninger, K. H. (1987). In preparation.

Schlosshauer, B. (1985). *Dev. Brain Res.* **19**, 237–244.

Schlosshauer, B., and Schwarz, U. (1983). *Neurosci. Lett.* **14**, 330.

Schlosshauer, B., Schwarz, U., and Rutishauser, U. (1984). *Nature (London)* **310**, 141–143.

Schwab, M., and Landis, S. (1981). *Dev. Biol.* **84**, 67–78.

Simkowitz, P., and Pfenninger, K. H. (1983). *Neurosci. Abstr.* **98**, 1422–1433.

Simkowitz, P., Ellis, L., and Pfenninger, K. H. (1986). Submitted.

Skene, J. H. P., and Willard, M. (1981). *J. Cell Biol.* **89**, 96–103.

Small, R. K., and Pfenninger, K. H. (1984). *J. Cell Biol.* **98**, 1422–1433.

Small, R. K., Blank, M., Ghez, R., and Pfenninger, K. H. (1984). *J. Cell Biol.* **98**, 1434–1443.

Spitzer, N. C. (1979). *Annu. Rev. Neurosci.* **2**, 363–397.

Strichartz, G. R., Small, R. K., and Pfenninger, K. H. (1984). *J. Cell Biol.* **98**, 1444–1452.

Wallis, I., Ellis, L., Suh, K., and Pfenninger, K. H. (1985). *J. Cell Biol.* **101**, 1990–1998.

Willard, M. B., Meiri, K., and Glicksman, M. (1985). *In* "Molecular Bases of Neural Development" (G. M. Edelman, W. E. Gall, and W. M. Cowan, eds.), pp. 341–361. Neurosciences Research Foundation, New York.

Zurn, A. D. (1982). *Dev. Biol.* **94**, 483–498.

CHAPTER 8

CARBONIC ANHYDRASE: THE FIRST MARKER OF GLIAL DEVELOPMENT

Ezio Giacobini

DEPARTMENT OF PHARMACOLOGY
SOUTHERN ILLINOIS UNIVERSITY
SCHOOL OF MEDICINE
SPRINGFIELD, ILLINOIS 62708

I. Introduction

Carbonic anhydrase (CA) is a zinc-containing enzyme which was first demonstrated in erythrocytes by Meldrum and Roughton (1934). The enzyme (E) catalyzes the reaction

$$CO_2 + H_2O \overset{E}{\rightleftharpoons} H_2CO_3 \rightleftharpoons H^+ + HCO_3^-.$$

Carbonic anhydrase is found in several tissues of vegetal and animal origin. Van Goor (1940) demonstrated the early appearance of CA activity in chick embryo retina. The presence of CA in the adult nervous system was first demonstrated in 1943 by Ashby who reported that, in several species, CA is highly localized to cortex rather than to white matter (1943, 1944a–c). Human brain, however, was an exception as higher CA activity was found in the white matter. Ashby (1943, 1944a–c) speculated that CA could play a role in "the capacity for speed of delivery of energy for the propagation of the nerve impulse." Subsequent studies, in which the inhibition of CA altered formation and electrolyte composition of the cerebrospinal fluid (CSF), implicated this enzyme in its production (Tschirgi *et al.*, 1954; Maren and Robinson, 1960). Although a method for the histochemical demonstration of CA activity was available as early as 1953 (Kurata), it was regarded as being too unspecific to allow for precise localization of enzyme activity in tissues. Lack of both specific histochemical techniques and sensitive quantitative methods did not permit reliable demonstration of CA in tissue sections and localization at the cellular level until 1961 (Giacobini, 1961). The development of a sensitive micromethod based on a modification of the Cartesian diver technique of Linderstrom-Lang (1937) allowed, for the first time, the determination of CA activity in single tissue elements (Giacobini, 1961, 1962).

CURRENT TOPICS IN
DEVELOPMENTAL BIOLOGY, VOL. 21

II. Carbonic Anhydrase: A Specific Glial Enzyme

The new microdiver technique was applied to samples of single nerve cells and surrounding glial cells (Giacobini, 1961, 1962). Single cell preparations (neurons and oligodendrocytes) were isolated under the dissection microscope from fresh unstained sections of the CNS of the rat. In the same investigation, the activity of single cell samples isolated from neural tissue, choroid plexus, and single erythrocytes was compared. This novel approach offered the possibilities of obtaining direct information on the localization of CA in CNS and exploring at a cellular level, the functional relationship between glia and neurons. Equivalent volumes of nerve and glial cells were dissected out from the lateral vestibular nucleus of Deiters and their CA activity measured by means of CO_2 evolution at 25°C from $NaHCO_3$ (final concentration, 1×10^{-4} M) in the presence of 0.1 M sodium phosphate buffer, pH 7.5 (Fig. 1). The uncatalyzed reaction and the activity curves of different known concentrations of semipurified CA preparations were determined for each experiment and compared with the curve from isolated cell preparation (Fig. 1). The mean enzyme concentration of a single cell preparation was calculated to be 6.4×10^{-20} mol of CA per neuron and 385×10^{-20} mol per corresponding volume of glia. The glial cells exhibited, therefore, a concentration of CA 126-fold higher than nerve cells. As a comparison, a single intact red cell showed a CA concentration of 2×10^{-20} mol. On the basis of the CA activity per unit volume, it was calculated that red cells had 670, choroid cells 250, and glial cells 126 times the activity of a nerve cell isolated from the nucleus of Deiter. The demonstration of a high and selective localization of CA in the glial elements of the CNS suggested these cells to be the site of the CO_2 hydration process (Fig. 2). The CO_2 produced by cell metabolism could rapidly diffuse within the neuron, and from it into the adjacent glial cells where it could be hydrated to carbonic acid (HCO_3^- at body pH) in the presence of CA (Fig. 2). A number of publications, mainly from Svaetichin's laboratory, have emphasized the importance of pCO_2 changes originating from CA activity of glial cells on neuronal excitability and function (Svaetichin *et al.*, 1965; Parthe, 1981). This functional–anatomical relationship between glia and neurons supports the postulated crucial role of glia in the regulation of the ionic environment of the neuron (Tschirgi, 1958; Laufer *et al.*, 1960; Svaetichin *et al.*, 1965). Based entirely on anatomical grounds, Ramon y Cajal (1928) speculated on a "mutually serviceable metabolic relationship, almost a symbiotic interaction, between neurons and glia in the mammalian nervous system." The demonstration of selective localization of a CO_2 regulation mechanism in glial

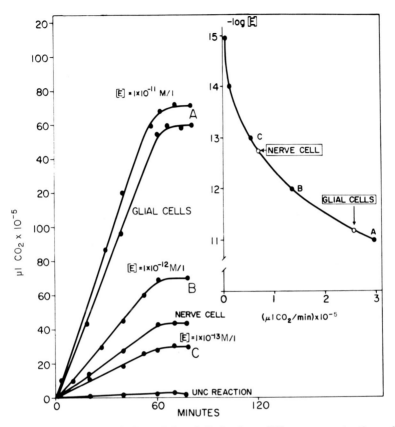

FIG. 1. Activity curves (A, B, and C at left) for three different concentrations of a semipurified carbonic anhydrase preparation and the curve for the uncatalyzed (UNC) reaction. These are compared (inset) with the activity of a single nerve cell and an equivalent volume of glial cells.

cells provided the first biochemical evidence for such a close neuronal–glial metabolic interrelation (Giacobini, 1964). Another important consequence of the demonstration of selective glial localization of a CA was the access to a reliable and specific marker of glial cells. The demonstration of a high activity for nonspecific cholinesterase (butyrylcholinesterase) in glial cells, reached with both histochemical (Koelle, 1951, 1955) and microchemical methods (Giacobini, 1959, 1964) in rat and cat, had provided a first possibility for selectively distinguishing glial cells from neurons. However, CA activity was found to be not only more specific but also more applicable as a marker as it could be used for glia identification in single cells, tissue sections, bulk preparations, and even tumors.

EZIO GIACOBINI

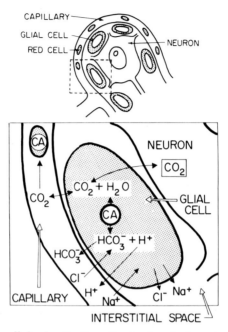

FIG. 2. (Top) The cellular localization of carbonic anhydrase in the central nervous system. (Bottom) A proposed mechanism for the transport of chloride and sodium. CA, Carbonic anhydrase.

III. Carbonic Anhydrase Activity in Oligodendroglia

The role of oligodendroglia in myelin formation has been discussed by Bunge (1968). The association of CA activity with oligodendroglia is important for the understanding of the process of myelination as this can be followed both *in vitro* and *in vivo*. Following the finding of a specific glia–CA association in dissected glial cell clusters (Giacobini, 1961), the enzyme was localized in white matter by histochemical techniques (Korhonen *et al.*, 1964). The development of methods for bulk separation of neurons, astrocytes, and oligodendrocytes from rat brain made it possible to study and compare activities of myelin- and glia-associated enzymes. The activities of three myelin-associated enzymes, CA, 5′ nucleotidase, and 2′,3′-cyclic nucleotide-3-phosphodiesterase (CNP) were measured in oligodendrocytes, neurons, and astrocytes isolated from the brain of rats of various ages (Snyder *et al.*, 1983). The CA-specific activity in oligodendrocytes was 3- to 5-fold higher than that in brain homogenates at each age, and at all ages. Only very low activities of this enzyme were measured in neuronal

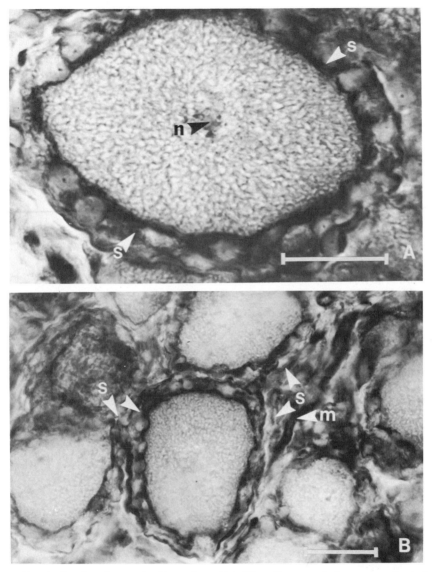

FIG. 3. Cat dorsal root ganglion cells. s, Satellite cells; n, nucleolus; m, myelinated axon. Bar, 20 μm. (A) Higher magnification showing a ganglion cell with surrounding satellite cells (courtesy of V. Parthe). The histochemical reaction for CA is negative in the cell body and its processes but positive in satellite glial cells and myelin.

and astrocytic preparations (Snyder *et al.*, 1983). Immunocytochemical techniques have demonstrated CA in oligodendrocyte perikarya in rat (Roussel *et al.*, 1979; Delaunoy *et al.*, 1980; Ghandour *et al.*, 1980), and biochemical studies have shown its presence in myelin preparations purified from rat brain (Cammer *et al.*, 1976; Yandrasitz *et al.*, 1976; Sapirstein *et al.*, 1978) and in primary astrocyte cultures (Church *et al.*, 1980). Myelinated fibers, as well as oligodendrocytes in human and mouse brain, bind anti-CA sera (Spicer *et al.*, 1979; Kumpulainein and Nystrom, 1981; Kumpulainein and Korhonen, 1982). Parthe (1981), using a refined histochemical method, has shown staining of CNS perineuronal oligodendrocytes throughout the myelin sheath cross section. In addition, in the dorsal root ganglion, a CA-positive reaction was found both in satellite glial cells and myelin (Fig. 3) (Parthe, 1981). The reaction was negative in the neuronal cell body and its processes (Fig. 3). The specific activity of CA increases 6- to 9-fold in oligodendrocytes in rats between the ages of 10 and 120 days (Snyder *et al.*, 1983). The formation of myelin during this period of development may account for the smaller change in relative specific activity related to homogenate activity. Peripheral nerves from younger animals show somewhat higher myelin CA-specific activities than nerves from adults (Cammer, 1979). This may depend on the fact that these nerves are still growing and the myelin sheath is thickening. However, the decrease in cytoplasmic volume of oligodendrocytes which occurs during maturation (Mori and Leblond, 1970) and the developmental increase of these cells seen by Snyder *et al.* (1983) suggest a real increase of CA concentration in the cytoplasm during development.

It is of interest to note that CA activity is histochemically negative in induced glial tumors and positive in the surrounding oligodendrocytes and white matter (Schiffer *et al.*, 1984). It is, therefore, possible that experimentally induced tumoral oligodendrocytes do not express CA.

IV. Carbonic Anhydrase Activity in Early Embryonic Development

Early studies of CA activity in embryos have dealt mainly with embryonic organs (retina) at the fourth day and after (Van Goor, 1940; Clark, 1951).

All organs which contain CA show an increase in their activity from the prenatal life to the adult. Van Goor (1940) was the first to demonstrate the early appearance of CA activity in chick embryos (retina). In the rat embryo, CA first appears at day 13, thereafter it triples on following days (Maren, 1967). In the amnion, CA activity is present at day 12. In the chick encephalon, Tautu and Voiculet (1958) found CA activity for hydration and dehydration on day 7 and day 11, respectively. Changes in tension of environmental CO_2 were shown to affect cellular differentiation (Shepard, 1962). The appearance of CA activity in early developing embryos which are producing CO_2 might lead to a reduced CO_2 tension ($H_2O + CO_2 \rightleftharpoons H_2CO_3$). Therefore, CA activity could control development by slowing down cellular differentiation (Shepard, 1962). Shepard (1962) found that the specific activity of CA of whole chick embryos increased abruptly between the fourth and fifth days. The increase occurred in both the anterior and posterior portions of the embryo and was not due solely to accumulation in the retina. It is interesting that the time of increasing CA activity coincides with the end of the embryonic period when differentiation is slowing down.

The relation between CA activity and the beginning of nervous function is puzzling. Young animals born in an immature condition (dog, cat, rabbit, and rat) show a late appearance of CA in brain (Ashley and Schuster, 1950). There is also a correlation between the appearance of CA activity in the CNS, the beginning of its functioning as judged by EEG (Ashley and Schuster, 1950), and its susceptibility to experimental seizures (Millichap, 1958; Nair and Bau, 1971). This relationship seems to be true for humans as well.

V. Concluding Remarks

Localizing of CA in cells is fundamental in understanding the function of this enzyme in the nervous tissue. Two qualitative techniques, histochemistry and immunocytochemistry, have independently confirmed our original finding of a selective localization of CA in glial cells as demonstrated by a quantitative cytochemical technique (Giacobini, 1961, 1962). All three approaches (histochemistry, immunocytochemistry, and cytochemistry) demonstrate a high concentration of CA activity in oligodendrocytes and its presence in myelin. The glial localization of CA holds true for both the CNS (including retina) and PNS, suggesting a general role of this enzyme in the conversion of carbonic acid formed from metabolic CO_2 back to CO_2. This product can then diffuse across the plasma membrane of neurons (into glia) or through the myelin sheath of nerves (Fig. 2). It is also likely that a relatively high concentration of CA activity is required during the period of rapid energy metabolism characteristic of growth and myelination. The possibility of demonstrating CA activity with several independent methods makes this enzyme a reliable probe of glial presence and development in various central and peripheral neural preparations, including *in vitro* cultures. This approach has proved to be useful whenever a detailed analysis of the enzyme distribution and localization is required or, more simply, when the presence of glial cells has to be ascertained.

ACKNOWLEDGMENTS

This chapter is dedicated to the memory of Dr. Gunnar Svaetichin (Instituto Venezolano de Investigaciones Cientificas, Caracas, Venezuela), who significantly contributed to our understanding of the function of CA in the nervous system.

REFERENCES

Ashby, W. (1943). *J. Biol. Chem.* **151,** 521.
Ashby, W. (1944a). *J. Biol. Chem.* **152,** 235.
Ashby, W. (1944b). *J. Biol. Chem.* **156,** 323.
Ashby, W. (1944c). *J. Biol. Chem.* **156,** 331.
Ashley, W., and Schuster, E. M. (1950). *J. Biol. Chem.* **184,** 109–116.
Bunge, R. P. (1968). *Physiol. Rev.* **48,** 197–251.
Cammer, W. (1979). *J. Neurochem.* **32,** 651–654.
Cammer, W., Fredman, T., Rose, A. L., and Norton, W. T. (1976). *J. Neurochem.* **27,** 165–171.
Church, G. A., Kimelberg, H. K., and Sapirstein, V. S. (1980). *J. Neurochem.* **34,** 873–879.
Clark, A. M. (1951). *J. Exp. Biol.* **28,** 332–43.
Delaunoy, J. P., Hog, F., Devilliers, G., Bansart, M., Mandel, P., and Sensenbrenner, M. (1980). *Cell. Mol. Biol.* **26,** 235–240.
Ghandour, M. S., Langley, O. K., Vincendon, G., and Gombos, G. (1980). *Neuroscience* **5,** 559–571.
Giacobini, E. (1959). *Acta Physiol. Scand.* **45,** 3–45.
Giacobini, E. (1961). *Science* **134,** 1524–1525.
Giacobini, E. (1962). *J. Neurochem.* **9,** 169–177.
Giacobini, E. (1964). *In* "Morphological and Biochemical Correlates of Neural Activity" (M. M. Cohen and R. S. Snider, eds.), pp. 15–38. Harper, New York.
Koelle, G. B. (1951). *J. Pharmacol. Exp. Ther.* **103,** 153.
Koelle, G. B. (1955). *J. Neuropathol. Exp. Neurol.* **14,** 23.
Korhonen, L. K., Naatanen, E., and Hyyppa, M. (1964). *Acta Histochem.* **18,** 336–347.
Kumpulainein, T., and Korhonen, L. K. (1982). *J. Histochem. Cytochem.* **30,** 283–292.
Kumpulainein, T., and Nystrom, S. H. M. (1981). *Brain Res.* **220,** 220–225.
Kurata, Y. (1953). *Stain. Technol.* **28,** 231.
Laufer, M., Svaetichin, G., Mitarai, G., Fatehchand, R., Vallecalle, E., and Villegas, J. (1960). *Neurophysiol. Psychophys. Symp.* p. 457.
Linderstrom-Lang, K. (1937). *Nature (London)* **140,** 108.
Maren, T. H. (1967). *Physiol. Rev.* **47,** 595–781.
Maren, T. H., and Robinson, B. (1960). *Bull. Johns Hopkins Hosp.* **106,** 1.
Meldrum, N. U., and Roughton, F. J. W. (1934). *J. Physiol. (London)* **80,** 113.
Millichap, G. J. (1958). *Proc. Soc. Exp. Biol.* **97,** 606–611.
Mori, S., and Lebond, C. P. (1970). *J. Comp. Neurol.* **139,** 1–30.
Nair, V., and Bau, D. (1971). *Brain Res.* **31,** 185–193.
Parthe, V. (1981). *J. Neurosci. Res.* **6,** 119–131.
Ramon y Cajal, S. (1928). *In* "Degeneration and Regeneration of the Nervous System." Oxford Univ. Press, London.

Roussel, G., Delaunoy, J-P., Nussbaum, J-L., and Mandel, P. (1979). *Brain Res.* **160,** 47–55

Sapirstein, V., Trachtenberg, M., Lees, M. B., and Koul, O. (1978). *In* "Advances in Experimental Medicine and Biology: Myelin and Demyelination" (J. Palo, ed.), pp. 55–70. Plenum, New York.

Schiffer, D., Bertolotto, A., Girodana, N. T., and Mauro, A. (1984). *In* "Induction of Brain Tumors by Transplacental E.N.U.: Correlation between Neurocytogenesis and Tumor Development" (F. Caciagli, E. Giacobini, and R. Paoletti, eds.). Elsevier, Amsterdam.

Shepard, T. H. (1962). *J. Embryol. Exp. Morphol.* **10,** 191–201.

Snyder, D. S., Zimmerman, T. R., Jr., Farooq, M., Norton, W. T., and Cammer, W. (1983). *J. Neurochem.* **40,** 120–127.

Spicer, S., Stoward, P. J., and Tashian, R. E. (1979). *J. Histochem. Cytochem.* **27,** 820–821.

Svaetichin, G., Neghishi, K., Fatehchand, R., Drujan, B. D., and Selvin de Testa, A. (1965). *In* "Biology of Neuroglia, Brain Research" (E. D. P. De Robertis and R. Carrea, eds.), pp. 243–266. Elsevier, Amsterdam.

Tautu, P., and Voiculet, N. (1958). *Comm. Acad. Rep. Pop. Romine* **8,** 233–239.

Tschirgi, R. D. (1958). *In* "Biology of Neurologia" (W. F. Windle, ed.), p. 130. Thomas, Springfield, Illinois.

Tschirgi, R. D., Frost, F. W., and Taylor, J. L. (1954). *Proc. Soc. Exp. Biol. Med.* **87,** 101.

Van Goor, H. (1940). *Acta Brev. Neerl. Physiol.* **10,** 37–39.

Yandrasitz, J. R., Ernst, S. A., and Salganicoff, L. (1976). *J. Neurochem.* **27,** 707–715.

CHAPTER 9

CHANGES IN AXONAL TRANSPORT AND GLIAL PROTEINS DURING OPTIC NERVE REGENERATION IN *XENOPUS LAEVIS*

Ben G. Szaro and Y. Peng Loh

LABORATORY OF NEUROCHEMISTRY AND NEUROIMMUNOLOGY
NATIONAL INSTITUTE OF CHILD HEALTH AND HUMAN DEVELOPMENT
NATIONAL INSTITUTES OF HEALTH
BETHESDA, MARYLAND 20892

I. Introduction

The formation of orderly nerve connections in the brain during development is a fundamental problem for developmental neurobiologists. Historically, the retinotectal system of the African clawed frog, *Xenopus laevis*, has been an important model system for studying the ontogeny of neural connectivity. As an experimental animal, *Xenopus* is hardy, can be raised in close quarters on a simple diet, and breeds on demand. Because *Xenopus* embryos develop externally, encased within an easily removed jelly coat, the *Xenopus* embryonic brain is easily accessible at all stages of development. In addition, the brain of *Xenopus*, relative to that of mammals and birds, is small and simple yet retains enough complexity for addressing those developmental problems unique to vertebrates.

Studies on the ontogeny of the *Xenopus* visual system include descriptive embryological studies which have identified cell populations present as early as the blastula and neural plate stages of development which contribute to eye formation (Jacobson and Hirose, 1978; Jacobson, 1983; Brun, 1981). Additional studies have described the morphogenetic movements of eye cup formation (Grant *et al.*, 1980; Holt, 1980), the differentiation of individual cell types in the retina and optic nerve (e.g., Sakaguchi *et al.*, 1984; Jacobson, 1968, 1976; Grant and Rubin, 1980; Cima and Grant, 1982a,b), the emergence of the primary visual projections to the optic tectum and thalamus (Gaze *et al.*, 1974; Hoskins and Grobstein, 1984), and the growth of the eye and tectum (Beach and Jacobson, 1979; Gaze *et al.*, 1979; Conway *et al.*, 1980). These studies have provided a rich data base and allowed work-

217

CURRENT TOPICS IN
DEVELOPMENTAL BIOLOGY, VOL. 21

ers in the field to speculate about the mechanisms involved in the formation of connections between the eye and optic tectum.

In addition to these descriptive studies, other experiments have revealed some of the intercellular mechanisms by which orderly projections are formed (for reviews see Hunt and Jacobson, 1974; Conway *et al.*, 1981; Fraser and Hunt, 1980; Meyer, 1982). In a growing axon, the process of neural connectivity can be broken down into a series of steps beginning with the initiation of axon elongation in the new postmitotic neuron. Next, the growing axon traverses inappropriate regions of the brain to the appropriate target area. Once there, it will initiate contacts with inappropriate target cells and, while competing with other ingrowing axons, will eventually find its appropriate target site. Once appropriate contact has been made, synapses must be formed and axon elongation slowed or terminated to prevent interference with normal physiological functions. The interplay of growth mechanics with the timing of differentiation, chemotaxis, contact guidance, differential cell adhesion, and synchronous firing of incoming neurons, has been proposed as an interdependent cellular mechanism which directs the process of neural connectivity. Investigators in the field are currently trying to resolve the relative contribution each process makes to each of the steps which comprise axon growth and the formation of orderly projections (e.g., see Holt, 1984; Harris, 1980, 1984; Rho and Hunt, 1980; Fraser, 1983; Meyer, 1983; Ide *et al.*, 1984; Schmidt and Edwards, 1983; Willshaw *et al.*, 1983). To date, the subtle interplay of each of these various mechanisms has precluded any understanding of the relationships between cellular mechanism and the developmental events leading to the formation of orderly neural projections. It is our belief that an understanding of the ontogeny of nerve connections would be greatly expedited if biochemical probes were to become available to study these phenomena.

II. Choice of Strategies for Studying the Biochemical Basis of Ontogeny of Neural Connections: General Screening or the Specific Biochemical Probe?

Conceptually, in beginning to look for molecular correlates to a process as complex as nerve growth and synaptogenesis, several different strategies can be followed. For instance, the appearance, regulation, pharmacology and control of known molecules can be examined for correlations with axon guidance and synaptogenesis during development. Specific inhibitors or antibodies which interfere with the function of such molecules might also interfere with axon growth and synaptogenesis, thus revealing relationships between such molecules and various cellular processes. For example, antibodies directed against extracellular matrix molecules, such as laminin, the L_1 antigen, and neural cell adhesion molecules on the surfaces of neurons and glia, have provided preliminary information on the possible roles of these antigens in the contact guidance of axons in several systems (e.g., see Agranoff *et al.*, 1984; Schachner *et al.*, 1983; Silver and Rutishauser, 1984).

Additional information on the cell biology of neural connectivity might be gained by referencing prior knowledge of the role of molecules known to be involved in motility and process formation in other systems. However, common proteins such as actin and tubulin are unlikely to play causal roles in the formation of topographic nerve connections. Since so few of the actual molecules involved in neural connectivity have been identified, simply focusing on a few known substances might overlook less understood molecules which are fundamental to the formation of neuronal projections.

Because so little is known about the biochemical basis of neural connectivity, it seems just as likely that general screening approaches might reveal new molecules involved in these processes. For instance, monoclonal antibodies have been produced which appear to recognize specific fasciculating nerves at particular times in the development of grasshopper embryos (see Goodman *et al.*, 1984). Such methods may be applied in *Xenopus*. Similarly, other more classical biochemical approaches such as gel electrophoresis may also be useful to identify macromolecules which correlate in time and place with various aspects of projection formation. Once these molecules are identified they can then, in turn, be used as specific markers to further probe synaptogenesis and axon guidance.

However, while classical biochemical separation procedures can provide broad surveys, efficient application of these techniques requires large amounts of starting materials and good control situations for comparison. Such situations are difficult to achieve in developing *Xenopus* embryos, where the distance over which the youngest optic axons grow is only a few hundred micrometers. Furthermore, new retinal ganglion cells are born continuously throughout development. Consequently, the formation of an orderly projection onto the tectum is a continuing process which takes months to complete. Thus, the different populations of axons involved in each stage of the growth process cannot be resolved adequately for conventional biochemical analyses. Fortunately in the *Xenopus* retinotectal system, optic nerve regeneration in the adult provides a physiological model for neural connectivity.

III. Nerve Regeneration and Its Relationship to the Ontogeny of Neural Connections in the *Xenopus laevis* Retinotectal System

In many respects, optic nerve regeneration recapitulates the basic processes of axon growth and synaptogenesis which occur during development. Moreover, during optic nerve regeneration neural connectivity is uncoupled from the more protracted developmental processes of differentiation and growth of the eye and tectum. Nevertheless, despite the apparent lack of any assistance from timing and growth mechanisms, the final regenerated retinotectal map, as assayed by current electrophysiological techniques, is indistinguishable from the original projection (Maturana *et al.*, 1959; Gaze, 1959). Thus, despite some differences between regeneration and ontogeny, optic nerve regeneration offers a practicable system for isolating some of the basic cellular mechanisms by which orderly neuronal projections are established.

The fact that during regeneration the entire population of retinal ganglion cell axons are induced to grow at once enables the biochemist to obtain sufficient material for analysis. Moreover, electrophysiological mapping techniques can be used to monitor the progress of the arrival of axons at the tectum and the refinement of topographic order in the projection. This allows the biochemist to correlate molecular changes with the time course of regeneration (e.g., when axon elongation or the refinement of ordered connectivity occurs). Thus, the regenerating visual system of lower vertebrates is a more suitable system for studying biochemical correlates to ordered synaptogenesis than other vertebrate regenerating systems (e.g., regenerating mammalian sciatic nerves) in which the point-to-point topographic relationships between populations of neurons are more difficult to assess.

While regenerating visual systems in other lower vertebrates have some advantages in size over the visual system of *X. laevis,* the chief attraction of the regenerating *X. laevis* retinotectal system is that the embryology of *Xenopus* is so well studied. Thus, biochemical changes in regenerating axons can subsequently be studied in the development of embryonic nerve projections in the same species.

IV. Axonally Transported Proteins: A Rational Paradigm for Studying the Biochemical Correlates of Nerve Regeneration

Because proteins serve many of the structural and functional needs of a cell, acting as enzymes, receptors, structural elements, membrane channels, and some classes of cell adhesion molecules, it seems logical to begin to study nerve regeneration at the biochemical level by examining proteins. However, this is not to say that proteins are necessarily the only candidates for molecules involved in neuronal outgrowth and synaptic recognition.

For example, the carbohydrate moieties of glycolipids and glycoproteins have been proposed to mediate some forms of cell adhesion, and thus neurite outgrowth, axon elongation, and cell recognition (e.g., see Fraser and Hunt, 1980, for a model of retinotectal specificity based on homophilic interactions between the glycocalyxes of glycolipids and glycoproteins). Carbohydrate residues similar to those found on gangliosides are also apparently involved in the gradient of cell adhesion between retinal and tectal cells *in vitro* (see Marchase, 1977; Pierce, 1982). Furthermore, the ganglioside GM1, differentially enhances neurite outgrowth and regeneration in conjunction with other serum factors in cultures of dorsal root, ciliary, and sympathetic ganglia. The exact combination of GM1, culture substrata, growth factors, and serum inhibitors required for maximal outgrowth differed for each ganglia tested (Skaper and Varon, 1984; Doherty *et al.*, 1985). Thus, gangliosides, as components of an *in vivo* substrata, could interact with growing axons from specific ganglia to effect outgrowth and sprouting along specific pathways or within unique target regions. These notions, along with observations that the axonal transport of gangliosides is increased during optic nerve regeneration (Sbaschnig-Agler *et al.*, 1984; Gammon *et al.*, 1985) and that antibodies to mixed gangliosides or to GM1 injected *in vivo* inhibit goldfish optic nerve regeneration (Sparrow *et al.*, 1984), suggest that gangliosides and possibly other carbohydrate-containing molecules play important roles during nerve regeneration. Nevertheless, it seems likely that, at some level, these other molecules will interact with specific proteins either as enzymatic substrates or ligands for cell surface receptors.

Because cell recognition in regenerating axons occurs at the axon tip, we focused on the proteins within the axons themselves. Except for some posttranslational modifications of proteins and a small amount of synthesis in mitochondria, protein synthesis does not occur in the axon. Instead, proteins are synthesized in the perikaryon and then transported down the axon. Thus, proteins involved in cell recognition, regulation of axon arborization, and basic axonal structural components must all be actively shipped to the growing axon terminals via axonal transport. Many of those proteins involved in general cellular housekeeping functions are thus likely to remain behind in the perikaryon. Consequently, in studies on regeneration, axonal transport serves as a filter to select molecules important for axon growth over those which might increase in the perikaryon as a generalized response to cell trauma.

As an additional advantage to the biochemist studying nerve regeneration, different axonally transported proteins move down the axon at different rates (e.g., Hoffman and Lasek, 1975a; Willard *et al.*, 1974; Willard and Hulebak, 1977). Moreover, the different rates appear to correspond with the movement of different organelles within the cell (Droz *et al.*, 1975; Lorenz and Willard, 1978; Lasek, 1980). The fastest moving compartment contains proteins associated with vesicles (mostly neurosecretory and membrane bound proteins). Subsequent phases contain mitochondria, followed next by proteins associated with the microtrabecular matrix and various soluble enzymes, and then by neurofilaments, microtubules, and associated proteins. Thus, by examining consecutive phases of transport one can potentially separate proteins on the basis of functionality. Proteins involved in synaptic recognition might be expected to be membrane bound, and, as such, to travel in the rapid phases; those proteins involved in regulating axon elongation through interactions with cytoskeletal components might be expected in the slower phases, associated with microtubules, microfilaments or neurofilaments.

More practically, studies on axonal transport of proteins are facilitated by the ease with which proteins can be labeled during synthesis, by injecting radioactive amino acids into the eye, and then subsequently tracked into the optic nerve. Also, autoradiographic techniques can be used to detect very subtle changes in the profiles of protein synthesis and cellular localization during regeneration.

V. Characterization of Phases of Axonally Transported Proteins in *Xenopus laevis* Optic Nerves

Axonal transport and its relationship to nerve function have been well studied in several systems (for reviews see for example Schwartz, 1979; Ochs, 1982; Grafstein and Forman, 1980; Forman, 1983; Baitinger *et al.*, 1983). Briefly, in mammalian optic axons, as in other neurons, there appear to be five individual phases of transport (Willard *et al.*, 1974, 1977) which are termed (from fastest to slowest) I, II, III, IV, and V. Phases IV and V appear to coincide in movement and rate to the slow component b (SCb) and slow component a (SCa) found in the peripheral nervous system (see Lasek, 1980, for overview). SCa is composed of the microtubule and neurofilament proteins and in rabbit optic nerves moves at 0.2–1 mm per day, while SCb consists of actin, among other proteins, and moves slightly faster at 2–4 mm per day. In addition, there are three faster phases, of which the most rapid phase, moving at more than 240 mm per day, consists of membrane bound and some secretory proteins associated with vesicles. Two additional intermediate components, including a mitochondrial component, also move through the nerve and are termed phase II (34–68 mm per day) and phase III (4–8 mm per day).

In general, axonal transport in the optic nerves of lower vertebrates like goldfish and the toad, *Bufo marinus,* is quite similar except that in these cold blooded animals the rates are somewhat slower (Elam and Agranoff, 1971a,b; Skene and Willard, 1981a). In *B. marinus,* the closest relative to *X. laevis* which has been studied to date, phase I moves at between 84 and 102 mm per day, phase II between 14 and 26 mm per day, phase III between 3.8 and 6.5 mm per day, phase IV (SCb) between 1.1 and 1.8 mm per day, and phase V (SCa) between 0.2 and 0.5 mm per day (Skene and Willard, 1981a).

Because axonal transport has been so well characterized in other systems, the basic aim of studying axonal transport in *Xenopus* optic nerve was to establish baseline data and methodologies for the regeneration studies. The short length of the *Xenopus* optic nerve (see Figs. 1 and 2) made it necessary to develop techniques for distinguishing axonally transported proteins in the optic nerve from those synthe-

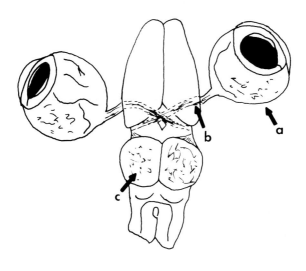

FIG. 1. A proportional schematic drawing of a dorsal view of the brain of *Xenopus laevis,* dissected free of the head (traced from a photograph provided by J. H. Rho). The optic fibers of the right eye (a) form the right optic nerve (b) and, after crossing at the optic chiasm on the ventral side of the brain, innervate the left optic tectum (c). Bar, 1 mm.

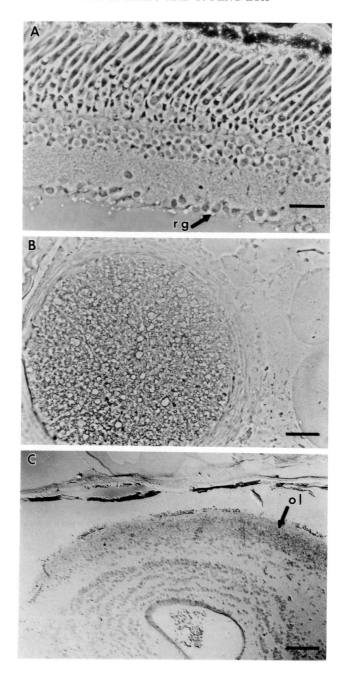

sized locally by glia from radiolabeled amino acids which had diffused the short distance from the site of injection in the eye.

To determine the contribution of labeled proteins from glial cells to the overall labeled protein patterns in the optic nerve, we, in collaboration with R. K. Hunt, blocked axonal transport by crushing the optic nerve immediately behind the eye. If done carefully, the nerve can be crushed without severing the nerve sheath and without interfering with the blood supply to the eye (see Fig. 3A and B which illustrates autoradiographically the lack of silver grains in the axons of a crushed optic nerve). In the crushed nerve, several hundred micrometers beyond the crush, only glial proteins would be present, and these can be subtracted from the nerve protein pattern to determine which proteins are axonally transported (Fig. 3C).

Using the crush control (a nerve crushed 5–30 minutes before injection of isotope) to identify bands which are axonally transported, it was possible (Table I) to identify when certain bands first appear in the nerve. On the basis of this table we were able to choose a set of 3 timepoints (2–4 hours, 18 hours, and 5–9 days) at which each of the phases of transport could be seen in the nerve.

Furthermore, by comparing protein patterns from optic tecta innervated by intact nerves to patterns from optic tecta innervated by crushed nerves, it was possible to separate the fastest wave (2–4 hours in the nerve) of transport in the *Xenopus* optic nerve into 3 separate waves, and thus estimate the transport rates of the 5 different phases

---

FIG. 2. Photomicrographs of cross sections of various regions of the visual system of *X. laevis*. (A) A cross section of the retina cut approximately from the level marked "a" in Fig. 1. The optic axons emanate from the retinal ganglion cells (rg) and course radially toward the optic nerve head along the inner vitreal surface of the eye (bar, 30 μm). (B) A cross section of the optic nerve cut at the level marked "b" in Fig. 1. The myelinated optic axons, visible here at the light level, are embedded in a matrix of glia; the entire nerve is surrounded by an epineural nerve sheath. A juvenile *X. laevis* froglet has from 3000 to 5000 myelinated optic axons, and an additional 30,000 to 50,000 unmyelinated axons (Gaze and Peters, 1961; Wilson, 1971) in a nerve which is about 1.5–2 mm long from the eye to the optic chiasm. The tissues shown in A and B were embedded in epon-araldite, sectioned at 3 μm, and photographed using phase-contrast optics (bar, 30 μm). (C) A parasagittal section of the optic tectum. The optic axons enter the tectum from the rostral (right) and lateral edges and innervate the tectum in the superficial optic layers of the tectum (ol), forming a topographic representation of the surface of the retina across the surface of the tectum. The optic nerve in this frog was labeled with intraocularly injected [³H]proline. The tissue was subsequently processed for paraffin sectioning and autoradiography, and finally stained with hematoxylin. Thus, the silver grains in the superficial layers of the optic tectum show the location of the terminal endings of the retinal ganglion cell axons (bar, 100 μm).

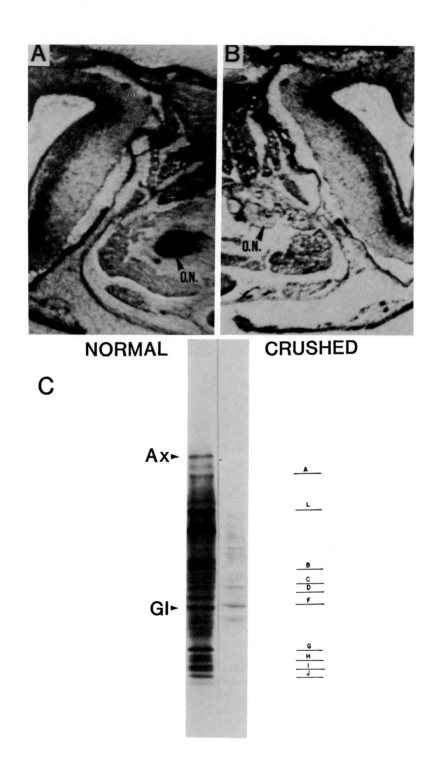

NORMAL   CRUSHED

C

Ax▸

Gl▸

A

L

B
C
D
F

G
H
I
J

of transport (see Szaro *et al.*, 1984, for further details). These phases moved at approximately 60–96, 30–48, 6–11, 1.6–2.8, and 0.2 mm per day, respectively. These velocities are not significantly different than those seen in the toad, *B. marinus.*

VI. Changes in Axonally Transported Protein Patterns in Regenerating *Xenopus* Optic Nerves: Surgical Paradigm and Time Course of Regeneration

During regeneration, the optic nerve distal to the site of trauma is not merely a passive structure through which the regenerating axons grow on their way back to the tectum. In the optic nerve, the glia appear to be involved in degrading degenerating axons and in preparing a pathway for the newly growing axons, and consequently may contribute to changes in the overall protein pattern during optic nerve regeneration (Sjostrand, 1965; Wolsburg, 1981; Reier and Webster, 1974). Thus, it was necessary to design a paradigm in which the contributions of the glia to protein patterns in regenerating nerves would be minimized. To accomplish this, the optic fibers were severed in the optic tract just prior to entering the optic tectum as shown in Fig. 4. This paradigm has been used previously in goldfish (Giulian *et al.*,

---

FIG. 3. A histological and biochemical illustration of the acute crush paradigm used to distinguish axonally transported proteins from those labeled locally by the support cells in the nerve. (A) A bright field autoradiograph of the intact optic nerve (o.n.) as it approaches the skull; (B) an optic nerve which was crushed at the orbit. Note that the crushed optic nerve (B) is devoid of label above background levels, while the intact nerve (A) is full of silver grains exposed by the [^{35}S]methionine proteins which have been transported through the nerve from the retinal ganglion cell bodies in the eye. Both eyes in this frog were injected with [^{35}S]methionine 30 minutes after the left optic nerve was crushed at the orbit. The frogs were sacrificed 1 hour after the injections. Slides were dipped in NTB2 emulsion, exposed for 6 days, and then counterstained in hematoxylin and metanil yellow. (C) The patterns of labeled proteins separated on an SDS 4–17% polyacrylamide gradient gel. Optic nerve segments (10) extending from the chiasm to approximately two-thirds of the distance to the site of the crush at the orbit were harvested 4 hours after intraocular injection of [^{35}S]methionine. Labeled proteins from gels run on crushed nerves are shown in the right lane. Proteins from the contralateral, intact nerve segments, which were injected and dissected similarly to the crushed nerves, are shown on the left. Proteins which were labeled in gels run on proteins from normal nerves, but were absent from gels run on crushed nerves, were considered axonally transported (e.g., the sample band labeled Ax). However, proteins which were present both in the normal and crushed nerves were considered glial (e.g., the sample band labeled Gl). The letters on the far right refer to internal molecular weight standards derived from the major Coomassie Blue-stained bands: A, 200 kDa; L, 125 kDa; B, 64 kDa; D, C, 55 kDa; E, 50 kDa; F, 46 kDa; G, 19.5 kDa; H, 15.5 kDa; I, 12 kDa; and J, 10.5 kDa. See Szaro *et al.* (1984) for further details.

TABLE I

AXONALLY TRANSPORTED PROTEINS PRESENT IN THE OPTIC NERVE AT VARIOUS TIMES AFTER INTRAOCULAR INJECTION OF [^{35}S]METHIONINE[a]

Time after intraocular injection

1 hour	4 hours	12 hours	18 hours	3 days	5 days	9 days	16 days
	350K						
					295K		
		290K					
					265K		
		260K					
240K							
					240K		
					215K		
	210K						
	200K						
			180K				
	170K						
	155/160K						
140K							
135K							
		120K					
103K							
98K							
92K							
82K							
		80K					
74K							
66K							
					65K		
64K							
	61K						
56K							
51K							
47K							
44K							
38K							
		37K					
		33K					
31K							
21.5K							
20K							
		19K					
			17K				
16K							
14K							
	12K						

[a]The numbers (K = 1000 Da) denote the molecular weight of the protein bands seen in the optic nerve. (Szaro *et al.*, 1984.)

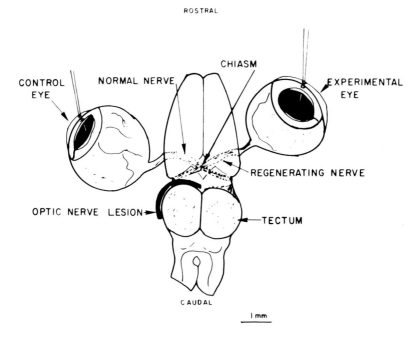

FIG. 4. A schematic diagram illustrating the surgical paradigm used in the studies on optic nerve regeneration in *X. laevis*. The optic fibers from the right eye, innervating the left tectum, were severed in the optic tract at the lateral and rostral margins of the optic tectum. This lesion preserved, intact, the optic nerve from the eye to the chiasm (the region of nerve analyzed for biochemical studies). The remaining, uninjured nerve on the left side of the animal was used as a control for protein synthesis in nonregenerating, sham-operated nerves. (From Szaro *et al.*, 1985.)

1980) and does not directly injure the region of the nerve harvested for biochemistry; thus, the nerve between the eye and chiasm served as a conduit through which proteins passing on their way to the growing nerve terminals were harvested.

It was first necessary to determine if optic axons injured in this way could regenerate an orderly retinotectal map. Also, in order to be able to resolve times during regeneration during which axon elongation might predominate or when synaptic refinement might occur, the time course of regeneration had to be determined. In part, these issues were addressed with autoradiography using [3H]proline as a label. These experiments showed that the lesion effectively denervated the tectum, and that regenerating fibers had covered over two-thirds of the denervated tectal space by 2 weeks after the lesion (see Szaro *et al.*, 1985, for further details).

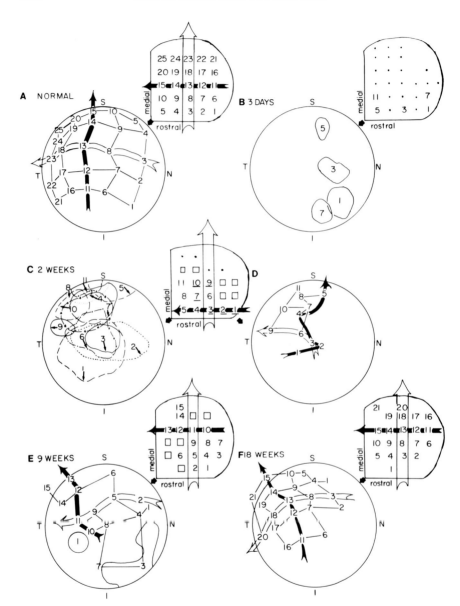

FIG. 5. Representative visuotectal projections assayed by electrophysiological mapping showing projections from a normal intact frog (**A**) and frogs at 3 days (**B**), 2 weeks (**C** and **D**), 9 weeks (**E**), and 18 weeks (**F**) following a tectal scratch lesion. The numbers on the tectal schematics show electrode recording positions at which a response to a small spot of light in the correspondingly numbered effective stimulus area of visual space was detected. Underlined numbers in C represent those regions of the tectum

This information was refined using extracellular recording electrodes to record from the terminal arborizations of the optic axons in the tectum after stimulation of these axons by a spot of light presented in the visual field of the contralateral eye innervating the injured tectum (for methods, see George and Marks, 1974; Hunt and Jacobson, 1974; Szaro et al., 1985). These recordings, as illustrated in Fig. 5, demonstrated that at 2 weeks after the lesion, much of the tectum gave only weak responses with abnormally large receptive fields. In addition, complete retinotopic order had not yet been reestablished. With this lesion, fiber sorting and refinements to the map continued for at least another 7 weeks. Thus, 2 weeks after the lesion some fiber growth was still actively in progress, and optic axons were sorting themselves into an orderly projection.

The appearance of synapses in the optic layers of the tecta of regenerating X. laevis frogs has been studied at the electron microscopic level (Ostberg and Norden, 1979). Synapses were seen forming simultaneously in the rostral and caudal tectum. This work suggested that synapses were formed by regenerating axons soon after arriving at the tectum, perhaps at inappropriate locations.

The conclusions of this ultrastructural study have received support from an electophysiological study done in frogs from the genus Rana (Adamson et al., 1984). In this study, the emergence of the indirect ipsilateral projection to the opposite tectum was observed in frogs whose optic nerve had been crushed at the orbit. Recordings from the tectum ipsilateral to a nerve crush require functional connections between regenerating optic axons and postsynaptic cells in the con-

---

which responded only to moving stimuli rather than to discrete stationary spots of light. These moving responses presumably represent either responses recorded postsynaptically from tectal cells or from overlapping presynaptic axons which had not yet formed a mature dense arbor. The boxes represent those regions of the tectum which could be weakly stimulated by gross changes in room illumination but which responded too weakly to spots of light to allow localization of the stimuli for these responses in visual space. The dots represent tectal recording loci which were unresponsive to visual stimulation. (A) A normal visuotectal projection. (B) A projection of a frog 3 days after a tectal scratch lesion. The tectum was unresponsive to visual stimulation except for a few loci, presumably spared by the surgery. (C) A partially recovered projection 2 weeks after injuring the nerve. (D) The same projection as in C, but with the centers of the effective stimulus areas indicated by numbers in order to illustrate the degree of retinotopy of the retinotectal map. (E) A projection at 9 weeks, mostly recovered except for a number of loci with weak responses, an abnormally shaped receptive field at locus 3, and one recording locus (1) which received input from 2 areas of visual field. (F) A fully regenerated retinotectal projection 18 weeks after injury. Frogs injured with this lesion typically took from 12 to 18 weeks to recover fully. (From Szaro et al., 1985.)

tralateral tectum. Recordings made in the tectum at early stages of regeneration from the *direct* contralateral projection showed abnormally large visual receptive fields. Similar recordings from the ipsilateral tectum initially showed no evidence of input from the contralateral tectum mediated by the indirect ipsilateral projection pathway, but by 20 days in some animals indirect ipsilateral recordings could be made which were in register with the direct contralateral projection. The fact that these earliest indirect ipsilateral receptive fields were similarly larger than normal, and at inappropriate positions relative to those in normal, intact frogs, showed that regenerating optic fibers make "test" synapses at inappropriate locations in the tectum, soon after their arrival and shortly after beginning to achieve some degree of retinotopy.

Thus, by comparison, our studies suggest that at 2 weeks postlesion, proteins necessary for synaptogenesis as well as those necessary for fiber growth would be enhanced. Consequently, to maximize the probability of detecting changes in proteins possibly involved in target recognition while minimizing the probability that proteins involved in fiber elongation would be overlooked, our subsequent biochemical studies were focused on regenerating optic nerves 2 weeks after injury.

VII. Quantitative Increases in Labeled Proteins during Regeneration

Historically, observations of increased incorporation of radioactive amino acids into retinal proteins during optic nerve regeneration provided some of the earliest direct evidence that axon growth involved new protein synthesis (e.g., see Murray and Grafstein, 1969; McQuarrie and Grafstein, 1982a). However, it is not always appropriate to infer increased protein synthesis directly from increased incorporation studies. In addition to increased levels of *de novo* protein synthesis, these observations could also reflect higher rates of uptake of amino acids or shifts in the intracellular pools of unlabeled amino acids resulting in more efficient incorporation of extracellularly applied radioactive amino acids into the proteins of regenerating retinal ganglion cells (Whitnall and Grafstein, 1981).

Several studies in other systems have shown that such general quantitative increases do occur (e.g., see Murray and Grafstein, 1969; McQuarrie and Grafstein, 1982a). While it is the tacit assumption that these increases reflect increases in protein synthesis, they could also reflect one or more additional processes: (1) a higher rate of uptake of extracellular amino acids into regenerating retinal ganglion cells; (2) a smaller, more rapidly turned over or bypassed intracellular pool of amino acids in regenerating cells which would result in more efficient (higher specific activity per mole of synthesized protein) incorporation of extracellularly applied radioactive amino acids into retinal ganglion cell proteins (Whitnall and Grafstein, 1981).

It has since been shown for regenerating optic nerves in goldfish that much of the increased incorporation of injected radioactive amino acids into retinal ganglion cell proteins is attributable to increased synthesis of proteins. In preparations in which retinal ganglion cells were isolated from the rest of the retina, a 4-fold increase in newly labeled proteins was observed in the retinal ganglion cells during regeneration (Giulian et al., 1980). Of the 4-fold increase, only a 20% increase could be attributed directly to residual radioactive amino acids taken up by cells but not incorporated into protein. Thus, the bulk of the increased labeling in regenerating retinal ganglion cells appeared to be due to an increase in protein synthesis rather than an increase in uptake of extracellular amino acids. However, the possibility of shifts in the intracellular pool size of unlabeled amino acids could have accounted for some of the remaining 380% increase.

Such shifts in intracellular pools have been shown in culture systems of regenerating goldfish retinas. Applying autoradiographic techniques to goldfish retinas in culture, Whitnall and Grafstein (1981) have shown that the level of incorporation into protein of extracellularly applied amino acids was more susceptible to fluctuations in the concentration and application of these labeled amino acids for retinas whose optic nerves had been regenerating than for normal, control retinas. These observations are consistent with increased uptake of extracellular amino acids in regenerating cells, smaller pool sizes of intracellular amino acids, and possibly more direct utilization of extracellular amino acids for protein synthesis without prior mixing with the intracellular pool (Whitnall and Grafstein, 1981). Thus, because of these shifts in pool sizes, absolute levels of increased protein synthesis are difficult to measure accurately.

However, more easily interpreted evidence has demonstrated increased levels of axonally transported proteins in regenerating nerves. For example, quantitative electron microscopy has shown, in regenerating goldfish retinal ganglion cells, that a higher percentage of newly synthesized proteins are routed into the Golgi apparatus, presumably for export through axonal transport (Whitnall and Grafstein, 1982). Thus, a greater fraction of the newly synthesized proteins are transported to axons in regenerating neurons. Because these measurements are normalized over general increases in protein synthesis, they are not subject to interference from artifacts derived from changes in amino acid pools and uptake, and substantiate the importance of increases in proteins supplied directly to regenerating axons.

Measurements made in collaboration with R. K. Hunt from regenerating optic nerves in *X. laevis* are consistent with these latter observations. After normalizing against the 1.5- to 3-fold increase in radioactivity incorporated into proteins in the eye during regeneration, the percentage of [^{35}S]methionine incorporated into protein in regenerating nerves increased over the contralateral control in the same animals by factors of 1.75 ± 0.40 (mean ± SEM) at 7 days post injection, 2.49 ± 0.33 at 2–4 hours postinjection, and 2.53 ± 0.25 at 18 hours postinjection (Szaro *et al.*, 1985). Thus, as in goldfish, increased levels of transport of newly synthesized proteins appear to occur during regeneration. However, these increases may be due to generalized responses to trauma rather than requirements for increased levels of specific proteins. Analysis of the patterns of axonally transported proteins on two-dimensional gel electrophoresis has shown in *Xenopus*, as in several other systems, that a small number of specific proteins increase significantly relative to the general increases observed for most other proteins, thus suggesting that at least some proteins play more specific roles in nerve regeneration.

VIII. Quantitative Changes in Specific Proteins during Regeneration

As described previously, the observations on specific protein changes during regeneration in *X. laevis* optic nerves were made 2 weeks after lesioning the nerve in the optic tract in order to maximize the probability of seeing proteins associated with synaptogenesis without eliminating proteins possibly associated with axon elongation (Szaro, 1982; Szaro *et al.*, 1985). Furthermore, the optic nerves were harvested either 2–4 hours, 18 hours, or 7 days later after injection of isotope, to sample each of the different waves of axonal transport. The changes in proteins during regeneration were normalized against the background increases observed for the majority of proteins in the system and thus probably underestimate the absolute levels of changes in these proteins.

The details and methodologies of these comparisons are presented in detail elsewhere (Szaro *et al.*, 1985). However, sample gel comparisons, with the measured spots indicated, are shown here in Figs. 6, 7, and 8. Spots which increased significantly (p < 0.05) are marked "△", while those which decreased are marked "▽".

In addition, the "crush" paradigm described previously for one-dimensional SDS gels was used to catalog, for the two-dimensional gels, a list of transported proteins at each of the different time points. Control nerves contained low amounts of those proteins found to increase during regeneration. This may reflect the fact that *Xenopus* eyes continue to grow throughout adult life and that normal optic nerves contain modest numbers of actively growing fibers (Straznicky and Gaze, 1971). Alternatively, even established optic nerve connections are tuned and refined in young postmetamorphic frogs, such that these oldest fibers present (which had first made synapses at very young tadpole stages) may continue to sprout new endings (Gaze *et al.*, 1979; Fraser, 1983; Scott and Lazar, 1976). Thus, the retinal ganglion cells may transport low levels of proteins involved in these processes even in uninjured nerves.

Table II summarizes the results of this study in regenerating *Xenopus* optic nerve. Nine transported proteins were seen to increase. These proteins had molecular weights of 240,000 (#1), 135,000 (#3), 65,000 (#11), 58,000 (#12), 54,000 (#17), 56,000 (#23), 64,000 (#8), 31,000 (#31), and 26,000 (#34). Of these nine, seven had similar isoelectric points and molecular weights as spots present in the optic tectum during regeneration. They were at 240,000 (#1), 135,000 (#3), 58,000 (#12), 54,000 (#17), 56,000 (#23), 31,000 (#31), and 26,000 (#34) Da. Thus, these proteins might be associated with growing terminals in the tectum.

Although we have not yet identified any of these specific proteins, it is possible that some may represent proteins already seen to be associated with growing axons in other systems. For example, one protein (#1, 240,000 Da) resembles, in acidity and molecular weight, the embryonic form of the non-calcium-dependent adhesion molecule N-CAM (in chicks, a broadly staining acidic band of between 200,000 and 250,000 Da: see Edelman, 1983) and is known to be present in *X. laevis* (see Fraser *et al.*, 1984). Protein #1 also has a similar molecular weight to the microtubule-associated protein, MAP-1, which is thought to play a role in axon growth (Calvert and Anderton, 1985). In addition, recent studies on GAP-43, another protein present in regenerating optic axons of *Bufo marinus,* have shown that, under electrophoretic conditions comparable to those employed in our studies, GAP-43 appears similar to protein #23 in *Xenopus* (Jacobson *et al.*, 1986). It will be interesting to further substantiate the identities of protein #1 and protein #23 with MAP-1 and GAP-43, respectively, using antibodies to these proteins.

4 HOUR NERVES

NORMAL

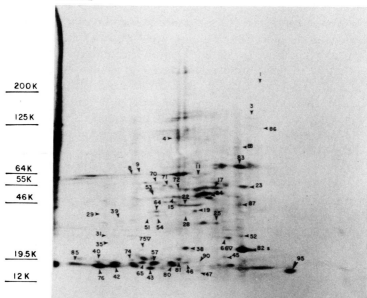

REGENERATING

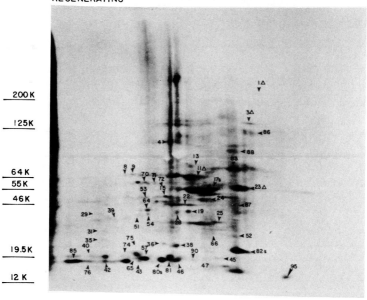

base   **acid**

Proteins which move in the slower phases and migrate with cytoskeletal components could be involved in axon elongation or its regulation (proteins #12, #17, #8, #31, and #34). One of these proteins, #17 at 54,000 Da with an isoelectric point of approximately 5, migrates similarly to tubulin (with molecular weight of 51,000 to 55,000, pI = 5.4; Strocchi et al., 1981). This cytoskeletal protein appears to increase in several regenerating systems (Hoffman and Lasek, 1975b; Heacock and Agranoff, 1976, 1982; Giulian et al., 1980; Skene and Willard, 1981c).

In addition to those proteins which increased during regeneration, 6 transported proteins decreased: 15,000 (#41), 17,000 (#42), 29,000 (#66), 49,000 (#53), 56,000 (#60), and 42,000 (#64) Da. The one protein associated with fast transport might be involved in normal synaptic function (#66, 29,000 Da) since levels of proteins involved in regulation of neurotransmitters have been known to decrease during regeneration (Frizell and Sjostrand, 1974; Heiwall et al., 1979; Boyle and Gillespie, 1970). In addition, perhaps those proteins associated with the slower phases may be involved in inhibiting axon sprouting and elongation in the normal nerve (#41, #42, #53, #60, and #64).

Several proteins which did not appear to be transported on the basis of crush controls also decreased [15,000 Da (#43), 72,000 Da (#59), and 22,000 Da (#75)]. Recently, it has been suggested that nonaxonal cells in normal nerves secrete growth inhibitors in goldfish optic nerves. Conditioned media from normal nerves inhibit axonal growth in goldfish retinal explants, while media conditioned by regenerating

---

FIG. 6. A visual comparison of fluorographs made from two-dimensional gels (for methods see Wilson et al., 1977; Neville, 1971; Szaro et al., 1985) of labeled proteins from optic nerves harvested 4 hours after intraocular injection of [^{35}S]methionine. The top gel consists of material from 10 labeled intact nerves, while the bottom gel was run on material from 10 optic nerves from animals injected with radioactive methionine 2 weeks after an optic nerve lesion. The X-ray film for each gel was exposed for 700,000 cpm-days. Molecular masses were obtained by comparing the migrations of prominent Coomassie blue-stained spots on both the left side of the gel and within the gel itself to prominent Coomassie blue-stained bands contained in one dimensional gels run on unlabeled optic nerves (see Szaro et al., 1984, for details). The acid end of the gel was at approximately pH 3.5; the basic end was at pH 8.5. The numbered spots represent those spots selected for quantitative measurements, either by computer or by direct scintillation counting of the physically excised spots (marked by "s" next to the number). This figure shows that, while most spots remained relatively constant (as a percentage of the total transported material in the nerve), a select few relatively either decreased or increased during regeneration. Those spots which increased during regeneration are designated by △ next to the appropriate number on the gel run on regenerating nerves (bottom). Those spots which were greater in gels run on normal nerves are designated by ▽ (top). K, kDa. (From Szaro et al., 1985.)

18 HOUR NERVES

NORMAL

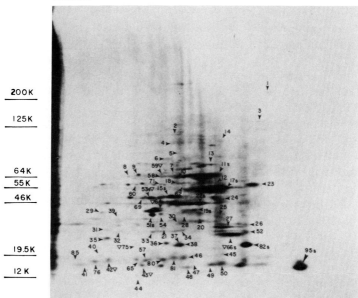

REGENERATING

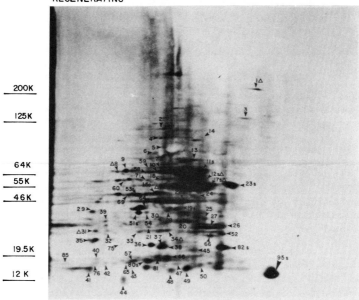

base   **acid**

nerves failed to inhibit growth. The media isolated from regenerating nerves lacked a prominent 13,000 Da component (Rachailovich and Schwartz, 1984). It would be interesting to determine if this protein is related to one of the proteins which decreased during regeneration.

However, at this stage we can only speculate on the roles these proteins play in regenerating *Xenopus* optic nerves. Now that a list of proteins which specifically increase during regeneration has been generated, it is necessary to begin to study and identify some of these individual proteins. Because of differences in preparations and species, some of these proteins might already be known in other developing and regenerating systems; we have suggested some similarities which might be identified using antibodies. For still others which do not bear any strong resemblance to any known proteins in other systems, several approaches may be taken to characterize and identify these proteins, some of which have already been taken in other systems.

For example, in the rapid phase of transport in regenerating optic nerves of *B. marinus,* a distant anuran cousin of *X. laevis,* three prominent growth associated proteins (GAPs) have been isolated: a basic protein at 50,000 Da, a 43,000 Da acidic protein, and a 24,000 Da protein which can only be resolved in the second dimension in the presence of urea (Skene and Willard, 1981b,c). The behavior of these proteins at different times during the course of regeneration has been examined. After sufficient time had elapsed for regeneration to be completed, the levels of these proteins returned to normal, thus strengthening the argument that these proteins were associated directly with axon regrowth rather than a nonspecific cellular response to trauma. In addition, the GAP-43 protein has been detected in neonatal rabbit optic nerves and regenerating rabbit hypoglossal nerves, while it is absent from injured adult optic nerves, which do not regenerate (Skene and Willard, 1981d). It has also been detected in the pyramidal tract axons in neonatal hamsters but not in adult axons (Skene and Kalil, 1984). GAP-43 is also prominent in neonatal rat cerebral cortex and cerebellum, but declines in mature brains (Jacobson *et al.,* 1986). Furthermore, polyclonal antibodies have been made to GAP-43 and have shown that GAP-43 is similar to the phosphoprotein pp46, a major component of mammalian growth cone membranes (Meiri *et al.,*

FIG. 7. A visual comparison of fluorographs made from two-dimensional gels of electrophoretically separated labeled proteins from 10 normal (top) and 10 regenerating (bottom) optic nerves harvested from the same frogs 18 hours after intraocular injection of [^{35}S]methionine. See Fig. 6 for further details. K, kDa. (From Szaro *et al.,* 1985.)

7 DAY NERVES

NORMAL

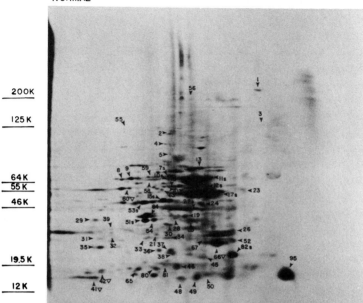

REGENERATING

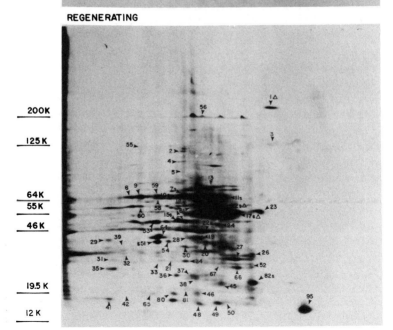

base   acid

1986; Skene *et al.*, 1986). These facts strengthen the argument that GAP-43 is associated directly with axon growth and is of general importance to growing vertebrate axons.

In addition, GAP-43 is similar to two other brain proteins, B-50 (Jacobson *et al.*, 1986), a major substrate of protein kinase C (Aloyo *et al.*, 1983) [studied because its phosphorylation was modulated by ACTH, dynorphin, and other behaviorally active neuropeptides (Zwiers et al., 1980; 1981], and F1 (Nelson et al., 1985) [a protein whose phosphorylation has been related to long term potentiation in rat hippocampus (Nelson and Routtenberg, 1985)]. That a protein associated with axon growth is identical to a substrate affected by neuropeptides and long-term potentiation suggests the possibility of intriguing interrelationships between development and plasticity.

The optic nerves of goldfish have also been used as a model for detecting changes in the axonal transport of proteins during regeneration. Using an optic tract lesion to maintain the integrity of the optic nerve during regeneration (an operation similar to the one we used in *Xenopus*), Giulian *et al.* (1980) found, 10–35 days after injury, changes in slowly transported proteins, including actin, tubulin, and additional components at 70,000, 90,000, and 300,000 [termed the large neuro polypeptide (LNP)] Da. They verified that these proteins were probably axonally transported by showing that these same proteins were seen to increase in isolated retinal ganglion cell bodies at the same stage of regeneration.

Other studies carried out on the rapid phases of transport in regenerating goldfish optic nerves (Benowitz *et al.*, 1981; Benowitz and Lewis, 1983) have found proteins at 44,000–49,000 Da, at 210,000 and 26,000 Da, and in the 110,000–140,000 Da range. These nerves were crushed at the back of the eye. The molecular weights of the heavier proteins are similar in molecular weight to the proteins we detected in *Xenopus* at 240,000 and 135,000 Da (#1 and #3). Furthermore, the appearance and disappearance of some of these proteins appeared to be correlated with the known arrival times of regrowing fibers at the edge of the tectum (Benowitz *et al.*, 1983). Specifically, proteins of molecular weights near 120,000 and 160,000 increased at times correlating to the arrival of regenerating fibers at the edge of the tectum, while a 26,000

Fɪɢ. 8. A visual comparison of fluorographs made from two-dimensional gels of electrophoretically separated labeled proteins from 10 normal (top) and 10 regenerating (bottom) optic nerves harvested from the same animals 7 days after intraocular injection of [³⁵S]methionine. See Fig. 6 for further details concerning the gel parameters. K, kDa. (From Szaro *et al.*, 1985.)

TABLE II

Axonally Transported Optic Nerve Proteins which Changed Significantly during Regeneration in X. laevis

Spot	Mass[a]	Ratio[b]	Transport phase[c]	n[d]
Increases				
$p < 0.01$				
1	240	2.08 ± 0.11	Fast	2
		(2.71 ± 0.62)	Intermediate–slow	3
3	135	1.87 ± 0.25	Fast	5
11	65	1.78 ± 0.19	Fast	5
12	58	1.78 ± 0.46	Intermediate–slow	3
17	54	1.92 ± 0.23	Intermediate–slow	3
23	56	2.07 ± 0.21	Fast	5
$0.01 < p < 0.05$				
8	64	1.77 ± 0.17	Intermediate	3
31	31	1.80 ± 0.32	Intermediate	3
34	26	1.80 ± 0.36	Intermediate	3
62[e]	58	2.54 ± 1.10	Intermediate	3
Decreases				
$p < 0.01$				
41	15	0.21 ± 0.80	Slow	2
42	17	0.18 ± 0.01	Slow	2
		(0.53 ± 0.21)	Intermediate	3
66	29	0.19 ± 0.03	Intermediate–slow	3
		(0.61 ± 0.11)	Fast	5
$0.01 < p < 0.05$				
43[e]	15	0.33 ± 0.09	Intermediate	2
53	49	0.42 ± 0.04	Intermediate	2
59[e]	72	0.42 ± 0.06	Intermediate	2
60	56	0.42 ± 0.06	Slow	2
64	42	0.40 ± 0.04	Intermediate	2
75[e]	22	0.38 ± 0.15	Fast	2
		0.45 ± 0.17	Intermediate	3

[a]In kDa; after Szaro et al., 1985.

[b]The ratios between regenerating and normal nerves are represented as the mean ± SE. Ratios expressed in parentheses indicate that the same protein is significant at the $0.01 \leq p \leq 0.05$ level at the indicated timepoint.

[c]Fast, intermediate, and slow phases of transport indicate proteins seen in the nerve 4 hours, 18 hours, and 7 days, respectively, after intraocular injection of isotope in the eye.

[d]n, Number of experiments. Each experiment compared the regenerating nerve to its contralateral control in 10 frogs.

[e]These proteins do not appear to be axonally transported as determined by crush control experiments.

Da protein decreased. Another acidic group of proteins in the range of 44,000–49,000 Da was unaffected (Benowitz *et al.*, 1983: Benowitz and Lewis, 1983). Thus, regenerating axons can modulate their biochemical state based on the interactions with neighboring tissue. Similar modulations have been observed in cultured explants of dorsal root ganglia in contact with their target tissues in coculture (Sonderegger *et al.*, 1983) and may be a general feature of regrowing axons.

IX. Other Protein Changes Observed in Injured Optic Nerves Not Directly Associated with Changes in Axonal Transport

So far, we have concentrated on describing changes in the axonal transport of retinal ganglion cell proteins with a view toward finding changes which might be investigated for possible roles in normal development. However, other changes, such as those made by the glia during regeneration, might influence the extracellular environment through which regenerating axons grow and possibly facilitate the regrowth of axons (Aguayo *et al.*, 1981). Such changes might produce an extracellular environment more like that seen by growing fibers during development. To date, little direct evidence exists to support the hypothesis that these changes mimic the extraaxonal environment during development, but some biochemical changes have been seen during regeneration in the glia and nerve sheath which could be investigated for possible correlations with developing systems.

For example, in *Xenopus* optic nerves crushed at the orbit, nerve segments dissected several hundred micrometers from the region of the crush showed fewer bands than the uncrushed nerves, but nerve segments taken distal to, but nearer, the crush showed not only the disappearance of bands (taken to be axonally transported) but also the appearance of additional bands not previously seen in uncrushed nerves (Szaro and Faulkner, 1982; Szaro, 1982; Szaro *et al.*, 1984). At least two of these (43,000 and 15,000 Da) appeared in crushed nerves whose eyes were injected as little as 5–30 minutes after crushing the nerve and sharply diminished within 1 day, while two additional bands at 54,000 and 48,000 Da appeared a day later (see Fig. 9). Autoradiography done on acutely injured frogs showed labeling in this region of the nerve occurred in the nerve sheath. Because these bands could also be labeled systemically by injecting the contralateral eye (Szaro *et al.*, 1984), it was thought that these bands represented a nonaxonal local response to the nerve injury.

While it cannot be stated unequivocably whether these proteins represent unique glial proteins responding to the injury or are a more generalized feature of cell trauma, it is tempting to hypothesize that

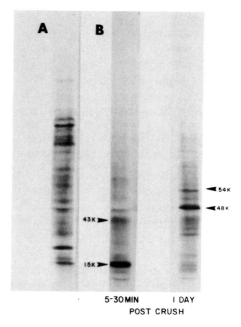

Fig. 9. Fluorographs of two gels containing labeled proteins from 10 optic nerve segments harvested from an area extending from the crush site near the eye to approximately one-third of the distance (about 0.5 mm) to the chiasm (B), and one gel showing proteins from a similar normal nerve segment for comparison (A). Nerves whose proteins are shown in the center lane were crushed 5–30 minutes before the injections and the segments harvested 1 hour after the injections, while the proteins shown in the far right lane came from nerves labeled 1 day after injury. In the region of the nerve near the crush, several proteins were greatly enhanced in the injured nerves. Two of these (arrows at 43 kDa and 15 kDa) were prominent in nerves labeled acutely, while two more were prominent in nerves labeled 1 day after the injury (arrows at 54 kDa and 48 kDa). Because these proteins appeared in nerve segments in which the normal transport of axonal proteins was blocked, and because they could also be labeled systemically, they are believed to be a support cell response to the trauma.

such proteins produced at the site of nerve injury may be taken up by injured nerve endings and stimulate regrowth. They may also be involved in the degradation of the disconnected axons downstream from the injury, and the laying down of new substrate for the axons to grow along, and thus may have developmental implications.

Similar proteins with molecular weights of 16,000, 30,000, and 42,000 have been seen as nonaxonal nerve constituents following orbital crushes in goldfish (Deaton and Freeman, 1983), and a fourth protein at 250,000 Da appears to be made in the glia surrounding the optic tract after a lesion near the tectum in goldfish (Giulian, 1984). In addition, a 37,000 Da protein has been reported to increase in injured and developing rat sciatic nerves and injured optic nerves. This protein appears to be specifically made by the epineural sheath cells and is similar to a plasma protein (Politis *et al.,* 1983; Skene and Shooter, 1983; Snipes and Freeman, 1984). Thus, in several systems, candidates for specific extraaxonal proteins which may play a role in nerve regeneration and ontogeny have been discovered.

X. Evidence Suggestive of Nerve Injury Factors Carried through the Bloodstream

In addition to a stimulus reaching the perikaryon from injured nerve endings by way of retrograde transport from the site of nerve injury, chemical signals which stimulate axon growth might travel indirectly through the bloodstream. Such a general signal would be expected to reach, and stimulate, an injury response in uninjured neurons as well as injured ones. In *Xenopus* retinal ganglion cells, we have evidence suggesting such a response in otherwise uninjured neurons, contralateral to a nerve lesion.

Uninjured optic nerves from frogs whose contralateral nerve has never been injured incorporate less radioactivity than similar uninjured optic nerves of frogs whose contralateral nerve has been lesioned 2 weeks prior to isotope injection. For example, the mean percentage of trichloroacetic acid (TCA)-precipitable radioactivity in a series of six sets (60 nerves in all) of optic nerves in animals which had never been injured contained 0.57% ± 0.06% (mean ± SEM) of the total precipitable material in the visual system at 18 hours after isotope injection. In a comparable series of three sets of nerves (30 nerves in all) which were contralateral to nerves which had been injured 2 weeks prior to injection contained 1.34% ± 0.33% of the total radioactivity in the visual system 18 hours after injection, a significant increase over the levels seen in untraumatized animals ($p < 0.05$, Student's two-tailed t test).

Similarly, the two-dimensional electrophoretic patterns of radiolabeled optic nerve proteins harvested 18 hours after [35S]methionine had been injected into the eyes of frogs whose nerves had never been injured were qualitatively different from analogous patterns taken 18 hours postinjection from the control nerves of frogs whose contralateral nerve had been injured 2 weeks prior to the injections (Fig. 10). When exposed for equal periods of time, the "control" nerves from previously injured frogs gave more intense protein patterns which

18 HOUR NERVES
Unoperated

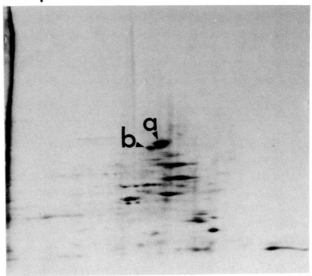

Control nerve in regenerating animal

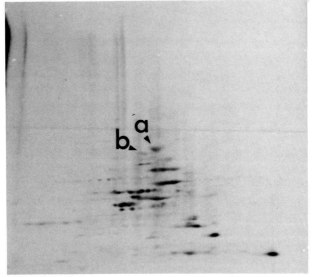

reflected the general increase in protein labeling. However, in contrast to the more intensely labeled proteins, two individual spots were barely labeled, although these spots were prominent in the patterns from uninjured frogs. While we do not at this time know the source of these changes, the fact that the protein patterns of the nerve contralateral to a nerve injury are different than similar samples taken from uninjured frogs suggests some indirect effect, possibly mediated through the bloodstream or as a result of secondary effects mediated across several synapses. A similar effect has been noted on the amount of radioactivity incorporated into retinal proteins (McQuarrie and Grafstein 1982b) and on neurite outgrowth and levels of ornithine decarboxylase in goldfish retinas (Kohsaka et al., 1981; Landreth and Agranoff, 1976). Such observations support the idea that humoral factors and secondary effects might be involved in normal development as well.

XI. Concluding Remarks

In this review, we have cited several systems, studied with a variety of surgical and biochemical paradigms, where potential key proteins correlated with optic nerve regeneration have been identified. Some of these same proteins are likely to be involved in axon outgrowth and target recognition during the ontogeny of the visual system.

In the lower vertebrate visual system, changes in specific proteins during regeneration occur in retinal ganglion cells in the eye, and also glial cells in the nerve and can be modulated by neuronal contacts with cells in the tectum. Whereas only a few years ago it was uncertain that any biochemical changes in specific proteins would be detected in re-

---

FIG. 10. Fluorographs of two-dimensional gel electrophoresis patterns of [35S]methionine labeled optic nerve proteins 18 hours after intraocular injection of the isotope. The top gel is the pattern obtained from 10 nerves pooled from the uninjured nerves of uninjured frogs, while the bottom gel has been pooled from 10 uninjured nerves contralateral to nerves which had been lesioned 2 weeks earlier. Note that spots labeled a and b were greatly diminished in control nerves from lesioned frogs but were prominent in the nerves of uninjured frogs. Thus, while the control nerves of injured frogs had significantly more labeling than similar nerve segments from uninjured frogs, the relative labeling of some proteins within them was diminished, suggesting a selective secondary effect on protein labeling in the uninjured nerves of injured frogs. The exposures have been balanced so that each gel received the same relative exposure based on trichloroacetic acid-precipitable counts of the total amount of labeled protein placed on the gel. The first dimension was an equilibrium isoelectric focusing gel (pH range approximately 3.5 to 8.5), while the second gel was a 4–17% polyacrylamide gradient SDS gel. The acid end of the isoelectric focusing gel is on the right. See text for further discussion.

TABLE III

Changes in Optic Nerve Protein Composition after Injury in Lower Vertebrates

Transported Proteins[a]

	Goldfish			X. laevis			B. marinus		
	Size	Phase	Modulation	Size	Phase	Modulation	Size	Phase	Modulation
	300 kDa	Slow	↑ [4,b]	240 kDa	Fast	↑ [10]	210 kDa	Slow	→ [8]
	210 kDa	Fast	↑ [1]				190 kDa	Fast	→ [8]
							185 kDa	Slow	→ [8]
	155 kDa	Fast	↑ [1]	135 kDa	Fast	↑ [10]	115 kDa	Fast	→ [8]
	110–140 kDa	Fast	↑ [1]						
	130 kDa	Fast	→ [1,2]						
	90 kDa	Slow	↑ [4]						
	76 kDa	Fast	↑ [1]						
	>70 kDa		↑ [4,c]						
	70 kDa	Slow	(2) ↑ [4,6]	65 kDa	Fast	↑ [10]	68 kDa	Fast	→ [8]
				64 kDa	Slow	↑ [10]			
	58 kDa	Fast	→ [1]	58 kDa	Slow	↑ [10]			
				56 kDa	Fast	↑ [10,h]			
				56 kDa	Slow	↑ [10]			
	51–55 kDa	Slow	↑ [1,4,6,d]	54 kDa	Slow	↑ [10,d]	50–55 kDa	Slow	↑ [8,d]
	54 kDa	Fast	↑ [1]						
	46–52 kDa	Slow	↑ [2]				50 kDa	Fast	↑ [8,e]
	50 kDa	Fast	→ [1]						

Size	Modulation	Size	Modulation	Size	Modulation
44–49 kDa	Fast ↑[1,2,6,f]				
45 kDa	Slow ↑[2,4,g]	49 kDa	Slow ↓[10]	43 kDa	Fast ↑[8,e]
				42–43 kDa	Slow ↑[8,g]
37 kDa	Slow ↑[2]	42 kDa	Slow ↓[10]		
36 kDa	Fast ↑[1]				
33 kDa	Fast ↓[1]			33 kDa	Fast ↑[8]
		31 kDa	Slow ↑[10]		
29 kDa	Fast ↓[1]	29 kDa	Fast ↓[10]		
28 kDa	Slow ↑[2]				
26 kDa	Fast ↑[1]	26 kDa	Slow ↑[10]	24 kDa	Fast ↑[8,e]
23 kDa	Slow ↑[2]				
		17 kDa	Slow ↓[10]		
		15 kDa	Slow ↓[10]		

SUPPORT CELL PROTEINS

Goldfish		X. laevis	
Size	Modulation	Size	Modulation
250 kDa	↑[5]	90 kDa	↑[9]
		72 kDa	↑[9]
		54 kDa	↑[9]
		48 kDa	↑[9]
42 kDa	↑[3]	43 kDa	↑[9]
30 kDa	↑[3]		

(continued)

249

TABLE III (Continued)

Support Cell Proteins

	Goldfish		X. laevis	
	Size	Modulation	Size	Modulation
			22 kDa	↓[9]
	16 kDa	↑[3]	15 kDa	↑[9]
			15 kDa	↓[9]
	13 kDa	↓[7]	13 kDa	↑[9]
			10 kDa	↑[9]

[a] References: [1]Benowitz et al. (1981); [2]Benowitz and Lewis (1983); [3]Deaton and Freeman (1983); [4]Giulian et al. (1980); [5]Giulian (1984); [6]Heacock and Agranoff (1982); [7]Rachailovich and Schwartz (1984); [8]Skene and Willard (1981c); [9]Szaro et al. (1984); [10]Szaro et al. (1985).

[b] This protein has been given the name "large neuropeptide polypeptide (LNP)."

[c] These proteins represent a group of fucosylated proteins which increase during regeneration.

[d] Based on their transport properties, abundance, and electrophoretic mobilities, these spots are thought to represent tubulin.

[e] Because these proteins appear to be associated with regenerating and growing axons in a variety of systems, they have been called growth-associated proteins (GAP-50, GAP-43, and GAP-24 respectively).

[f] Thought possibly to be homologous with GAP-43.

[g] Based on their transport properties and electrophoretic mobilities, these proteins are thought to represent actin.

[h] Although under the solubilization and electrophoretic conditions employed in Xenopus this protein has a higher molecular weight than 44–49 kDa, because of its acidity and abundance in fast-transported proteins during regeneration this protein might be similar to the 44–49 kDa group of proteins reported in goldfish and the GAP-43 protein reported in B. marinus.

generating nerves (Hall *et al.*, 1978), it is now clear that a number of changes do occur (see the summary in Table III). While these changes may not necessarily represent new proteins, unique to growing nerves (Szaro *et al.*, 1985; Benowitz and Lewis, 1983; Perry and Wilson, 1981), they nevertheless confirm at the biochemical level that a subset of specific proteins are modulated during regeneration and thus are likely to play important roles in nerve regrowth and response to trauma.

Thus, the beginning foundations for a biochemical understanding have been laid, and preparations for finding roles for these macromolecules can be made. As new separation techniques and detection criteria are brought to bear on these and other growing neuronal systems, still other molecules may be discovered. Recent advances in the production of monoclonal antibodies from small amounts of protein (Van Ness *et al.*, 1984) and gene cloning techniques will be essential in aiding the identification of these molecules and exploring their functional roles in nerve regeneration and neural connectivity during development. Meanwhile, the progress made so far has renewed the hope that eventually the biochemical basis for the regulation of axon elongation and the recognition of appropriate target sites may be in the near, rather than distant, future.

ACKNOWLEDGMENTS

Most of the work discussed here on the axonal transport of proteins in regenerating *X. laevis* optic nerves was done in collaboration with R. Kevin Hunt and supported in part by National Science Foundation Grant PCM-26987 and National Institutes of Health Grant NS-14807.

REFERENCES

Adamson, J., Burke, J., and Grobstein, P. (1984). *J. Neurosci.* **4**, 2635–2650.
Agranoff, B. W., Hopkins, J. M., Davis, R. E., Ford-Holevinski, T. S., and McCoy, J. P. (1984). *Soc. Neurosci. Abstr.* 10, 282.
Aguayo, A. J., David, S., and Bray, G. M. (1981). *J. Exp. Biol.* **95**, 231–240.
Aloyo, V. J., Zwiers, H., and Gispen, W. H. (1983). *J. Neurochem.* **41**, 649–653.
Baitinger, C., Cheney, R., Clements, D., Glicksman, M., Hirokawa, N., Levine, J., Meiri, K., Simon, C., Skene, P., and Willard, M. (1983). *Cold Spring Harbor Symp. Quant. Biol.* **48**, 791–802.
Beach, D. H., and Jacobson, M. (1979). *J. Comp. Neurol.* **183**, 603–614.
Benowitz, L. I., and Lewis, E. R. (1983). *J. Neurosci.* **3**, 2153–2163.
Benowitz, L. I., Shashoua, V. E., and Yoon, M. G. (1981). *J. Neurosci.* **1**, 300–307.
Benowitz, L. I., Yoon, M. G., and Lewis, E. R. (1983). *Science* **222**, 185–188.
Boyle, F. C., and Gillespie, J. S. (1970). *Eur. J. Pharmacol.* **12**, 77–84.
Brun, R. B. (1981). *Dev. Biol.* **88**, 192–199.
Calvert, R., and Anderton, B. H. (1985). *EMBO J.* **4**, 1171–1176.
Cima, C., and Grant, P. (1982a). *J. Embryol. Exp. Morphol.* **72**, 225–249.
Cima, C., and Grant, P. (1982b). *J. Embryol. Exp. Morphol.* **72**, 251–267.
Conway, K., Feiock, K., and Hunt, R. K. (1981). *Curr. Top. Dev. Biol.* **15**, 217–317.

Deaton, M. A., and Freeman, J. A. (1983). *Soc. Neurosci. Abstr.* **9**, 693.
Doherty, P., Dickson, J. G., Flanigan, T. P., and Walsh, F. S. (1985). *J. Neurochem.* **44**, 1259–1265.
Droz, B., Rambourg, D., and Koenig, H. L. (1975). *Brain Res.* **93**, 1–13.
Edelman, G. M. (1983). *Science* **219**, 450–457.
Elam, J. S., and Agranoff, B. W. (1971a). *J. Neurochem.* **18**, 375–387.
Elam, J. S., and Agranoff, B. W. (1971b). *J. Neurobiol.* **2**, 379–390.
Forman, D. S. (1983). *In* "Progress in Neurology: Spinal Cord Reconstruction" (C. C. Kao, R. P. Bunge, and P. J. Reier, eds.) Raven, New York.
Fraser, S. E. (1983). *Dev. Biol.* **95**, 505–511.
Fraser, S. E., and Hunt, R. K. (1980). *Annu. Rev. Neurosci.* **3**, 319–352.
Fraser, S. E., Murray, B. A., Chuong, C.-M., and Edelman, G. M. (1984). *Proc. Natl. Acad. Sci. U.S.A.* **81**, 4222–4226.
Frizell, M., and Sjostrand, J. (1974). *J. Neurochem.* **22**, 845–850.
Gammon, C. M., Goodrum, J. F., Toews, A. D., Okabe, A., and Morell, P. (1985). *J. Neurochem.* **44**, 376–387.
Gaze, R. M. (1959). *Q. J. Exp. Physiol.* **44**, 209–308.
Gaze, R. M., and Peters, A. (1961). *Q. J. Exp. Physiol.* **46**, 299–309.
Gaze, R. M., Keating, M. J., and Chung, S.-H. (1974). *Proc. R. Soc. London Ser. B* **185**, 301–330.
Gaze, R. M., Keating, M. J., Ostberg, A., and Chung, S.-H. (1979). *J. Embryol. Exp. Morphol.* **53**, 103–143.
George, S. A., and Marks, W. B. (1974). *Exp. Neurol.* **42**, 467–482.
Giulian, D. (1984). *Proc. Natl. Acad. Sci. U.S.A.* **81**, 3567–3571.
Giulian, D., Des Ruisseaux, H., and Cowburn, D. (1980). *J. Biol. Chem.* **255**, 6494–6501.
Goodman, C. S., Bastiani, M. J., Doe, C. Q., du Lac, S., Helfand, S. L., Kuwada, J. Y., and Thomas, J. B. (1984). *Science* **225**, 1271–1279.
Grafstein, B., and Alpert, R. (1976). *Neurosci. Abstr.* **2**, 46.
Grafstein, B., and Forman, D. S. (1980). *Physiol. Rev.* **60**, 1167–1283.
Grant, P., and Rubin, E. (1980). *J. Comp. Neurol.* **189**, 671–698.
Grant, P., Rubin, E., and Cima, C. (1980). *J. Comp. Neurol.* **189**, 593–613.
Hall, M. E., Wilson, D. L., and Stone, G. C. (1978). *J. Neurobiol.* **9**, 353–366.
Harris, W. A. (1980). *J. Comp. Neurol.* **194**, 303–317.
Harris, W. A. (1984). *J. Neurosci.* **4**, 1153–1162.
Heacock, A. M., and Agranoff, B. W. (1976). *Proc. Natl. Acad. Sci. U.S.A.* **73**, 828–832.
Heacock, A. M., and Agranoff, B. W. (1982). *Neurochem. Res.* **7**, 771–788.
Heiwall, P.-O., Dahlstrom, A., Larsson, P.-A., and Booj, S. (1979). *J. Neurobiol.* **10**, 119–136.
Hoffman, P. N., and Lasek, R. J. (1975a). *J. Cell Biol.* **66**, 351–366.
Hoffman, P. N., and Lasek, R. J. (1975b). *Soc. Neurosci. Abstr.* **5**, 1237.
Holt, C. (1980). *Nature (London)* **287**, 850–852.
Holt, C. (1984). *J. Neurosci.* **4**, 1130–1152.
Hoskins, S. G., and Grobstein, P. (1984). *Nature (London)* **307**, 730–733.
Hunt, R. K., and Jacobson, M. (1974). *Curr. Top. Dev. Biol.* **8**, 203–259.
Ide, C. F., Reynolds, P., and Tompkins, R. (1984). *J. Exp. Zool.* **230**, 71–80.
Jacobson, M. (1968). *Dev. Biol.* **17**, 219–232.
Jacobson, M. (1976). *Brain Res.* **103**, 541–545.
Jacobson, M. (1983). *J. Neurosci.* **3**, 1019–1038.
Jacobson, M., and Hirose, G. (1978). *Science* **202**, 637–639.
Jacobson, R. D., Virag, I., and Skene, J. H. P. (1986). *J. Neurosci.* **6**, 1843–1855.
Kohsaka, S., Schwartz, M., and Agranoff, B. W. (1981). *Brain Res.* **227**, 391–401.

Landreth, G. E., and Agranoff, B. W. (1976). *Brain Res.* **118**, 299–303.

Lasek, R. J. (1980). *Trends Neurosci.* **3**, 87–91.

Lorenz, T., and Willard, M. (1978). *Proc. Natl. Acad. Sci. U.S.A.* **75**, 505–509.

McQuarrie, I. G., and Grafstein, B. (1982a). *Brain Res.* **235**, 213–223.

McQuarrie, I. G., and Grafstein, B. (1982b). *Brain Res.* **251**, 25–37.

Marchase, R. B. (1977). *J. Cell Biol.* **75**, 237–257.

Maturana, H. R., Lettvin, J. Y., McCulloch, W. S., and Pitts, W. H. (1959). *Science* **130**, 1709–1710.

Meiri, K. F., Pfenninger, K. H., and Willard, M. B. (1986). *Proc. Natl. Acad. Sci., U.S.A.* **83**, 3537–3541.

Meyer, R. L. (1982). *Curr. Top. Dev. Biol.* **17**, 101–146.

Meyer, R. L. (1983). *Dev. Brain Res.* **6**, 293–298.

Murray, M., and Grafstein, B. (1969). *Exp. Neurol.* **23**, 544–560.

Nelson, R. B., and Routtenberg, A. (1984). *Exp. Neurol.* **89**, 213–224.

Nelson, R. B., Routtenberg, A., Hyman, C., and Pfenninger, K. H. (1985). *Soc. Neurosci. Abstr.* **11**, 927.

Neville, D. (1971). *J. Biol. Chem.* **246**, 6328–6334.

Ochs, S. (1982). "Axoplasmic Transport and its Relation to Other Nerve Functions." Wiley, New York.

Ostberg, A., and Norden, J. (1979). *Brain Res.* **168**, 441–456.

Perry, G. W., and Wilson, D. L. (1981). *J. Neurochem.* **32**, 1203–1217.

Pierce, M. (1982). *J. Cell Biol.* **75**, 237–257.

Politis, M. J., Pellegrino, R. G., Oadlander, A. L., and Murdock, R. (1983). *Brain Res.* **273**, 392–395.

Rachailovich, I., and Schwartz, M. (1984). *Brain Res.* **106**, 149–155.

Reier, P. J., and Webster, H. deF. (1974). *J. Neurocytol.* **3**, 591–618.

Rho, J.-H., and Hunt, R. K. (1980). *Dev. Biol.* **80**, 436–453.

Sakaguchi, D. S., Murphey, R. K., Hunt, R. K., and Tompkins, R. (1984). *J. Comp. Neurol.* **224**, 231–251.

Sbaschnig-Agler, M., Ledeen, R. W., Grafstein, B., and Alpert, R. M. (1984). *J. Neurosci. Res.* **12**, 221–232.

Schachner, M., Faissner, A., Kruse, J., Lindner, J., Meier, D. H., Rathgen, F. G., and Wernecke, H. (1983). *Cold Spring Harbor Symp. Quant. Biol.* **48**, 557–568.

Schmidt, J. R., and Edwards, D. L. (1983). *Brain Res.* **269**, 29–39.

Schwartz, J. H. (1979). *Annu. Rev. Biophys. Bioeng.* **8**, 27–45.

Scott, T. M., and Lazar, G. (1976). *J. Anat.* **121**, 485–496.

Silver, J., and Rutishauser, U. (1984). *Dev. Biol.* **106**, 485–499.

Sjostrand, J. O. (1965). *Z. Zellforsch. Histochem.* **68**, 481–493.

Skaper, S. D., and Varon, S. (1985). *Int. J. Devl. Neurosci.* **3**, 187–185.

Skene, J. H. P., and Kalil, K. (1984). *Soc. Neurosci. Abstr.* **10**, 1030.

Skene, J. H. P., and Shooter, E. (1983). *Proc. Natl. Acad. Sci. U.S.A.* **80**, 4169–4173.

Skene, J. H. P., and Willard, M. (1981a). *J. Neurochem.* **37**, 79–87.

Skene, J. H. P., and Willard, M. (1981b). *J. Neurosci.* **1**, 419–426.

Skene, J. H. P., and Willard, M. (1981c). *J. Cell Biol.* **89**, 86–95.

Skene, J. H. P., and Willard, M. (1981d). *J. Cell Biol.* **89**, 96–103.

Skene, J. H. P., Jacobson, R. D., Snipes, G. J., McGuire, C. B., Norden, J. J., and Freeman, J. A. (1986). *Science* **233**, 783–785.

Snipes, G. J., and Freeman, J. A. (1984). *Soc. Neurosci. Abstr.* **10**, 1029.

Sonderegger, P., Fishman, M. C., Bokoum, M., Bauer, H. C., and Nelson, P. G. (1983). *Science* **221**, 1294–1297.

Sparrow, J. R., McGuinness, C., Schwartz, M., and Grafstein, B. (1984). *J. Neurosci. Res.* **12**, 233–243.

Straznicky, K., and Gaze, R. M. (1971). *J. Embryol. Exp. Morphol.* **26**, 67–79.

Strocchi, P., Brown, B. A., Young, J. D., Bonventre, J. A., and Gilbert, J. M. (1981). *J. Neurochem.* **37**, 1295–1307.

Szaro, B. G. (1982). Ph. D. thesis. The Johns Hopkins University.

Szaro, B. G., and Faulkner, L. A. (1982). *Biophys. J.* **37**, 59a.

Szaro, B. G., Faulkner, L. A., Hunt, R. K., and Loh, Y. P. (1984). *Brain Res.* **297**, 337–355.

Szaro, B. G., Loh, Y. P., and Hunt, R. K. (1985). *J. Neurosci.* **5**, 192–208.

Van Ness, J., Laemmli, U. K., and Pettijohn, D. E. (1984). *Proc. Natl. Acad. Sci. U.S.A.* **81**, 7897–7901.

Whitnall, M. H., and Grafstein, B. (1981). *Brain Res.* **220**, 362–366.

Whitnall, M. H., and Grafstein, B. (1982). *Brain Res.* **239**, 41–56.

Willard, M. B., and Hulebak, K. L. (1977). *Brain Res.* **136**, 289–306.

Willard, M., Cowan, W. M., and Vagelos, P. R. (1974). *Proc. Natl. Acad. Sci. U.S.A.* **71**, 2183–2187.

Willshaw, D. J., Fawcett, J. W., and Gaze, R. M. (1983). *J. Embryol. Exp. Morphol.* **74**, 29–45.

Wilson, M. A. (1971). *Q. J. Exp. Physiol.* **56**, 83–91.

Wilson, D. L., Hall, M. E., Stone, G. C., and Rubin, R. W. (1977). *Anal. Biochem.* **83**, 33–44.

Wolsburg, H. (1981). *Adv. Anat. Embryol. Cell Biol.* **67**, 1–94.

Zwiers, H., Schotman, P., and Gispen, W. H. (1980). *J. Neurochem.* **34**, 1689–1699.

Zwiers, H., Aloyo, V. J., and Gispen, W. H. (1981). *Life Sci.* **28**, 2545–2551.

CHAPTER 10

MONOCLONAL ANTIBODY APPROACHES TO NEUROGENESIS

Shinobu C. Fujita

DEPARTMENT OF PHARMACOLOGY
GUNMA UNIVERSITY SCHOOL OF MEDICINE
SHOWA-MACHI, MAEBASHI 371, JAPAN

I. Introduction

The nervous tissue is an extremely intricate and precise network. This network is a physical as well as functional entity, comprising cells of a large number of subtypes with complex and often greatly extended morphologies. And, in sharp contrast to the most sophisticated computing machines, this network develops "automatically" in the embryo.

Since the earliest days of neuroscience, an immense amount of work has been expended in the analyses of development of the nervous system. In spite of the considerable body of phenomenological knowledge gained (Jacobson, 1978; Patterson and Purves, 1982), we still fall short of answering many fundamental questions. What regulates the division, differentiation, and migration of neural cells? What directs the growing neuronal processes? What specifies the formation and maintenance of synapses? It appears that new methods and strategies need to be incorporated.

Monoclonal antibody technique was introduced into neuroscience around 1980 (McKay *et al.,* 1981). It has now proven its effectiveness in providing reagents for distinguishing a large number of cell types (Zipser and McKay, 1981; Fujita *et al.,* 1982), showing the existence of a hypothetical molecule (Trisler *et al.,* 1981), and for revealing unexpected cellular organizations in the mature brain (Hendry *et al.,* 1984; Hawkes *et al.,* 1984; Fujita *et al.,* 1985; Imamura *et al.,* 1985).

The usefulness of this technique derives from the fact that the immunogen preparation need not be pure to obtain specific monoclonal antibodies (mAbs), and in fact the chemical nature of the antigen molecules need not even be known as long as certain criteria, such as histological localization, are available for the screening of the anti-

CURRENT TOPICS IN
DEVELOPMENTAL BIOLOGY, VOL. 21

bodies. One other important aspect of the technique is that the antibody obtained can serve as a bridge between the histochemistry and biochemistry of the antigen molecule. In this respect mAbs are fundamentally different from histochemical dyes and silver, which can sometimes bring out subtle cellular differences. These aspects of the monoclonal antibody technique allow adoption of entirely new strategies that were previously beyond access with conventional immunological or histological methodologies.

We and others have sought to apply this technique to the analysis of neurogenesis by constructing libraries of mAbs that are useful in immunological dissection of the developing nervous system and, possibly, in uncovering the molecules directly involved in specific neurogenetic processes (Ciment and Weston, 1982; McKay et al., 1983; Fujita and Obata, 1984a; Tanaka and Obata, 1984; Hockfield and McKay, 1985). Such an approach has been only recently initiated. The purpose of this contribution is not to summarize or survey the fruits of this immature field but rather to discuss certain emerging directions in applications of the mAb technique to the analysis of neurogenesis. For illustration I will have to draw mainly from the works I am better acquainted with. First, I shall consider several uses of mAbs obtained against the chick embryonic central nervous system. These mAbs serve as markers revealing interesting features of the developing vertebrate nervous system. mAbs against *Drosophila* nervous system will then be considered. The antigen molecules for some of these mAbs have been identified and characterized (Zipursky et al., 1984, 1985). A new approach of identifying interesting molecules through mAbs, and strategies for studying their function, will be discussed. For a review on application of mAbs in neuroscience in general, readers are referred to McKay (1983). Technical aspects of the method have been amply discussed by Cuello et al. (1983).

II. mAbs as Markers

The chick embryo is very accessible and has the best studied developing nervous system. A library of about 60 mAbs has been constructed by immunizing mice with homogenates of embryonic optic nerves and searching for the antibodies with characteristic staining patterns in immunohistochemical screening using sections of fixed embryonic neural tissue (Fujita and Obata, 1984a). Some are specific to the nervous system, while others are not, yet do show differential binding between certain regions within the nervous system. A few are specific to certain nonneuronal structures. In this section examples will be taken from this library to illustrate several directions in their use.

A. DIFFERENTIATION OF NEURAL TUBE

Formation of the neural tube occurs well before any sign of neuronal differentiation. Is there any function specifically associated with the cells of neural plate that gives rise to the neural tube?

mAb C94A2[1] labeled the walls of both the central canal of the spinal cord (Fig. 1A) and the ventricles in the brain (Fig. 1B) of the chick embryo. In the retina, it bound to the region between the pigment epithelium and outer limiting membrane (Fig. 1C). These are, of course, embryologically homologous structures. In early embryos, the antibody labeled the wall of the neural groove (Fig. 1D) and, even earlier, the outer surface of the neural plate (Fig. 1E).

Thus, this mAb is virtually specific to a homologous structure of the neural tissue, presumably a specialization of the neuroepithelial and ependymal cells. It is possible that the antigen molecule is involved in a primarily neural function. The mAb shows that the expression of this molecule precedes the appearance of differentiated neurons and is not consequent to the formation of the closed neural tube.

It is known that certain treatments of early vertebrate embryos, such as exposure to a high concentration of vitamin A, cause morphological defects in the neural tube (Morris, 1980). An injection of retinoic acid into fertilized chicken eggs incubated for 1 day caused the neural tube defect as examined the following day (Fig. 1H vs F) (Obata and Fujita, 1985). When such specimens were tested with mAb C94A2, Fig. 1I and G was obtained. The figures show, by virtue of the specificity of the antibody, that the defect is not a complete loss of cellular order, but rather the consequence of the formation of a number of small "neural tubes" with their own central canals.

Thus, mAbs can be obtained that serve both to mark developmentally related structures and to detect the onset of a particular cytodifferentiation. Such mAbs can also identify cellular specializations in abnormal development or in experimentally manipulated situations.

B. DEVELOPMENT OF SPINAL TRACTS

Longitudinal neuronal fibers of the spinal white matter are of four major categories: sensory, long ascending, long descending, and propriospinal (interneuronal). In the mature spinal cord each of these comprise a number of subtypes of fibers with distinct origins and destinations, and therefore presumably with specialized functions. Fibers of a given subtype tend to run in a defined region in a cross section of

---

[1] This antibody was originally designated as 94A2. To distinguish the chick antibodies from those against *Drosophila*, the former will be referred to with the prefix C in this article.

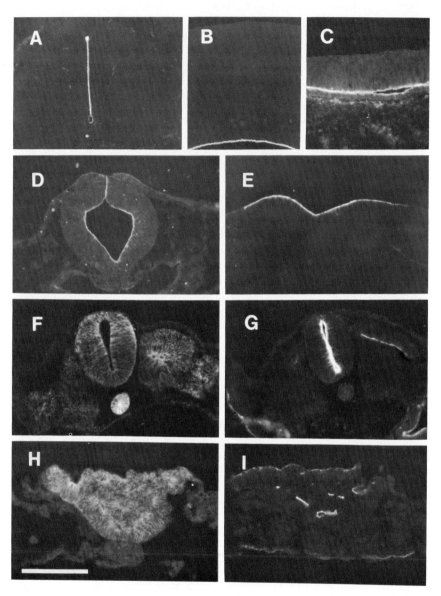

Fig. 1. Specificity of mAb C94A2 and the development of the neural tube in the chick embryo. A, Transverse section of the spinal cord (6-day embryo), dorsal toward top; B, optic tectum (8-day embryo), radial section, pial toward top; C, retina (4-day embryo), radial section, vitreal toward top; D, neural tube; E, neural plate; F and G, normal embryo at 48 hours, transverse section, as controls to H and I; H and I, abnormal neural tube in an embryo given retinoic acid, G and I were stained with mAb C94A2 while F and H were stained with mAb C86F7 to illustrate the overall cellular order. Indirect immunofluorescence on cryostat sections from formaldehyde-fixed embryos (same for Figs. 2, 4–7). Bar, 400 μm for A and B, 200 μm for C–I. (Obata and Fujita, 1986.)

the cord and thus constitute a spinal tract. It is not known why individual growing fibers do not get lost amid such heavy traffic.

Formation of the spinal tracts is only beginning to be analyzed (Nornes *et al.*, 1980; Schreyer and Jones, 1982; Okado and Oppenheim, 1985). The small size of the tissue in early embryos, however, has precluded the study of the earliest tracts. When the chick mAbs were tested on the embryonic spinal cord, a majority were found to bind differentially between certain regions of spinal funiculi, later to become the white matter (Fig. 2). Cross sections of the spinal cord of 6-day embryos were particularly useful in comparing the specificity of the mAbs, because the spinal tracts then are presumably simple in number and disposition, and the glia aside from the radial glia are not even in existence at this stage.

The examples of Fig. 2 were taken from 23 mAbs reported (Fujita and Obata, 1984a). A group of mAbs (Fig. 2A–C) shared a striking character; an outer layer of the ventral funiculus stained intensely while the inner layer was associated with much less stain. (See Fig. 3 for anatomical nomenclature.) In the lateral funiculus, more stain was seen toward the lateral edge. Six mAbs exhibited such characteristics and were named the α family. There was an antibody entirely specific to the dorsal funiculus (Fig. 2E). Others were not specific to a single region but showed diversity with respect to the funicular region most intensely stained (Compare Fig. 2D,F,G, and H).

These staining patterns were consistent from one embryo to another. They were also consistent along the length of the spinal cord as far as was studied. Since glial cells are not evident on day 6 and the staining had a fibrous appearance for most of the antibodies, the diversity of mAb staining is likely to reflect molecular diversity of neuronal fibers.

Based on the observation with 23 mAbs, the spinal funiculi of 6-day chick embryo could be divided into at least 6 regions of immunochemically distinct fiber groups as diagramed in Fig. 3. It is possible that this division can be further refined with a more extensive analysis. The 6 regions probably represent developing spinal tracts. It is not possible, however, to identify each region with specific tracts, as the tract organization in early chick embryo is not established. It is expected that the combination of the mAb approach with the classical anatomical methods will be fruitful here.

The fiber tracts in embryonic spinal cord can show immunochemical diversity for two reasons: (1) the tracts may be at different maturational stages so that maturation-dependent antigenic molecules are expressed differently among the tracts; and (2) the tracts

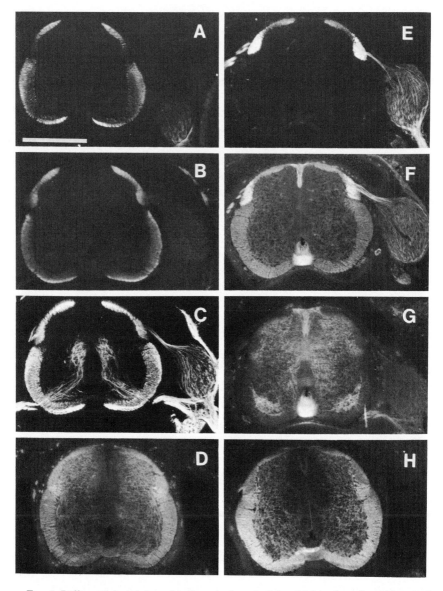

FIG. 2. Differential staining of early spinal cord of the chick embryo by a library of mAbs. A, mAb C94C2; B, mAb C93E11; C, mAb C82E10; D, mAb C81H11; E, mAb C87D10; F, mAb C87C7; G, mAb C85B12; H, mAb C83D4. Dorsal root ganglia shown toward right except in D and H. Bar, 400 μm.

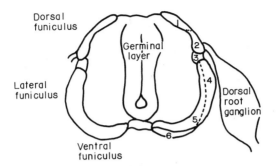

FIG. 3. Schematic diagram of cross section of the spinal cord of 6-day chick embryo. The numbers represent subregions of the spinal funiculi distinguished by the mAbs.

may contain different or even unique subclasses of fibers, each with distinct composition of antigenic molecules.

For all but one of the α family mAbs discussed above, regional localization of the immunoreactivity in the 6-day embryo was replaced by a more-or-less homogeneous distribution of positive fibers in a nearly mature spinal cord of an 18-day embryo. Expression of the antigens involved is thus likely to be related to maturation of fibers. Most of the other mAbs, however, stained the 18-day spinal cord sections with varying degrees of localization of the immunoreactivity, which was particularly clear in the dorsal horn (Fig. 4). Since histological fiber selectivity of many of the mAbs thus appears to persist to maturity, these antibodies are likely to reflect intrinsic molecular heterogeneity among certain subtypes of neuronal fibers. It is interesting to note that about half of these antibodies was not specific to the nervous system. Polypeptide bands have been detected in immunoblot analysis for nine of the antibodies.

The use of mAbs discussed in this section illustrates the power of these immunological reagents in differentiating subtypes of histological structures in embryonic nervous systems that are otherwise uniform in appearance.

C. CEREBELLAR CORTICOGENESIS

Many regions of the vertebrate brain involved in simultaneous processing of a large amount of neural information have cortical organization. The mechanisms of formation of cortical cytoarchitecture and functional neural network have been the subject of numerous studies. Works using radiolabeled thymidine, radiological lesions, or genetic mutations have revealed the birthdates and migration of neurons. To understand the control of cell migration and differentiation

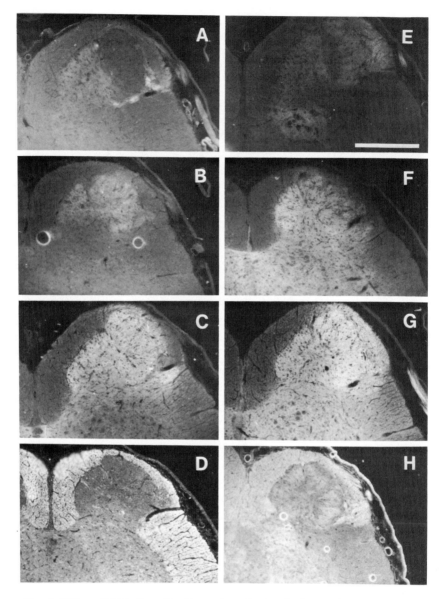

FIG. 4. Differential binding of mAbs to the spinal cord of an 18-day embryo, transverse sections. Only dorsal spinal cord is shown for comparison. Midline toward left, dorsal toward top. A, mAb C81H9; B, mAb C81F3; C, mAb C82C12; D, mAb C96D2; E, mAb C87C7; F, mAb C85B12; G, mAb C82B6; H, mAb C83C2. Bar, 400 μm.

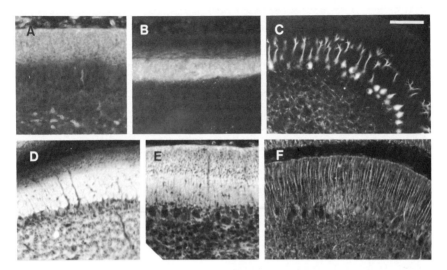

FIG. 5. Developing cerebellum, mAbs characterize layers. A, mAb C83C2 (external granular layer); B, mAb C86D10 (molecular layer); C, mAb C94C2 (Purkinje cells); D, mAb C84H3 (molecular and internal granular layers); E, mAb C95F1 (primarily external granular and molecular layers); F, mAb C96D2 (Bergmann glias). A, B, and D from 13-, 15-, and 18-day chick embryos, respectively; others from 16-day embryos. Bar, 100 μm.

into respective neuronal types, it is important to be able to distinguish the subtypes of interacting neurons and to monitor the maturational stages of the cells involved.

Many of the chick mAbs discussed in the previous section indeed distinguished between cortical layers in the developing cerebellum as illustrated in Fig. 5 (Obata and Fujita, 1984). Certain antibodies characterize the external granular layer (Fig. 5A), molecular layer (Fig. 5B), or Purkinje cells (Fig. 5C). Such differential stainings corroborate the interpretation that many of these antibodies distinguish between subtypes of neurons. These antibodies will also be useful in identifying the cell types and layers in cultures of cerebellar slices or dissociated cells.

Another interesting observation was that, with some antibodies, staining of the Purkinje cells was dependent upon maturational stages. mAb C82E10 binds to the medium and heavy subunits of neurofilament (Fujita & Obata, 1986). This mAb stained the Purkinje cells between embryonic days 11 and 18 (Fig. 6A and B), but the staining was gradually lost during the perinatal period (Fig. 6C). In the mature tissue the fibers of basket cells were intensely stained (Fig. 6D). This antibody belongs to the α family. Such a change in staining was ob-

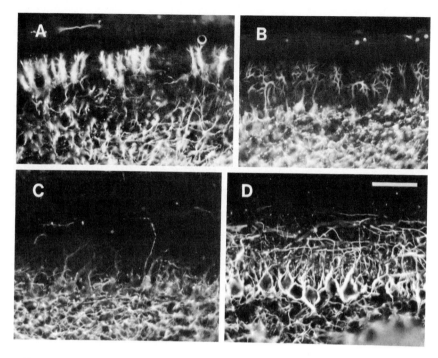

Fig. 6. Developmental changes in expression of mAb C82E10 immunoreactivity in the chick cerebellum. A, 15-day embryo; B, 18-day embryo; C, 2 days after hatching; D, 29 days after hatching. Bar, 100 μm.

served with two other mAbs of this family. It is to be noted that this observation was facilitated by the fact that the Purkinje cells are large and readily identifiable neurons in the tissue.

This phenomenon raises the possibility of using mAbs as monitors of certain dynamic, physiological neuronal states at the time of tissue fixation. Consonant with such an idea, it was reported that antibodies to heavy neurofilament subunit labeled mouse retinal ganglion cells after axotomy (Dräger and Hofbauer, 1984). More recently, mAbs were found that stain the retinal tissue dependent on light or dark adaptation (Balkema and Dräger, 1985). Further, it may be possible to obtain mAbs that can mark growing processes, such as from brain transplants, or that can detect cells undergoing redeployment of synapses after certain experimental manipulation.

D. Developing Optic Nerve

For processing of certain sensory information, such as visual or somatosensory, central cortical areas are organized so as to have one-to-one spatial representation of the stimulus as it impinges upon the two-dimensional array of the peripheral sensors. Mechanisms for achieving such topographical specificity of neuronal connections have been a central issue in neurobiology. The visual system has received particularly intense attention. Despite a large number of interesting experiments on the formation of retinotopic retinotectal projection, however, debate over mechanisms at work, with chemoaffinity and contact-guidance theories at the extremes, has not been settled (Fraser and Hunt, 1980).

Some of the earliest applications of the mAb technique in neuroscience were indeed made in this field with encouraging results (Trisler et al., 1981; Lemmon and Gottlieb, 1982; Henke-Fahle and Bonhoeffer, 1983). Their direct relevance to retinotopy, however, is yet to be demonstrated. In an attempt to devise a penetrating approach to this problem, we planned to reexamine the embryonic development of the visual system with a distinction of subtypes of retinal ganglion cells. The occurrence of such subtypes has long been known on morphological and physiological grounds (Ramón y Cajal, 1893; Stone, 1983). It is possible that only certain subtypes may be relevant in establishing the retinotopic organization. Anatomical distinction of subtypes of optic fibers was not feasible previously, but the mAb technique now offers a possibility. Thus, we generated a panel of mAbs by immunizing mice with homogenates of chicken embryonic optic nerves. As discussed above, many of these mAbs differentiate between certain subtypes of neuronal fibers in the developing spinal cord and cerebellum.

On the eighth day of incubation, only a fraction of the optic nerve fibers has reached the optic tectum (Goldberg, 1974). Molecules encoding the retinotopic information, if they occur, are most likely to be present at this stage. Some mAbs, such as C95H2, bound to the cross section of the optic nerve uniformly (Fig. 7A). A majority of the mAbs, however, showed clear regional variation in the intensity of stain, and this variation occurred roughly dorsoventrally (Fujita and Obata, 1984b). For example, mAb C82E10 stained fibers of the middle to ventral region with the dorsal region largely negative (Fig. 7B), while mAb C87D10-positive fibers were observed in the middle region, but not toward the dorsal or ventral margins (Fig. 7C). mAb C86D10 stained the dorsal region better than the ventral (Fig. 7D).

These observations demonstrate the occurrence of chemically distinct classes of nerve fibers in the embryonic optic nerve. Three possibilities can be considered in interpreting such a result. (1) It may reflect a heterogeneity in maturational stages among the optic fibers; retinal ganglion cells are known to differentiate in a central to pe-

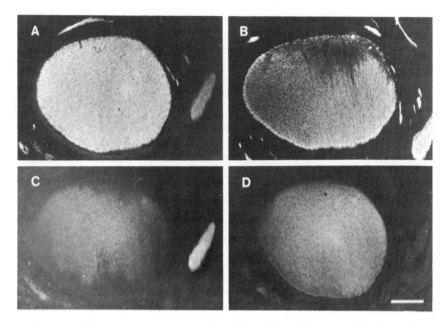

Fig. 7. Differential distribution of mAb immunoreactivities across developing optic nerve of 8-day chick embryo. Dorsal toward upper right, roughly nasal toward upper left. A, mAb C95H2; B, mAb C82E10; C, mAb C87D10; D, mAb C86D10. Bar, 200 μm.

ripheral sequence (Kahn, 1974). (2) Different subtypes of optic fibers may have unique sets of antigens, and the fibers of a given subtype may be enriched in certain regions of the nerve; regional functional specialization in avian retina has been inferred from an analysis of ganglion cell density variation (Ehrlich, 1981). (3) Optic fibers may carry molecules relating to the retinal position of the ganglion cell somata; a retinal antigen distributed in a dorsoventral gradient of density has been described (Trisler *et al.*, 1981). It should be possible eventually to know which of these possibilities actually obtains for each of the mAbs through analyses of developmental changes of immunoreactivity and its distribution in the visual centers as well as in the retina.

Availability of molecular markers of subtypes of nerve fibers, or molecular monitors of fiber maturation, will open new approaches to the long-standing question of the formation of retinotopic retinotectal connection.

III. From Antibody via Antigen to Function

Uses of mAbs as markers discussed in the previous section were effective without knowledge of the respective antigen molecules. However, the feasibility of identifying the antigen also allows the use of the mAb technique as a means to discover interesting molecules. Thus, certain attributes of an "interesting" molecule that can be used in screening for mAbs are first set up. For example, developmentally transient expression or tissue localization can be used in a variety of screening procedures. When specific mAbs are obtained, it is, in principle, a straightforward task to identify a protein (Burnette, 1981) or glycolipid antigens (Levine *et al.*, 1984).

When an interesting molecule is found, its function must be established. As mAbs themselves are often without biological activity, this step is not simple. As an alternative, polyclonal antisera can be raised against the antigen purified using mAbs. Antisera are more likely to block biological function of the antigen. Such an approach will be effective in tissue culture or other *in vitro* situations in which permeability of the antibody is less of a problem.

The most elegant way, however, would be to manipulate or alter the antigen molecules themselves through genetic techniques. To do so, the DNA region that codes or controls the antigen is identified, then mutated sequences are put into appropriate systems to express mutant phenotypes which are studied. Such an approach can be carried out to best advantage using *Drosophila*. This section will review a pioneering attempt along such a strategy (Fujita *et al.*, 1982; Zipursky *et al.*, 1984, 1985).

A. DISCOVERY OF A PHOTORECEPTOR ANTIGEN

The number of neurons in *Drosophila* is estimated at 10^5 (compared with 10^{10} for man). Yet the fly can perform a wide range of complex behavior (Benzer, 1971; Quinn and Greenspan, 1984). The whole fly weighs about 1 mg. This small size has been a formidable obstacle to molecular approaches to its nervous system (Hotta, 1979). When the mAb technique became available, a library of mAbs was generated against the nervous tissue of the fruit fly to circumvent the above difficulty in obtaining chemical handles on specific components of the nervous system (Fujita *et al.*, 1982).

Mice were immunized with homogenates of *Drosophila* brains or whole heads. Hybridoma antibodies were screened by indirect immunofluorescence histochemistry using frozen sections of fly heads. As might be expected of such an experiment, hybridoma antibodies of wide range of specificity were observed. From 4 fusion experiments, 148 hybridoma lines were cloned. Fig. 8 compares some of the mAbs as they stain the fly brains in frontal sections of the head.

mAb 2G4B (Fig. 8A) stained the inner neuropil of the brain, while mAb D12A (Fig. 8C) strongly stained the cellular cortex. Neuronal

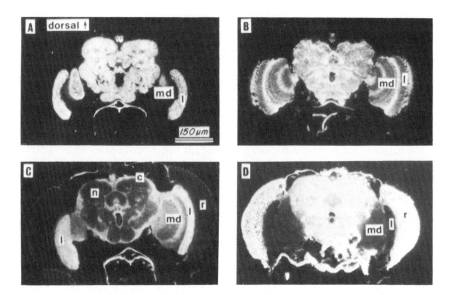

FIG. 8. mAbs characteristically stain the *Drosophila* brain. Transverse sections of the fly heads. A, mAb 2G4B; B, mAb 4C6; C, mAb D12A; D, mAb 2E6. l, Lamina, the first optic ganglion; md, medulla, the second optic ganglion; c, cortex; n, neuropil; r, retina. Indirect immunofluorescence on cryostat sections formaldehyde fixed after sectioning fresh (same for Fig. 9). (From Fujita *et al.*, 1982.)

fibers, fiber tracts in particular, were stained by mAb 4C6 (Fig. 8B). mAb 2E6 was not specific to a tissue, but it stained the brain in an interesting way (Fig. 8D). The lamina, the first optic ganglion beneath the retina, was devoid of stain, with the medulla only weakly stained, while a strong reaction was seen in the midbrain. Such a specific absence of antigen could play a role in the neurogenetic specification. Several other antibodies showed clear regional differences within the brain in the intensity of stain. Among the antibodies obtained, none was specific to a small subset of neurons except for those that were specific to the photoreceptors.

Figure 9 compares one such antibody (mAb 24B10, Fig. 9D and E) with four others as they stain radial sections of *Drosophila* compound eye. While mAb 21A6 (Fig. 9F and G) was specific to the retina, mAb 24B10 stained regularly arranged photoreceptor axons in the lamina and medulla of the brain optic lobe in addition to the photoreceptor cell bodies within the retina. Thus, this antigen molecule is likely to be involved in some neural function other than phototransduction itself. Could it perhaps be the formation and maintenance of specific connections?

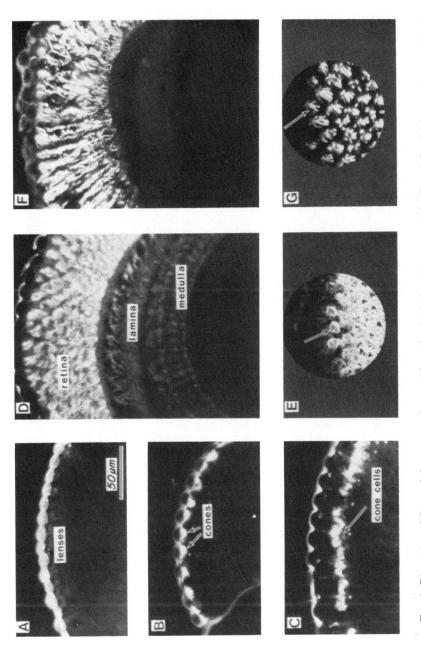

Fig. 9. Dissection of *Drosophila* compound eye with mAbs. Radial sections except for E and G, which were tangential. A, mAb 3F12 (lenses); B, mAb 22G8 (cones); C, mAb 3F11 (cone cells); D and E, mAb 24B10 (photoreceptor cells); F and G, mAb 21A6. (From Fujita *et al.*, 1982.)

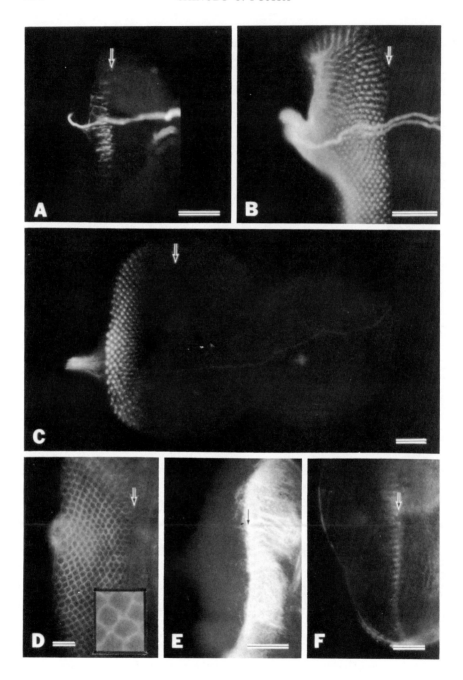

When whole-mount preparations of the eye–antennal disc, the anlage of the compound eye, from third instar larvae were studied with a set of mAbs, various aspects of the morphogenetic process were highlighted by each antibody (Fig. 10) (Zipursky et al., 1984). mAb 22C10 revealed the neurofibrillar differentiation (Fig. 10A and B) that starts immediately after a zone of cellular activity (morphogenetic furrow) sweeps the tissue. mAb 24B10 labeled the photoreceptor axons some distance behind the furrow. Thus, the antigen appears early in the photoreceptor axons as they grow toward the brain centers. mAb 6D6 showed the partitioning of the ommatidia (Fig. 10D), while two others marked transient immunoreactivities associated with the furrow (Fig. 10E and F). Such a set of mAbs will be useful in reexamining the histogenesis of the compound eye (Ready et al., 1976) and in analysis of the defects in the morphogenetic mutations of the compound eye (Meyerowitz and Kankel, 1978).

In view of its specificity, mAb 24B10 was chosen for further study of the antigen molecule.

B. From Antibody to the Gene

As discussed earlier, identification of the gene coding or controlling the immunoreactivity is an important step in establishing the biological function of the molecules initially defined by mAbs. In the case of polypeptide antigens, there are at least four ways to achieve this: (1) the antigen molecule is purified by immunoaffinity chromatography, partial amino acid sequence is determined, and, with appropriate oligonucleotide sequence probes synthesized, cloned DNA libraries are screened for the structural gene; (2) a cDNA expression vector library is screened directly with the mAbs for expression of the epitopes (Helfman et al., 1983; Young and Davis, 1983); (3) mRNA is taken from polysomes purified on immunoaffinity columns, and specific cDNA probes are made; and (4) if an electrophoretic variant of the antigen is available that can be detected on immunoblots, the locus can be mapped with appropriate genetic markers.

The first strategy was successfully used for mAb 24B10 by the Caltech group. (Zipursky et al., 1984). Figure 11 (lane A) shows the

---

FIG. 10. Patterns in the eye anlage revealed by mAbs. Dissected eye-antennal discs were immunostained as whole mounts. A and B, mAb 22C10. Discs from early (A) and late (B) third instar larvae. Note advancing array. C, mAb 24B10. The immunoreactivity appears far behind the morphogenetic furrow (arrow) sweeping toward right. D, mAb 6D6; E, mAb 3E1; F, mAb 22G8. Bars, 50 μm. Arrows indicate the position of the morphogenetic furrow. (From Zipursky et al., 1984.)

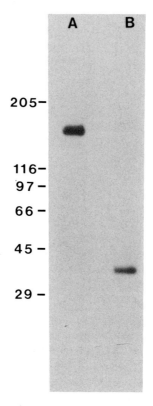

Fɪɢ. 11. Purified antigen molecules radiolabeled and electrophoresed. A, The photoreceptor antigen 24B10; B, a control antigen 8C3. The numbers are molecular weight calibrations. (From Zipursky *et al.*, 1984.)

autoradiogram of the electrophoretic band of the purified and radiolabeled antigen 24B10. It has a molecular weight of 160,000. Lane B is a control experiment using a different mAb. Two-dimensional electrophoretic analysis identified the antigen 24B10 with a spot previously shown to be specific to the retina by manual dissection. The antigen was also shown to be glycosylated. This suggests that the molecule is expressed on the cell surface, and independent experiments yielded results in accord with this possibility.

Partial N-terminal amino acid sequence of the antigen was determined using 8 μg of the purified material (Fig. 12). Oligonucleotides corresponding to two suitable stretches of the amino acid sequence were synthesized. These probes were used to screen a *Drosophila* genomic library, and a homologous λ genomic clone was isolated. Further-

NH$_2$-X-MET-PHE-ASP-ARG-GLU-MET-GLU-GLU-THR-HIS-TYR-PRO-PRO-HIS-X-TYR-ASN-VAL-MET-HIS

DNA PROBES:

(I) TAC-CT^T_C-CT^T_C-TG^T_C-GT^G_A-AT^G_A-GG 5′

(III) AT^G_A-TT^G_A-CA$^{G}_{A,T,C}$-TAC-GT 5′

(II) TAC-CT^T_C-CT^T_C-TG^G_A-GT^G_A-AT^G_A-GG 5′

FIG. 12. Partial amino-terminal amino acid sequence of the photoreceptor antigen, 24B10, and corresponding DNA probes synthesized. X denotes uncertain residues. (From Zipursky *et al.*, 1984.)

more, an appropriate fragment of the clone was used for *in situ* hybridization to the giant chromosomes, and the gene was localized to band 100B (Zipursky *et al.*, 1985).

Now that the genetic locus is defined, various classical genetic techniques can be mobilized to study the gene dosage effect and to collect mutations of the locus, or even to delete the gene. Then artificially altered genes can be introduced into the fly using transposon vehicles to study systematically the possible range of mutant phenotypes that may be manifested in the histogenesis of the retina, in retinal physiology, or in visually triggered behaviors. It will also give an opportunity to study temporal and tissue control of expression of a gene whose product is restricted to a defined subset of neurons. Such a molecular genetic approach will, of course, be complemented with analyses using organ cultures of eye–antennal disks in which polyclonal antisera raised against the purified antigen are injected to test for possible blockage of normal retinofugal projection.

IV. Concluding Remarks

The challenge of the nervous system is the diversity of neurons. Diversity with respect to cellular morphology has long been appreciated. By contrast, not much is known about the molecular diversity of neurons. Several neurotransmitters have been recognized for some time, but current knowledge does not permit even an estimate of the total number of neurotransmitters and neuromodulators. We know even less about the neuronal surface molecules which are often invoked to play key roles in neuronal recognition (Barondes, 1970; Edelman, 1983).

With the mAb technique, it became possible for the first time to explore systematically the molecular diversity of neurons. This point was illustrated in Section II by a set of mAbs that bind differentially to groups of nerve fibers in the developing spinal cord, cerebellum, and optic nerve. It is possible that the embryonic nervous tissue is even better suited than mature ones for this type of analysis because of simpler cytoarchitecture and less-developed dendritic arborization. An mAb that labeled a subcellular specialization of ependyma was also mentioned to illustrate the use of mAbs as differentiation markers. In relation to maturation-dependent changes in the immunoreactivity of Purkinje cells, the possibility of using mAbs as monitors of dynamic cellular physiological states was discussed.

The usefulness of mAbs actually goes beyond the markers and monitors discussed above. The antigen molecules can be identified, and in favorable circumstances their functions can be experimentally questioned. In Section III I discussed genetic strategies for achieving this and used, for illustration, an attempt being made to study the function of a photoreceptor-specific *Drosophila* glycoprotein initially identified by an mAb.

When a truly innovative technique is introduced, it takes years of experience by a number of practitioners before its potential is fully realized. Many more incisive uses of mAbs will certainly emerge in near future. Nevertheless, the uses discussed here, both as markers and monitors and as the means to new molecules of interest, will continue to provide main avenues of insights into the molecular processes in neurogenesis.

ACKNOWLEDGMENTS

I wish to thank Dr. K. Obata for stimulating collaboration for the works discussed in Section II and for supplying the data for Fig. 1 (prior to publication) and Figs. 5 and 6. I am also deeply grateful to Dr. S. Benzer and colleagues for numerous rewarding discussions and for preparing the figures in Section III. I thank Mrs. Y. Roppongi and Mrs. Y. Aoki for skillful assistance. I acknowledge helpful support by grants-in-aid from the Ministry of Education, Science and Culture of Japan.

REFERENCES

Balkema, G. W., and Dräger, U. D. (1985). *Nature (London)* **316**, 630–633.
Barondes, S. H. (1970). *In* "The Neurosciences, Second Study Program" (F. O. Schmitt, ed.), pp. 747–760. Rockefeller Univ. Press, New York.
Benzer, S. (1971). *J. Am. Med. Assoc.* **218**, 1015–1022.
Burnette, W. N. (1981). *Anal. Biochem.* **112**, 195–203.
Ciment, G., and Weston, J. A. (1982). *Dev. Biol.* **93**, 355–367.
Cuello, A. C., Milstein, C., and Galfre, G. (1983). *In* "Immunohistochemistry" (A. C. Cuello, ed.). Wiley, New York.
Dräger, U. C., and Hofbauer, A. (1984). *Nature (London)* **309**, 624–626.
Edelman, G. M. (1983). *Science* **219**, 450–457.
Ehrlich, D. (1981). *J. Comp. Neurol.* **195**, 643–657.
Fraser, S. E., and Hunt, R. K. (1980). *Annu. Rev. Neurosci.* **3**, 319–352.
Fujita, S. C., and Obata, K. (1984a). *Neurosci. Res.* **1**, 131–148.
Fujita, S. C., and Obata, K. (1987). In preparation.
Fujita, S. C., and Obata, K. (1984b). *Soc. Neurosci. Abstr.* **10**, 669.

Fujita, S. C., Zipursky, S. L., Benzer, S., Ferrus, A., and Shotwell, S. L. (1982). *Proc. Natl. Acad. Sci. U.S.A.* **79,** 7929–7933.

Fujita, S. C., Mori, K., Imamura, K., and Obata, K. (1985). *Brain Res.* **326,** 192–196.

Goldberg, S. (1974). *Dev. Biol.* **36,** 24–43.

Hawkes, R., Leclerc, N., and Colonier, M. (1984). *Soc. Neurosci. Abstr.* **10,** 44.

Helfman, D. M., Feramiso, J. R., Fiddes, J. C., Thomas, G. P., and Hughes, S. H. (1983). *Proc. Natl. Acad. Sci. U.S.A.* **80,** 31–35.

Hendry, S. H., Hookfield, S., Jones, E. G., and McKay, R. (1984). *Nature (London)* **307,** 267–269.

Henke-Fahle, S., and Bonhoeffer, F. (1983). *Nature (London)* **303,** 65–67.

Hockfield, S., and McKay, R. D. G. (1985). *J. Neurosci.* **5,** 3310–3328.

Hotta, Y. (1979). *In* "Mechanisms of Cell Change" (J. D. Ebert and T. S. Okada, eds.), pp. 169–182. Wiley, New York.

Imamura, K., Mori, K., Fujita, S. C., and Obata, K. (1985). *Brain Res.* **328,** 362–366.

Jacobson, M. (1978). "Developmental Neurobiology." Plenum, New York.

Kahn, A. J. (1974). *Dev. Biol.* **38,** 30–40.

Lemmon, V., and Gottlieb, D. I. (1982). *J. Neurosci.* **2,** 531–535.

Levine, J. M., Beasley, L., and Stallcup, W. B. (1984). *J. Neurosci.* **4,** 820–831.

McKay, R. D. G. (1983). *Annu. Rev. Neurosci.* **6,** 527–546.

McKay, R., Raff, M. C., and Reichardt, L. F., eds. (1981). "Monoclonal Antibodies to Neural Antigens." Cold Spring Harbor Laboratory, Cold Spring Harbor, New York.

McKay, R. D. G., Hockfield, S., Johansen, J., Thompson, I., and Frederiksen, K. (1983). *Science* **222,** 788–794.

Meyerowitz, E. M., and Kankel, D. R. (1978). *Dev. Biol.* **62,** 112–142.

Morriss, G. (1980). *Nature (London)* **284,** 121–123.

Nornes, H. O., Hart, H., and Carry, M. (1980). *J. Comp. Neurol.* **192,** 119–132.

Obata, K., and Fujita, S. C. (1984). *Neurosci. Res.* **1,** 117–129.

Obata, K., and Fujita, S. C. (1985). *Neurosci. Res.,* supplement 1, S34 (Abstr).

Obata, K., and Fujita, S. C. (1987). In preparation.

Okado, N., and Oppenheim, R. W. (1985). *J. Comp. Neurol.* **232,** 143–161.

Patterson, P. H., and Purves, D., eds. (1982). "Readings in Developmental Neurobiology." Cold Spring Harbor Laboratory, Cold Spring Harbor, New York.

Quinn, W. G., and Greenspan, R. J. (1984). *Annu. Rev. Neurosci.* **7,** 67–93.

Ramón y Cajal, S. (1893). "The Vertebrate Retina," trans. in Rodieck (1973). "The Vertebrate Retina, Principles of Structure and Function." Freeman, San Francisco.

Ready, D., Hanson, T. E., and Benzer, S. (1976). *Dev. Biol.* **53,** 217–240.

Schreyer, D. J., and Jones, E. G. (1982). *Neuroscience* **7,** 1837–1853.

Stone, J. (1983). "Parallel Processing in the Visual System." Plenum, New York.

Tanaka, H., and Obata, K. (1984). *Dev. Biol.* **106,** 26–37.

Trisler, G. D., Schneider, M. D., and Nirenberg, M. (1981). *Proc. Natl. Acad. Sci. U.S.A.* **78,** 2145–2149.

Young, R. A., and Davis, R. W. (1984). *Proc. Natl. Acad. Sci. U.S.A.* **80,** 1194–1198.

Zipser, B., and McKay, R. (1981). *Nature (London)* **289,** 549–554.

Zipursky, S. L., Venkatesh, T. R., Teplow, D. B., and Benzer, S. (1984). *Cell* **36,** 15–26.

Zipursky, S. L., Venkatesh, T. R., and Benzer, S. (1985). *Proc. Natl. Acad. Sci. U.S.A.* **82,** 1855–1859.

CHAPTER 11

SYNAPSE FORMATION IN RETINA IS INFLUENCED BY MOLECULES THAT IDENTIFY CELL POSITION

David Trisler

LABORATORY OF BIOCHEMICAL GENETICS
NATIONAL HEART, LUNG AND BLOOD INSTITUTE
NATIONAL INSTITUTES OF HEALTH
BETHESDA, MARYLAND 20892

I. Introduction

The development and function of the nervous system involves the formation of complex networks of cellular organization. Highly specific interconnections are formed between diverse neuron types in various and distant locations. In the human, for example, this complexity and specificity involves perhaps 10^{12} neurons and as many as 10^{15} synapses. One model proposed to account for the orderly establishment of neuronal connections in the assembly of the nervous system is a molecular system of neuronal recognition that identifies cell type and position (Sperry, 1963; Hood *et al.*, 1977; for review, Trisler, 1982).

The retina is a favorable system for the study of the molecular bases of mechanisms that impart positional information and identity to cells in the nervous system because of its relatively simple three-dimensional organization. During development the neural retina grows from a single layer of epithelial cells to a stratified epithelium, and finally cells segregate on the basis of cell type forming multilayered sheets of cells (Fig. 1). There are five major functional classes of neurons and one glial cell type in chicken retina. The outermost layer of cell bodies in the neural retina contains the photoreceptors. Photoreceptor cells synapse with horizontal and bipolar cells in the outer synaptic layer of retina. The inner nuclear layer is made up of the cell bodies of interneurons—horizontal, bipolar, and amacrine cells, as well as the nuclei of Müller cells, the retina glial cells that span the retina from the outer limiting membrane to the inner limiting membrane. Bipolar and amacrine cells synapse with ganglion cells in the inner synaptic layer. The innermost layer of cell bodies is the ganglion cell layer. And finally the ganglion cell axons course along the vitreal margin of the retina to the optic nerve head, form the

277

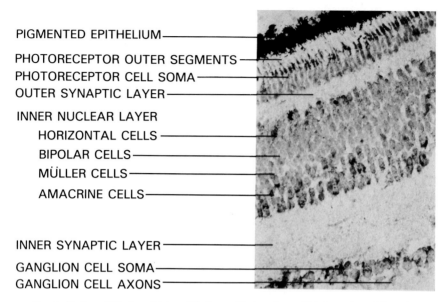

PIGMENTED EPITHELIUM————————

PHOTORECEPTOR OUTER SEGMENTS——

PHOTORECEPTOR CELL SOMA————

OUTER SYNAPTIC LAYER————

INNER NUCLEAR LAYER

   HORIZONTAL CELLS ————

   BIPOLAR CELLS————

   MÜLLER CELLS————

   AMACRINE CELLS————

INNER SYNAPTIC LAYER————

GANGLION CELL SOMA————

GANGLION CELL AXONS————

FIG. 1. Retina of 19-day chicken (*Gallus gallus*) embryo. Thick sections (10-μm) were cut from unfixed frozen retina and stained with 1.0% toluidine blue in water.

optic nerve, and project to the optic tectum and other centers in the brain. Thus, cellular organization and position across the thickness of the retina are important during development.

Similarly, cell position in the plane of the retina is important. Interneurons receive information from fields of photoreceptors, in some cases sending long processes along the retinal plane to form a dendritic field at some distance from the cell body and transmit the information to the ganglion cells that also receive from fields of interneurons (Ramón y Cajal, 1972 translation). In addition, topographic relationships of neurons in the plane of retina are conserved when retina ganglion cell axons synapse with neurons of the optic tectum and other higher centers in the brain forming point-to-point maps of the retina in these brain regions (Sperry, 1963). Therefore, cell position is ordered in three dimensions during development of the retina.

Cell surface molecules that identify cell position both across the thickness of retina and in the plane of retina have been detected with monoclonal antibodies to embryonic chicken retina. These surface molecules can be used to form the basis of a neuronal recognition system. Antibodies to these molecules were used to examine the role of such molecules in development of the nervous system.

II. Molecules That Identify Cell Type and Position in the Nervous System

Two of the models for conferring identity on neurons have been proposed by Sperry (1963) and Hood *et al.* (1977). Sperry's model of molecular gradients states that ganglion cell position in the retina can be determined by two molecules distributed in orthogonal gradients. This molecular system could be used to distinguish one ganglion cell from another. The second model will distinguish functional classes of cells from one another in the nervous system, e.g., ganglion cells from photoreceptors. This model is the "area-code" hypothesis of Hood *et al.* (1977), in which cells bear a combinatorial set of molecules that identify the location and function of a cell by reflecting its lineage of differentiation, e.g., nervous system, retina, ganglion cell. Molecular patterns consistent with these two models have been identified.

A. TOPOGRAPHIC GRADIENT MOLECULES

Molecules that identify cell position in the plane of the retina were detected with a monoclonal antibody generated by the fusion of spleen cells from mice immunized with a small portion of dorsoposterior retina with P3X63 Ag8 mouse myeloma cells (Fig. 2). One antibody to a cell surface molecule bound more to cells from dorsoposterior retina than to cells from the remainder of the retina. This molecule, termed TOP for toponymic (i.e., a marker of position), is distributed in a topographic gradient in retina (Fig. 3). A bilaterally symmetrical gradient of TOP from the dorsoposterior margin to the ventroanterior margin of retina was found in both right and left eyes. TOP molecules are present on most or all cells in the retina. The number of TOP molecules detected per cell varies continuously along the axis of the antigen gradient. Thus, TOP can be used to identify cell position in the plane of retina along that axis and thereby distinguish cells within a functional class.

B. MOLECULES THAT IDENTIFY CELL TYPE AND STRATUM IN RETINA

Antibodies obtained from 1752 other hybridoma colonies generated from mice immunized with dorsal, ventral, anterior, and posterior portions of retina and with clonal retinal hybrid cell lines bound equally in the plane of retina. However, some of these antibodies detected molecules that identify cell type and position across the thickness of retina from the vitreal margin to the pigmented epithelium (Fig. 4). Two molecules present in most or all cells of retina are 57D8, a cell-surface molecule, and 56D11, an intracellular molecule that may be associated with the Golgi apparatus, although further work is needed

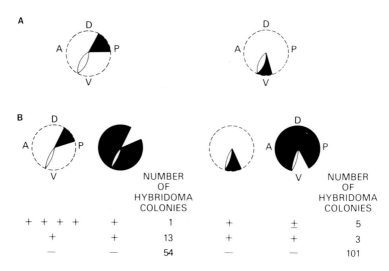

Fig. 2. Strategy for detecting region-specific retina surface-membrane antigens. A, Mouse immunization with chick retina segments; B, hybridoma antibody specificity. Lymphocyte hybridomas were derived from mice immunized with cells from dorsoposterior (left) or ventral (right) retina; hybridoma antibody binding to paired cultured-cell monolayers from the portions of 14-day chick embryo retina shown in black was determined (bottom). A, D, P, and V correspond to anterior, dorsal, posterior, and ventral, respectively. The choroid fissure, through which axons exit and enter the retina, shown extending from the ventroanterior margin of retina, was used as a landmark for dissection. (From Trisler *et al.*, 1981a.)

to confirm this point. Two molecules that distinguish cell bodies from processes in retina are 95D4 and 94C2. Antibody 95D4 binds to molecules associated with photoreceptor outer segments and with cell processes. In contrast, antibody 94C2 recognizes molecules restricted to cell bodies in the retinas. Antigen 94C2 has also been detected on the surface of quail retina neurons in culture (Pessac *et al.*, 1983). An antigen present on synapsing pairs of cells, i.e., photoreceptors and bipolar cells, was identified with antibody 107H11. A cell-type-specific molecule, 96B11, is present in Müller cells. Antibody 66E1 binds abundantly to a double row of punctate structures in the outer synaptic layer. Oblique sections show that the structures are beaded rings.

C. Categories of Molecular Patterns in the Nervous System

A summary of distributions of molecules across the retina is shown in Table I. Hybridoma antibodies were obtained that are specific for photoreceptor outer segments, photoreceptor cell bodies, the outer synaptic layer of retina, the inner nuclear layer, Müller cells, the inner

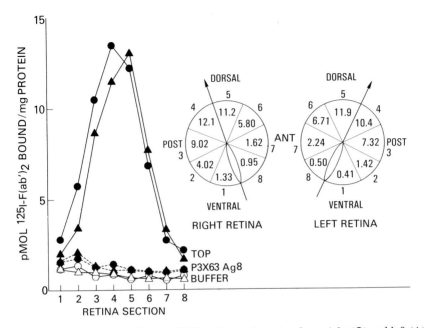

Fig. 3. Topographic gradients of TOP molecules in retina from right (●) and left (▲) eyes of 14-day chicken embryos. ●—● and ▲—▲, anti-TOP monoclonal antibody from mouse; ●- - -● and ▲- - -▲, P3X63 Ag8 mouse myeloma antibody; ○ △, buffer without antibody to TOP. Values shown (right) within appropriate sections of retina are picomoles of ^{125}I-labeled F(ab′)$_2$ fragment of rabbit IgG directed against mouse IgG specifically bound per mg protein (i.e., picomoles of ^{125}I-labeled F(ab′)$_2$ bound in the presence of anti-TOP antibody minus picomoles of ^{125}I-labeled F(ab′)$_2$ bound in the presence of P3X63 Ag8 antibody). (From Trisler et al., 1981a.)

synaptic layer, and ganglion neurons. Some antibodies stained two or more layers of retina. The categories of antigen distribution found in the retina are shown in Table II:

1. Nervous system compartment-specific molecules are present on all cells of retina but are not found in other areas of the nervous system;

2. Retina stratum-specific molecules were detected that are restricted to the inner nuclear layer which contains horizontal, bipolar, and amacrine cell bodies or molecules restricted to the inner synaptic layer;

3. Synapsing-pair specific molecules present in photoreceptor cells and bipolar cells;

4. Cell-type-specific molecules that are photoreceptor specific, Müller cell specific, or ganglion cell specific;

TABLE I

DISTRIBUTION OF ANTIGENS IN RETINA RECOGNIZED BY MONOCLONAL ANTIBODIES

Number of hybridomas	Photoreceptors		Outer synaptic layer	Inner nuclear layer	Müller cells	Inner synaptic layer	Ganglion neurons
	Outer segments	Soma					
4	+						
3		+					
3			+				
2				+			
5					+		
1						+	
3							+
2	+		+				
1			+			+	
1				+		+	+
2		+		+			+

^aFrom Trisler *et al.*, 1984.

5. Functional class subset on a subset of ganglion cells;
6. Molecules specific for cell body elements, i.e., cell soma or cell processes; and
7. Molecules that identify topographic position.

Thus, the retina contains molecules that distinguish cell types and that can be used to mark position both across the thickness of retina and in the plane of retina.

Additional categories of molecular specificities have been reported that identify other stages of differentiation in the lineage of cells in the nervous system (Table III). Molecules that distinguish neurons from glial cells are N-CAM and NS-4, which are present on all central and peripheral neurons studied but are not present on glial cells. Conversely, RET-G1 is glial cell specific. Among neurons, antibody A4 binds to all central nervous system neurons but not to those of the peripheral nervous system, and 38/D7 binds to peripheral but not to central neurons. Cells of one functional system, the limbic system, can

FIG. 4. Indirect immunofluorescence analysis of the distribution of molecules in frozen sections of unfixed 19-day chick embryo retina detected by monoclonal antibodies. A, 57D8; B, 56D11; C, 95D4; D, 94C2; E, 107H11; F, 96B11; G, 66E1 on retina in cross-section; H, 66E1 on retina in oblique section. Bar, 30 μm. (B,C,D, and G from Trisler *et al.*, 1984.)

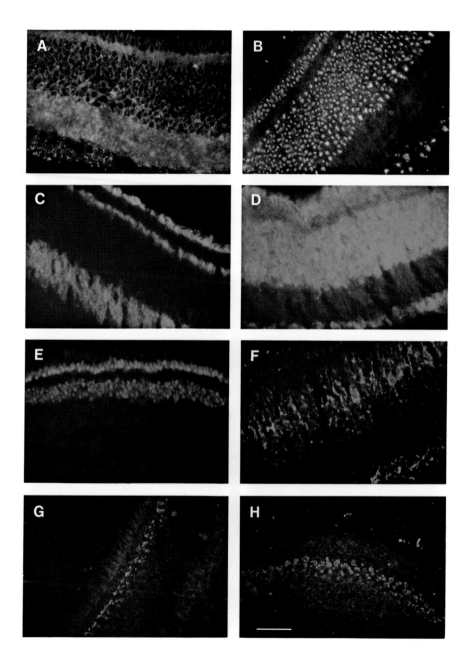

TABLE II

MOLECULAR SPECIFICITIES IN RETINA[a]

Specificity	Antibody
Compartment	
All cells of retina, no other cells of the nervous system	13H9
Stratum	
All cells of inner nuclear layer, no other cells of retina	39C12
Inner synaptic layer only	37A11
Synapsing pairs	
Photoreceptor + bipolar cells	107H11
Cell class	
Photoreceptor cell specific	37F12
Müller cell specific	96B11
Ganglion cell specific	41A4
Cell class subset	
Ganglion cell subset	41C5
Cell body elements	
All cell soma	94C2
All process layers	95D4
Topographic position	
All cells, graded expression	TOP (14H3)

[a]From Grunwald et al., 1983; Trisler et al., 1984; Hilt et al., 1985.

be distinguished from those of other functional systems with antibody 2G9 that was generated to cells of the hippocampus. This antigen is shared by cells of the hippocampus as well as the afferent or efferent cells that are thought to constitute the limbic system. The *Drosophila* nervous system is derived from compartments or cellular polyclones of the embryo that form precisely bounded regions in the adult nervous system (Lawrence, 1978). Antibody DK1A4 recognizes an antigen that is present in all cells of anterior retina but not posterior retina in late larvae. Similarly, antigens have been reported that are present on all cells of the chicken retina but are not found elsewhere in the nervous system. These molecules mark the retina as a possible compartment of the vertebrate nervous system. Cell surface molecules that are markers of functional classes of neurons identify, for example, rat rod photoreceptor cells, *Drosophila* photoreceptor cells, horizontal cells in carp retina, and rat cerebellar granular cells; those that mark glial cell

TABLE III

Positional Distribution of Cell-Surface Molecules in the Nervous System

Specificity	Antigen/antibody	Animal[a]	Reference
Neuron/glia			
Neuron specific	N-CAM	Chicken	Thiery et al. (1977)
	NS 4	Mouse	Schnitzer and Schachner (1981)
Glia cell specific	RET-G1	Rat	Barnstable (1980)
CNS/PNS neurons			
All CNS neurons	A4	Rat	Cohen and Selvendran (1981)
All PNS neurons	38/D7	Rat	Vulliamy et al. (1981)
Functional system			
Limbic system	2G9	Rat	Levitt (1984)
Compartment			
Retina specific	Cognin	Chicken	Hausman and Moscona (1976)
	13H9	Chicken	Grunwald et al. (1983)
	C1H3	Chicken	Cole and Glaser (1984)
Anterior retina	DK1A4	Drosophila	Wilcox et al. (1981)
Cell class			
Schwann cells	Ran-1	Rat	Brockes et al. (1977)
Rod photoreceptor cells	RET-P1	Rat	Barnstable (1980)
Oligodendroglia	01-04	Mouse	Sommer and Schachner (1981)
Photoreceptor cells	Mab 24B10	Drosophila	Fujita et al. (1982)
Cerebellar granular cells	7-8D2	Rat	Webb and Woodhams (1984)
Horizontal cells	HC-I	Fish	Young and Dowling (1984)
Cell class subset			
Mechanosensory neurons	Lan 3-2	Leech	Zipser and McKay (1981)
Lateral mechanosensory neurons	Lan 4-2	Leech	
Horizontal (cone type I) cells	HC-II.7	Fish	Young and Dowling (1984)
Dorsal root ganglion cell set	SSEA-3,4	Rat	Dodd et al. (1984)

(continued)

TABLE III *(Continued)*

Motorneuron subset	Mes-2	Grasshopper	Kotrla and Goodman (1984)
Topographic gradient			
Retina gradient	TOP	Chicken	Trisler *et al.* (1981)

a Animal in which the molecular specificity was described.

classes specify mouse oligodendroglia and rat Schwann cells. Subsets of neuronal cells within the general functional classes also have been reported: mechanosensory and lateral mechanosensory neuron subsets in leech, cone type I horizontal cells in carp retina, a rat dorsal root ganglion cell subset, and a subset of motorneurons in the grasshopper.

More complex combinations of neurons that bear a common cell-surface marker have been reported. For example, subsets of cells in the ocular dominance columns of monkey striate cortex and subsets of cells in cat spinal cord share antigen Cat 301 (McKay and Hockfield, 1982). Also, retina ganglion cells and the cells of the inner half of the inner nuclear layer of chicken retina, perhaps amacrine cells only, express RET4 (Lemmon and Gottlieb, 1982).

Many of these molecular patterns, recently defined with monoclonal antibodies, do not fit known distributions of cell-surface molecules, e.g., cytoarchitectural molecules, cell-junctional proteins, membrane enzymes, transport proteins, ion channels and pumps, cell adhesion molecules, neurotransmitter receptors, hormone or trophic/tropic molecule receptors, histocompatibility molecules, cellular oncogene proteins, or extracellular matrix molecules. The patterns, however, do fit proposed cell recognition systems.

A hypothetical molecular recognition system that incorporates the molecular patterns discussed combines the gradient and area-code models of Sperry (1963) and Hood *et al.* (1977) in Fig. 5. In this model an individual retina ganglion cell can be identified as unique by expressing a combination of molecules that mark it as (1) a neuron, (2) of the central nervous system, (3) of the visual system, (4) of retina, (5) of the ganglion cell class, (6) of a ganglion cell subset, and (7) at a particular topographical position. However, a strict area-code system, i.e., a unique molecule for each functional set and subset of cells in the nervous system, is not necessary. Cells of distinct noninteracting classes in different regions of the nervous system could be identified by

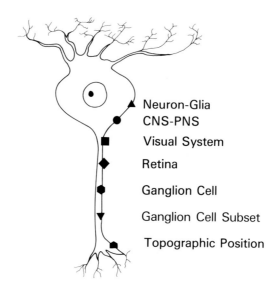

FIG. 5. Retina ganglion cell model of a neuronal recognition system proposed from categories of molecular patterns detected in the nervous system.

the same marker. For example, retina ganglion cells and cerebellar Purkinje cells could be marked with the same molecule, or topographic position in retina and cerebellum might utilize the same set of molecules. Also, the same marker could be interpreted differently by distinct sets of cells. Such marker molecules may be expressed progressively as they are needed by cells and may not persist throughout development and in the adult, although they do in some of the cases discussed.

The model suggests that positional information and organization of the nervous system can be encoded by both ordered and graded mechanisms. A combination of the two reflects both qualitative and quantitative differences among neurons. Further work is needed to demonstrate that such a system of molecules in the nervous system functions in neuronal recognition.

III. Characteristics of TOP, a Marker of Cell Position in Retina

A. GEOMETRY OF THE TOP GRADIENT

The retina grows by accretion of concentric rings of neurons at the periphery; thus, central retina is the oldest portion of the retina and peripheral retina is the youngest. To determine whether the bilaterally symmetrical gradient of TOP is a polar gradient that rotates around the center of the retina with uniform antigen concentration along any radius or is a circumferential gradient extending from dorsoposterior to ventroanterior retina, retinas were cut into eight central and eight peripheral sections (Fig. 6A) and assayed for TOP. A 35-fold gradient of antigen was found extending from the dorsoposterior to ventroanterior margins of the retina aligned parallel to the long axis of the choroid fissure.

As shown in Fig. 6B, strips of retina extending from the dorsoposterior to ventroanterior margins, or perpendicular to this axis from anterior to posterior margins, were removed and each was cut into nine pieces which were assayed for antigen. The concentration of TOP molecules detected varied continuously and logarithmically with the logarithm of distance along the circumference of the retina from the ventroanterior pole of the gradient to the dorsoposterior pole, with

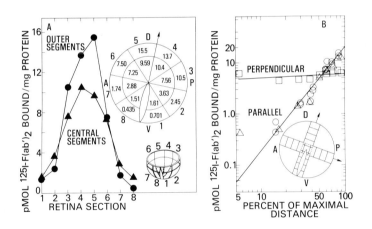

FIG. 6. Geometry of the TOP gradient in 14-day chicken embryo retina. Specifically bound ^{125}I-labeled $F(ab')_2$ is shown on the ordinate in A and B and within the appropriate segment of retina tested in A. (A) Each retina (left eye) was cut into eight 45° sections (7.25 mm in length from the center to periphery of retina) and each was divided into central (4.9 mm) and outer (2.35 mm) segments. (B) Demonstration that TOP concentration detected depends on the square of distance from the ventroanterior margin of the retina. Percentage of maximal circumferential distance is shown on the abscissa; 100% corresponds to 14.5 mm. △, Strips of retina from ventroanterior (0%) to dorsoposterior (100%) retina margins, 14.5 × 2.5 mm, were removed from eight retinas (left eyes), and each was cut into nine segments (1.6 × 2.5 mm) as shown; each segment was assayed for TOP antigen. □, Strips of retina from dorsoanterior (0%) to posterior (100%) margins of retina perpendicular to the choroid fissure were prepared and assayed as above. ○, Data from A. The length of the arc from the ventral pole of the gradient to the center of each segment was calculated by assuming the retina to be a hemisphere and using equations for spherical triangles. (From Trisler et al., 1981a.)

a slope of 2. The data extrapolate to a 400-fold gradient of TOP molecules between the two poles. In contrast, little or no change was detected in retina cells along a perpendicular axis from anterior to posterior margins of the retina.

The concentration of TOP detected (B_x) is a function of the square of the circumferential distance (D_x) from the ventroanterior pole of the gradient to the dorsoposterior pole. Thus, TOP molecules can be used as a marker of cell position along the axis of the gradient, i.e.,

$$D_x = D_{max} \left[\frac{B_x}{B_{max}} \right]^{0.5}$$

where D_{max} and B_{max} are maximal values for retina at the dorsoposterior margin, i.e., 14.5 mm and 20 pmol of ^{125}I-labeled $F(ab')_2$ specifically bound per milligram protein, respectively.

B. CELLULAR DISTRIBUTION OF TOP

Autoradiography revealed that most TOP antigen is associated with the outer and inner synaptic layers and the ganglion cell axon layer of retina (Fig. 7). The antigen was detected by autoradiography and indirect immunofluorescence on most, if not all, cell types in 14-day chick retina. All cells mechanically dissociated from 8-day chick embryo dorsoposterior retina exhibited punctate ring fluorescence; all cells from middle retina were also fluorescent, but to a lesser extent than dorsoposterior cells. No fluorescent cells were detected at the ventroanterior margin of retina, although low levels of TOP were detected by ^{125}I-labeled $F(ab')_2$ binding. No obvious heterogeneity was

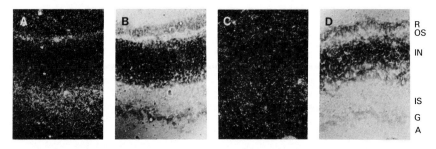

FIG. 7. Autoradiographs of 14-day chicken embryo retina. (A and B) Dark-field (A) and phase-contrast (B) views of dorsal retina. In A, some silver grains over cell soma in the inner nuclear layer appear dim due to staining of cells by toluidine blue. (C and D) Dark-field (C) and phase-contrast (D) views of ventral retina. R, Photoreceptor layer; OS, outer synaptic layer; IN, inner nuclear layer; IS, inner synaptic layer; G, ganglion cell layer; A, ganglion cell axon layer. ×630. (From Trisler *et al.*, 1981a.)

observed in the cell population from each location. No evidence was found for topographically distributed differences in antigen affinity for antibody; however, variation in antigen accessibility has not been excluded. These results suggest that the gradient of antigen results from differences in the number of antigen molecules per cell rather than in the proportion of cells bearing the antigen.

C. TOP EXPRESSION DURING DEVELOPMENT

The concentration of TOP antigen detected was higher in dorsoposterior retina than in ventroanterior retina at every age tested from the 4-day embryo through the adult (Fig. 8A, inset), and the axis and polarity of the gradient did not change during development (Fig. 8A). The concentration of antigen detected in the dorsal half of the retina increased 3-fold between the 4th and 12th days of embryonic

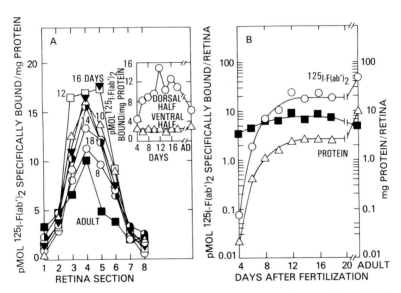

FIG. 8. TOP antigen in chick retina as a function of developmental age. (A) Topographic gradient vs developmental age. Each retina (left eye) was cut into eight sections as shown in Fig. 3. Symbols and days *in ovo:* ○, 8; △, 10; □, 12; ○, 14; ▽, 16; ◑, 18; and ■, adult. (Inset) ^{125}I-labeled F(ab')$_2$ bound specifically is shown on the ordinate; days *in ovo* and adult (AD) are shown on the abscissa. ○, dorsal half of retina (sections 3–6); △, ventral half of retina (sections 1,2,7, and 8). Data for retina from 4- and 6-day embryos are shown only in the inset. (B) Antigen and protein/retina. ○, ^{125}I-labeled F(ab')$_2$ bound specifically per retina; △, protein per retina; ■, ordinate represents picomoles ^{125}I-labeled F(ab')$_2$ bound specifically per milligram of protein. (From Trisler *et al.,* 1981a.)

development and then decreased slightly in the adult. The amount of protein per retina and TOP antigen detected per retina increased 470- and 620-fold, respectively, between the 4-day embryo and the adult; the amount of antigen detected per milligram of protein remained relatively constant (Fig. 8B). These results show that a gradient of TOP molecules is formed early in retina development, during the period of active neuroblast proliferation and neuron genesis, and that the gradient is maintained after neuron genesis ceases.

D. Tissue Specificity

Highest concentrations of TOP antigen detected were in regions of the nervous system derived from prosencephalon (forebrain): retina > cerebrum > thalamus (Table IV). Low levels of antigen were found in dorsal and ventral retina pigment epithelium, optic nerve, optic tectum, and cerebellum; little or no antigen was detected in heart, liver, kidney, or cells from blood.

E. Species Specificity

Gradients of TOP molecules with similar orientation and symmetry were detected in turkey, quail, and duck embryo retina, 17, 15, and 16 days after fertilization, respectively (Fig. 9). The antigen was not

TABLE IV

Distribution of Top Antigen in Chicken Tissues[a]

	^{125}I-labeled $F(ab')_2$ specifically bound (pmol/mg protein)	
Tissue	14-Day embryo	Adult
Dorsal neural retina	12.0	7.70
Ventral neural retina	1.08	2.70
Cerebrum	3.30	3.12
Thalamus	2.13	—
Optic nerve	—	0.58
Optic tectum	0.15	—
Cerebellum	0.23	0.46
Dorsal retina pigment epithelium	0.26	—
Ventral retina pigment epithelium	0.25	—
Heart, liver, kidney, or blood cells	0.002–0.055	—

[a]From Trisler et al., 1981a.

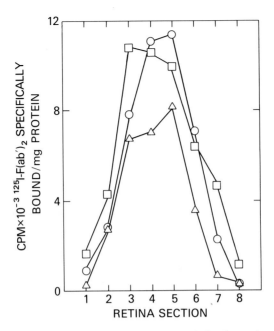

FIG. 9. TOP gradients in retina from Japanese quail (○), *Coturnix coturnix japonica* (15-day embryo); White Pekin duck (△), *Anus platyrhynchos* (16-day embryo); and turkey (□), *Meleagris gallopavo* (17-day embryo). The eggs hatch 17, 28, and 28 days after fertilization, respectively. (From Trisler *et al.*, 1981a.)

TABLE V

EFFECT OF TRYPSIN OR HEAT ON TOP ANTIGENICITY[a]

	Treatment of retina cells		^{125}I-labeled F(ab′)$_2$ bound specifically to retina cells	
Experiment	0–30 minutes	30–40 minutes	cpm	Percentage
1	Control	+ Trypsin inhibitor	1700	100
	Trypsin	+ Trypsin inhibitor	93	6
	Trypsin + trypsin inhibitor	—	1803	106
2	4°C, 30 minutes		1607	100
	100°C, 30 minutes		132	8

[a]Assay conditions in Trisler *et al.* (1981a).

detected in retina of goldfish, *Xenopus laevis, Rana pipiens,* turtle, or Fisher rats (not shown).

F. Properties of TOP Antigen

TOP antigenicity was lost after incubation at 100°C or by incubation with trypsin (Table V). All TOP antigenicity in retina cell homogenates was recovered from the 100,000 × g particulate fraction; soluble antigen was not detected (not shown). Bovine gangliosides (10–10,000 μM) did not inhibit the binding of anti-TOP antibody to retina. Rabbit antisera to mono-, di-, and trisialogangliosides bound to retina; however, a ganglioside gradient was not detected. These results suggest that TOP is a protein or is associated with a protein in cell membranes.

IV. Regulation of TOP Expression

Molecular gradients and cell position are thought to play a role in several aspects of embryonic development (Child, 1941; Turing, 1952; Crick, 1970; Wolpert, 1971; French *et al.,* 1976; Malacinski and Bryant, 1984). Among the mechanisms proposed for establishing and maintaining positional information are (1) a linear concentration gradient of molecules generated by a source–sink mechanism (Crick, 1970) in which a molecule is synthesized at a site, i.e., the source, in the tissue and diffuses through the tissue to a sink, a site where it is destroyed; (2) a threshold point in the gradient (Wilcox *et al.,* 1973) where properties may change abruptly as a result of the gradient; and (3) accumulation of events and a permanent record of the time cells spend in a progress zone, e.g., the region of growing cells at the·tip of the chick limb bud (Summerbell *et al.,* 1973), which in retina would be the circular peripheral margin. The results presented suggest that a combination of these three models may be involved in regulation of TOP expression.

A. TOP Gradient Increases with Retinal Growth

The retina grows by adding rings of cells in the proliferative zone at the peripheral margin (Coulombre, 1955; Kahn, 1974) (Fig. 10B). Expression of the gradient during retina growth was examined by comparison of the amount of TOP in the proliferative zone at the poles of the gradient in day 4–10 embryo retinas with the amount of TOP at the corresponding distances along the gradient axis in day 12 retina (Fig. 10). The cells in the proliferative zone at the dorsoposterior margin of retina expressed progressively more TOP with retinal growth, while those in the proliferative zone at the ventroanterior margin

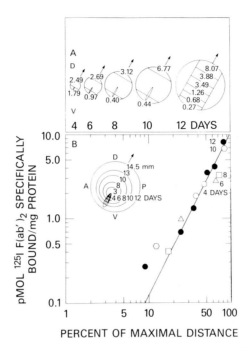

FIG. 10. Increase in TOP gradient with retinal growth. (A) TOP antigen detected [picomoles ^{125}I-labeled F(ab′)$_2$ specifically bound per milligram of protein] at the poles of the gradient of days 4–10 and along the axis of the gradient of day 12 embryonic retina. (B) TOP values at these positions for each age as a function of distance along the axis of the gradient in day 12 retina. Circumferential distances from dorsoposterior to ventroanterior pole for each age are shown: ○, day 4; △, day 6; □, day 8; ○, day 10; ●, day 12.

expressed progressively less TOP thereby increasing the magnitude of the gradient. There was a corresponding increase in TOP levels of dorsal-half retina and a decrease in ventral-half retina, indicating an increase in magnitude of the gradient during development and not an expansion of a complete gradient that is present in day 4 retina.

There was close agreement in the amount of TOP detected on cells at a given distance along the gradient axis from the fundus, the oldest portion of the retina, throughout the developmental period tested. Thus, each position along the axis has a constant TOP value all through these developmental ages.

B. CELL POSITION DETERMINES THE AMOUNT OF TOP IN PROGENY

The level of TOP expressed in the progeny of cells dividing at various angles to the axis of the antigen gradient was determined by comparing TOP expression in cells of the proliferation zone of day 4 and day 12 retinas at different positions around the peripheral margin (Fig. 11). Both the magnitude and the sign of change in TOP expression during development varied depending on the position of the parental cells and therefore the direction of cell division. Progeny cells in dorsoposterior retina expressed more TOP than parental cells while those in ventroanterior retina expressed less. The greatest magnitude of change in expression was along the axis of the gradient. Little or no change in TOP expression occurred in cells with retinal growth along the perpendicular axis (90°). Intermediate rates of change in TOP expression were found during cell division at 45° to the gradient axis.

C. TOP Gradient Is Expressed *in Vitro*

Cells from dorsoposterior, middle, and ventroanterior day 8 embryo retina were cultured separately *in vitro* after dissociation with trypsin to remove TOP antigen. TOP was reexpressed *in vitro* by cells from all areas of the retina, but the level of TOP in each culture was different depending on the position of origin of the cells in retina (Fig. 12). TOP was in highest abundance in cell cultures from dorsoposterior retina, intermediate abundance from middle retina, and lowest abundance from ventroanterior retina. Thus, the gradient was reexpressed *in vitro*. We believe the gradient is not maintained by a "source–sink"

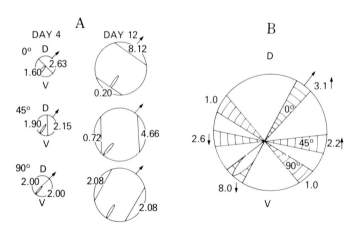

Fig. 11. TOP antigen expression in cell progeny during embryonic development from day 4–12 as a function of the angle of growth from the axis of the gradient in retina. (A) TOP antigen detected [picomoles ^{125}I-labeled F(ab')$_2$ specifically bound per milligram of protein] at the peripheral margin of day 4 and day 12 retina along the axis at 0, 45, and 90° to the axis. (B) Fold and sign of change in TOP level at each angle from day 4 to day 12 of embryonic growth.

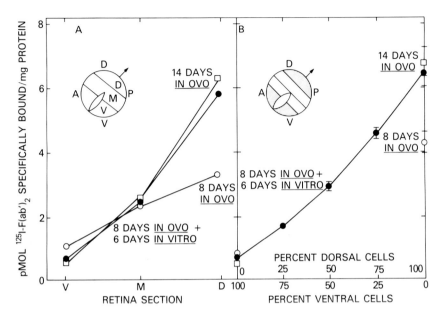

FIG. 12. Gradient increases and antigen expression in retinal cultures. (A) TOP antigen in monolayer cultures containing cells dissociated with trypsin from dorsal, middle, or ventral 8-day chick embryo retina and cultured for 6 days. (B) Cells from dorsal and ventral retina cultured in varying proportions. ○, cells from 8-day *in ovo* retina; □, cells from 14-day *in ovo* retina; ●, cells from 8-day *in ovo* retina that were grown *in vitro* 6 days. [(B) From Trisler *et al.*, 1981b.]

mechanism after day 8 of embryonic development because TOP was reexpressed in middle and ventral retina cell culture in the absence of the most positive cells, those that would be the likely "source." The gradient is also reexpressed in the absence of cells from surrounding tissue, e.g., pigmented epithelium and lens, suggesting that interactions with these tissues are not required for reexpression and maintenance of the gradient.

There was also no evidence of the influence of cells from dorsal retina on the expression of TOP in cells from ventral retina and vice versa when cells from these regions were cocultured (Fig. 12). Cells from dorsal and ventral day 8 retina that were mixed in different proportions and cocultured for 6 days contained antigen levels that were nearly additive (linear correlation coefficient, r^2, 0.98). No evidence of induction or suppression of antigen was obtained under the conditions tested.

The magnitude of the TOP gradient increased with retinal growth *in ovo* from day 4 to day 12 postfertilization. This increase also occurred *in vitro* (Fig. 12). After six days *in vitro* cells from day 8 embryonic retina expressed a gradient similar in magnitude to day 14 embryos *in vivo* and greater than day 8 embryos. Cells from dorsal retina increased TOP expression in culture. Those from ventral retina decreased and there was little change in the cells from middle retina. Thus, the mechanism for generating an increased gradient during retinal growth is determined in the retina cells of day 8 embryos. It is present at day 6, the earliest time tested (not shown).

The gradient was maintained at least 10 days in culture and is reexpressed in cells after a second exposure to trypsin and transfer to fresh culture dishes (Fig. 13). Thus, retina cells from day 6 and 8 embryos grown *in vitro* up to 10 days behave as *in vivo* cells in expression of the TOP gradient.

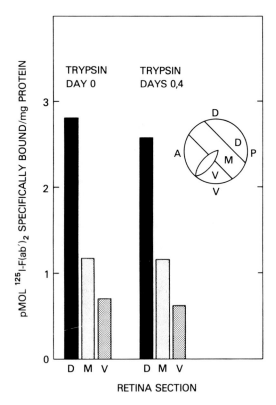

FIG. 13. TOP antigen in monolayer cultures containing cells dissociated with trypsin from dorsal, middle, or ventral 8-day chick embryo retina and cultured for 10 days. Where indicated, cells were dissociated again on the 4th day of culture, recovered, and cultured in fresh plates for 6 additional days. (From Trisler *et al.*, 1981b.)

Our hypothesis is that a gradient is established in very young retina by diffusion of molecules from a point source that becomes one pole of the gradient to a sink that becomes the opposite pole. This gradient gives a TOP value to all cells of the young retina. Each cell in the progress (growth) zone at the margin of the hemiretina delineated by the axis of the gradient bears a different TOP value (bilateral cells at a given position in both hemiretinae bear the same value). A threshold near the middle of the gradient at the retina fundus determines the polarity, i.e., sign (+, −) of change, of TOP expression in cell progeny. After the gradient is established in the young retina, it is no longer maintained by diffusion of molecules from a point source, but the gradient of TOP increases by cell progeny expressing progressively more TOP in dorsal retina and less in ventral as a function of the length of time blast cells are in the progress zone at the retinal margin. Intermediate rates of increase and decrease in TOP expression in progeny of cells dividing at an angle to the gradient axis are a result of the position of the early parental cell at the time the gradient was established and not a result of an intermediate amount of time spent in the progress zone. Since the retina grows by addition of rings of cells at the peripheral margin, the amount of time blast cells spend in the progress zone is nearly equal around the entire retinal margin. Those blast cells dividing along the orthogonal to the gradient axis at the threshold or polarization point are unaffected by the amount of time spent in the progress zone and express constant TOP levels in their progeny. Thus, the behavior of cells in regulating the level of TOP in progeny after the gradient is initially established is independent of a diffusible molecule. Further work is required to determine the nature of the early gradient, to identify events (e.g., dilution of molecule, elapsed time, number of cell divisions) that increase the magnitude of the TOP gradient during retinal growth, and to understand the molecular basis of regulation of TOP expression, e.g., DNA–protein complex modification by changes in local ion concentration (Groudine and Weintraub, 1982) or local protein concentrations (Brown, 1984), DNA supercoiling (review Weintraub, 1985), DNA methylation (Vanyushin et al., 1970; review Razin et al., 1984), or, less probably, gene amplification or gene rearrangement (review Brown, 1981).

V. Function of TOP in Retinal Development

The role of molecular markers of cell position in the development of the nervous system was examined using a monoclonal antibody to TOP. Antibodies provide a means of blocking molecular function (Levi-Montalcini and Booker, 1960; Crawford et al., 1982; Warner et *al.,* 1984; Strauss and Nirenberg, 1985), and antibodies to neuronal surface molecules have been used to disturb growth cone behavior and disrupt neurite outgrowth (Schwartz and Spirma, 1982; Henke-Fahle and Bonhoeffer, 1983; Trisler, 1983; Trisler and Nirenberg, 1983; Leifer *et al.,* 1984; Thanos *et al.,* 1984; Fraser *et al.,* 1984).

Our objectives were to determine the accessibility of TOP in the *in vivo* retina to antibody, to determine the persistence of antibody in the retina after injection into the embryo, and to identify changes in the development of retina that is continuously exposed to anti-TOP antibody.

A. Ab · TOP Distribution after Extraembryonic Injection

Antibody to TOP (Ab·TOP) injected into the amniotic cavity of *in ovo* chick embryos 2–4 days after fertilization was detected on retina cells one day after injection. The concentration of Ab·TOP complexes detected was higher in dorsal retina than in ventral retina (Fig. 14). Antibody injected into the amniotic cavity of embryos older than 4 days and antibody injected into the yolk, a rich source of maternal antibody for the chick (Malkinson, 1965), of 3-day embryos was not detected in retina (not shown). Thus, TOP antigen in retinas of 4-day embryos and younger is accessible to antibody from an extraembryonic source.

B. Ab · TOP Gradient after Intraocular Injection

Anti-TOP antibody was injected intraocularly into the vitreal space of day 7–19 embryos. Retinas from the injected eyes were cut into

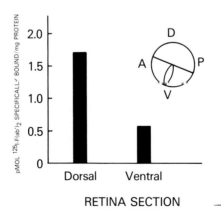

FIG. 14. Anti-TOP antibody distribution in retinas of 3-day chicken embryos one day after injection of the antibody into the amniotic cavity. (From Trisler *et al.,* 1986.)

eight sections as shown in Fig. 15, and cells from each section were assayed for Ab·TOP complexes. A dorsoposterior → ventroanterior gradient of Ab·TOP complexes of the same magnitude and orientation detected by *in vitro* binding studies was present in retina 24 hours after intraocular injection of mouse ascites fluid containing anti-TOP antibody into day 11 chick embryos (Fig. 15). Little or no antibody was detected bound to retina cells from the contralateral noninjected eye and from eyes injected with P3X63 Ag8 antibody. A gradient of anti-TOP antibody was present in retina 24 hours after intraocular injection for all embryonic ages tested up to day 18. Anti-TOP antibody was not detected in retina 24 hours after intraocular injection of day 18 and older embryos (not shown).

C. Distribution of Ab · TOP in Retina Cell Strata

Distribution of Ab·TOP complexes in the cell soma and cell process strata of retina after intraocular injection was determined in day 17 embryo retina 6 days after injection. Indirect immunofluorescence revealed that most anti-TOP antibody in chick embryo retina was distributed in the outer and inner synaptic layers and the ganglion cell axon layer of retinas after injection (Fig. 16A). Little antibody was detected in retinas from eyes injected with P3X63 Ag8 antibody (Fig.

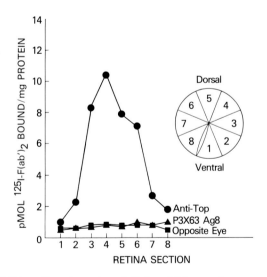

FIG. 15. Ab·TOP gradient in retina of 12-day chicken embryos one day after intraocular injection of antibody. ●, Retinas from eyes injected with anti-TOP antibody; ▲, retinas from eyes injected with P3X63 Ag8 antibody; ■, retinas from noninjected eyes opposite the anti-TOP injected eyes. (From Trisler *et al.*, 1986.)

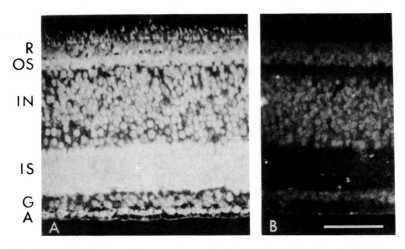

FIG. 16. Indirect immunofluorescence analysis of the distribution of antibody in 17-day chicken embryo dorsoposterior retina six days after intraocular injection of anti-TOP antibody producing hybridoma cells (A) and P3X63 Ag8 antibody producing myeloma cells (B). R, Photoreceptor cell layer; OS, outer synaptic layer; IN, inner nuclear layer; IS, inner synaptic layer; G, ganglion cell layer; A, ganglion cell axon layer. Bar, 50 μm. (From Trisler *et al.*, 1986.)

16B). The nuclear stain was seen in the absence of antibody with rhodamine-conjugated avidin alone (not shown). These results show that anti-TOP antibody diffuses from the vitreal space, penetrates the entire thickness of retina, and binds most abundantly in the cell process layers of retina.

D. DURATION OF Ab · TOP GRADIENT

A gradient of Ab·TOP complexes was detected in retina 24 hours after intraocular injection of mouse ascites fluid containing anti-TOP IgG$_1$ and persisted for 3–4 days (Fig. 17A). The Ab·TOP gradient was maintained for 9–10 days when hybridoma cells that synthesize anti-TOP antibody were injected (Fig. 17B). The hybridoma cells provide a continuous source of antibody for long-term maintenance of Ab·TOP complexes in the retina.

E. STABILITY OF THE TOP GRADIENT IN THE PRESENCE OF ANTIBODY

The stability of TOP expression in retina after intraocular injection of anti-TOP hybridoma cells and P3X63 Ag8 myeloma cells was determined by measuring both the amount of antibody from the injected source that was bound to the retina cells 5–11 days after injection and

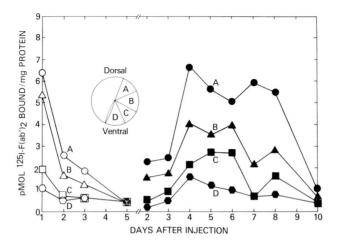

Fig. 17. Duration of Ab·TOP gradient in retina after intraocular injection of mouse ascites fluid containing anti-TOP antibody (A open symbols) and hybridoma cells producing anti-TOP antibody (B, closed symbols) into 11-day chicken embryos. One-half of each retina was prepared for electron microscopy (Fig. 19) and synapse counts (Fig. 20), and the other half retina was cut into four sections as shown and assayed for anti-TOP antibody. ○ and ●, section A, dorsoposterior retina; △ and ▲, section B, posterior retina; □ and ■, section C, ventro-posterior retina; ○ and ◗ , section D, ventral retina. (From Trisler *et al.*, 1986.)

the total amount of TOP detectable in the same retina by *in vitro* assay with additional exogenous antibody. A gradient of TOP molecules was detected at all ages tested after injection (Fig. 18). Eleven days after injection, when anti-TOP antibody no longer was present in retina, a normal gradient of TOP was detected by *in vitro* assay with exogenous antibody (Fig. 18A). The difference in amount of anti-TOP antibody bound to retina cells in the presence of injected antibody alone and injected antibody plus exogenously added antibody may reflect the loss of injected antibody from the cells during the course of incubation of duplicate samples of cells with exogenous antibody or may represent a compartment of TOP molecules in retina that is not accessible to injected antibody. The reduction in binding of anti-TOP antibody per milligram of retina protein between embryonic days 15 and 21 (days 5 and 11 after injection) occurs normally in development (Fig. 8). A normal gradient of TOP was present in retina after intraocular injection of P3X63 Ag8 cells at all ages tested from day 15 to day 21 embryos, days 5 to 11 postinjection (Fig. 18B), and in day 7 posthatch chicken retina 19 days after intraocular injection of anti-TOP anti-

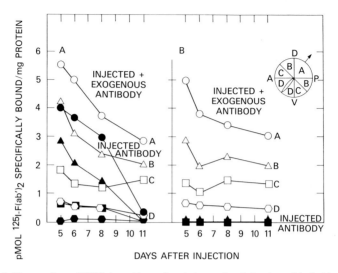

Fig. 18. Expression of TOP in retina after intraocular injection of hybridoma cells producing anti-TOP antibody (A) and myeloma cells producing P3X63 Ag8 antibody (B) into 10-day chicken embryos. Five days after intraocular injection, and on each subsequent day through 11 days, retinas were removed from the embryos, cut into sections as shown, and assayed for TOP and Ab·TOP complexes. Cells from corresponding sections of retina were pooled into A, B, C, and D. Cells in half of each pool were assayed for TOP by incubation with exogenously added anti-TOP antibody, and half were assayed for antibody bound to the retina cells from the injected source in the vitreous. Half of each pool of cells was incubated 1 hour at 4°C in buffer containing anti-TOP antibody and the other half was incubated in buffer without antibody. The total antibody bound to the retina cells was then detected with ^{125}I-labeled F(ab')$_2$. ○ and ●, Retina section A; △ and ▲, retina section B; □ and ■, retina section C; ○ and ●, retina section D. The closed symbols represent specific binding of ^{125}I-labeled F(ab')$_2$ to retina cells in the presence of anti-TOP antibody from the intraocular injected source. The open symbols represent the total specific binding of ^{125}I-labeled F(ab')$_2$ in the presence of both injected and exogenously added anti-TOP antibody (A) and specific binding of ^{125}I-labeled F(ab')$_2$ in the presence of injected P3X63 Ag8 antibody and exogenously added anti-TOP antibody (B). (From Trisler et al., 1986.)

body (not shown). Thus, the presence of anti-TOP and P3X63 Ag8 antibody in retina does not prevent expression of the TOP antigen gradient.

F. Time Course of Synapse Formation
in the Presence of Antibody

Antibody was injected into eyes of day 11 embryos at the time neuron process layers are forming and two days before the first structurally identifiable synapses appear (Sheffield and Fishman, 1970;

Hughes and LaVelle, 1974; Daniels and Vogel, 1980). Two components of the process of synapse formation—the disappearance of growth cones and the appearance of synapses—were studied in retina from embryos exposed to anti-TOP antibody, control antibodies P3X63 Ag8 and 57D8, and no antibody. For this study growth cones were defined as bodies containing large, irregular membrane cisternae or vesicles (Del Cerro and Snider, 1968; Kawana et al., 1971). These bodies were larger in section than most neurites and sometimes were seen in continuity with neurites and filopodia. Anti-TOP antibody from the injected hybridoma source reached maximum binding in the retina four days after injection (day 15 embryo). At this time, retinal development appeared normal. Electron microscopic analysis revealed that the number of growth cones and synapses and the amount of extracellular space between neurites was similar in retinas exposed to anti-TOP antibody or to no antibody (Fig. 19A and B; Fig. 20). However, with continued exposure to anti-TOP antibody retinal development was altered.

In normal retinal development from embryonic day 15 to day 18, the number of growth cones decreased 50% (Figs. 19C and 20A), whereas the number of growth cones increased 50% during the same developmental period in retinas exposed to anti-TOP antibody (Figs. 19D and 20A) and then decreased, from 7 to 10 days after injection, to 3 growth cones/100 μm^2 near control levels (not shown after day 19). The number of synapses increased 150% in normal retinas from embryonic days 15 to 18, then plateaued at 39–40 synapses/100 μm^2 between days 18 and 21 as previously described (Daniels and Vogel, 1980); however, in the presence of anti-TOP antibody the number of synapses increased minimally between days 15 and 18 (Figs. 19D and 20B). Then from day 18 to day 21 the number increased to 35/100 μm^2, approaching control levels.

Extracellular space between neurites also was affected by anti-TOP antibody. The amount of extracellular space between neurites remained unchanged from 4 to 10 days after injection of anti-TOP antibody when neurite packing normally increases. Maintenance of a large extracellular space in retinas exposed to anti-TOP antibody was restricted to the neuron process layers. No enlargement of extracellular space was detected in the cell soma layer of retina at these ages (not shown). The width of the retina cell body layers and neurite layers was normal after intraocular injection of hybridoma cells. No cell death was detected in retina from 4 to 10 days after injection of anti-TOP hybridoma cells. There was no evidence that inhibition of synapse formation was restricted to any particular cell type. Both

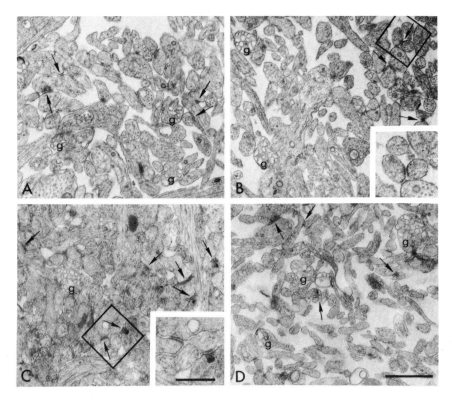

Fɪɢ. 19. Electron micrographs of the inner synaptic layer of 15-day (A,B) and 18-day (C,D) chicken embryo retina 4 and 7 days, respectively, after intraocular injection of hybridoma cells producing anti-TOP antibody (B,D) and myeloma cells producing P3X63 Ag8 antibody (A,C). Growth cones (g) and synapses (arrows) can be seen. Inset in B is an enlargement of an area containing a synapse in day 15 retina, and the inset in C is an enlargement of an area containing two synapses, one a ribbon synapse of a bipolar cell. Bar in D, 1.0 μm. Bar in inset, 0.5 μm. (From Trisler et al., 1986.)

conventional synapses and bipolar cell ribbon synapses were present from days 5 to 7 after injection of anti-TOP antibody. Synapse formation appeared to be interrupted across the entire inner synaptic layer in both dorsal and ventral retina. No difference in development was detected between retinas from noninjected eyes and from eyes injected with myeloma cells that produce P3X63 Ag8 antibody, which binds uniformly in all regions of retina in low abundance, and 57D8 antibody, which binds in high abundance to the surface of cell soma and neurites in all regions of the retina (Fig. 2). Retinal development, thus, appeared normal in the presence of control antibodies, and the changes

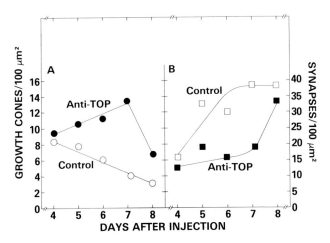

Fig. 20. Time course of synapse formation in the inner synaptic layer of retina after intraocular injection of hybridoma cells producing anti-TOP antibody, 57D8 antibody, and myeloma cells producing P3x63 Ag8 antibody into 11-day embryos. Growth cones and synapses were counted on electron micrographs of retinas (e.g., Fig. 19) from embryos in which the duration of Ab·TOP complexes was measured in Fig. 17. ●, Growth cones, and ■, synapses in the presence of anti-TOP antibody. ○, Growth cones, and □, synapses in the presence of control and no antibody. The number of growth cones/100 μm^2 and the number of synapses/100 μm^2 is shown on the ordinate of A and B, respectively. Each point represents 55–300 growth cones (total counted 1469) and 204–868 synapses (total counted 4523) from 1–8 retinas. Controls include regions of retina comparable to anti-TOP injected retinas from eyes injected with P3X63 Ag8 myeloma cells, the noninjected opposite eyes and noninjected eyes opposite those injected with anti-TOP hybridomas, and eyes injected with 57D8. The number of growth cones in experimental and control retinas was significantly different at days 5, 6, and 7 (p values, Student's t test: day 5, $p < 0.05$; day 6, $p < 0.01$; day 7, $p < 0.01$), and the number of synapses was significantly different at days 6 and 7 (day 6, $p < 0.05$; day 7, $p < 0.01$). (From Trisler et al., 1986.)

detected in the presence of anti-TOP antibody were inhibition neurite development and synapse formation.

Several groups have demonstrated effects of antibodies against nervous system molecules on the development of nerve cells and tissues in vivo and in vitro. Antibodies against chicken cognin and N-CAM have been shown to inhibit cell–cell adhesion (Hausman and Moscona, 1979, and Thiery et al., 1977, respectively) and to disrupt axonal fasciculation and the projection of most retinal ganglion cell axons on the optic tectum (Thanos et al., 1984; Fraser et al., 1984), although some axons in direct contact with tectal cells correct the disruption (Thanos et al., 1984). Antibody T61/3/12 blocks neurite outgrowth of chick retina cells *in vitro* (Henke-Fahle and Bonhoeffer, 1983), whereas antibody to Thy-1 stimulates neurite outgrowth of rat retina ganglion cells *in vitro* (Leifer *et al.*, 1984). Antibody L1 inhibits granular cell migration in rat cerebellum explants (Lindner *et al.*, 1983).

We could not demonstrate inhibition of cell–cell adhesion with anti-TOP antibody or inhibition of neurite outgrowth *in vitro* in cocultures of retina cells and hybridoma cells (not shown). Retina cells were not killed during *in vitro* coculture with hybridoma cells. No evidence of cell death was seen by electron microscopy in retinas exposed to anti-TOP antibody for 11 days after injection, the longest time tested. A normal gradient of TOP molecules was detected in retinas as Ab·TOP complexes for 9 to 10 days after injection. After Ab·TOP complexes no longer were detected, a normal gradient of TOP molecules was present. The decrease in Ab·TOP complexes in retina after day 18 represents the loss in accessibility of TOP to the intraocular hybridoma antibody source and not down-regulation of TOP antigen. Thus, retinas appeared normal after intraocular injection of anti-TOP antibody except for extracellular space between neurites, an increased number of growth cones, and inhibition of synapse formation, suggesting that TOP is involved in synapse formation.

Our working hypothesis is that TOP molecules mark cell position in the retina and that, in the presence of Ab·TOP complexes, growing neurites fail to detect the TOP gradient. This could result in the prolonged presence of growth cones, since neurites fail to find their targets, and delayed synapse formation. Near-normal numbers of synapses, however, are formed eventually after exposure to anti-TOP antibody. Further work is required to determine whether the synaptic connections which occur in retina after intraocular injection of anti-TOP antibody are positionally correct.

VI. Summary

Molecules that identify cell type and position in the nervous system were detected by monoclonal antibodies. One molecule, TOP, is distributed in a 35-fold topographic gradient from the dorsoposterior margin to the ventroanterior margin of avian retina. The gradient is present in young embryos, increases with retinal growth, and persists in the adult. TOP molecules are present on most or all cells of retina. The number of TOP molecules detected per cell varies continuously along the axis of the antigen gradient. Thus, TOP can be used to identify position in the plane of retina along that axis. Other antigens that identify cell type and position across the thickness of retina also were detected. Molecules that mark such cellular organization may represent a neuronal recognition system.

Antibodies were used to examine the role of markers of cell position in development of the nervous system. Antibody to TOP from hybridoma cells that were injected into *in vivo* embryo eyes diffused into the retina and bound in a topographic gradient of Ab·TOP complexes. Synapse formation in retina was inhibited in the presence of anti-TOP antibody. This suggests that TOP is involved in synapse formation and that recognition of position by neurons is necessary for normal synapse formation.

REFERENCES

Barnstable, C. (1980). *Nature (London)* **286**, 231.

Barnstable, C., Akagawa, K., Hofstein, R., and Horn, J. P. (1983). *Cold Spring Harbor Symp. Quant. Biol.* **48**, 863.

Brockes, J., Fields, K. L., and Raff, M. C. (1977). *Nature (London)* **266**, 364.

Brown, D. D. (1981). *Science* **211**, 667.

Brown, D. D. (1984). *Cell* **37**, 359.

Child, C. M. (1941). "Patterns and Problems of Development" Univ. of Chicago Press, Chicago.

Cohen, J., and Selvendran, S. (1981). *Nature (London)* **291**, 421.

Cole, G., and Glaser, L. (1984). *Proc. Natl. Acad. Sci. U.S.A.* **81**, 2260.

Coulombre, A. J. (1955). *Am. J. Anat.* **96**, 153.

Crawford, G., Slemmon, J. R., and Salvaterra, P. M. (1982). *J. Biol. Chem.* **257**, 3853.

Crick, F. (1970). *Nature (London)* **225**, 420.

Daniels, M. P., and Vogel, Z. (1980). *Brain Res.* **201**, 45.

Del Cerro, M. P., and Snider, R. S. (1968). *J. Comp. Neurol.* **133**, 341.

Dodd, J., Salter, D., and Jessell, T. M. (1984). *Nature (London)* **311**, 469.

Fraser, S., Murray, B. A., Chuong, C.-M., and Edelman, G. M. (1984). *Proc. Natl. Acad. Sci. U.S.A.* **81**, 4222.

French, V., Bryant, P. J., and Bryant, S. (1976). *Science* **193**, 969.

Fujita, S., Zipursky, S. L., Benzer, S., Ferrus, A., and Shotwell, S. L. (1982). *Proc. Natl. Acad. Sci. U.S.A.* **79**, 7929.

Groudine, M., and Weintraub, H. (1982). *Cell* **30**, 131.

Grunwald, G. B., Trisler, D., and Nirenberg, M. (1983). *J. Neurosci.* **9**, 692.

Hausman, R., and Moscona, A. (1976). *Proc. Natl. Acad. Sci. U.S.A.* **73**, 3594.

Hausman, R., and Moscona, A. (1979). *Exp. Cell Res.* **119**, 191–204.

Henke-Fahle, S., and Bonhoeffer, F. (1983). *Nature (London)* **303**, 65.

Hilt, D. C., Strauss, W. L., Trisler, D., Darveniza, P., and Nirenberg, M. (1985). *J. Cell Biol.* **101**, 223.

Hood, L., Huang, H., and Dreyer, W. (1977). *J. Supramol. Struct.* **7**, 531.

Hughes, W. F., and LaVelle, A. (1974). *Anat. Rec.* **179**, 297.

Kahn, A. J. (1974). *Dev. Biol.* **38**, 30.

Kawana, E., Sandri, C., and Akert, K. (1971). *Z. Zellforsch. Mikrosk. Anat.* **115**, 284.

Kotrla, K., and Goodman, C. (1984). *Nature (London)* **311**, 151.

Lawrence, P. (1978). *Zoon* **6**, 157.

Leifer, D., Lipton, S. A., Barnstable, C. J., and Mashland, R. H. (1984). *Science* **224**, 303.

Lemmon, V., and Gottlieb, D. I. (1982). *J. Neurosci.* **2**, 531.

Levi-Montalcini, R. M., and Booker, B. (1960). *Proc. Natl. Acad. Sci. U.S.A.* **46**, 384.

Levitt, P. (1984). *Science* **223**, 299.

Lindner, J., Rathjen, F. G., and Schachner, M. (1983). *Nature (London)* **305**, 427.

McKay, R., and Hockfield, S. (1982). *Proc. Natl. Acad. Sci. U.S.A.* **79**, 6747.

Malacinski, G. M., and Bryant, S. V. eds. (1984). "Pattern Formation." Macmillan, New York.

Malkinson, M. (1965). *Immunology* **9**, 311.

Pessac, B., Girard, A., Romez, G., Crisanti, P., Lorinet, A.-M., and Calothy, G. (1983). *Nature (London)* **302**, 616.

Ramón y Cajal, S. (1972). "The Structure of the Retina." Thomas, Springfield, Illinois.

Razin, A., Cedar, H., and Riggs, A. D., eds. (1984). "DNA Methylation." Springer-Verlag, New York.

Schnitzer, J., and Schachner, M. (1981). *J. Neuroimmunol.* **1**, 457.

Schwartz, M., and Spirma, N. (1982). *Proc. Natl. Acad. Sci. U.S.A.* **79**, 6080.

Sheffield, J. B., and Fishman, D. A. (1970). *Z. Zellforsch. Mikrosk. Anat.* **104**, 405.

Sommer, I., and Schachner, M. (1981). *Dev. Biol.* **83**, 311.

Sperry, R. (1963). *Proc. Natl. Acad. Sci. U.S.A.* **50**, 703.

Strauss, W. L., and Nirenberg, M. (1985). *J. Neurosci.* **5**, 175.

Summerbell, D., Lewis, J. H., and Wolpert, L. (1973). *Nature (London)* **244**, 492.

Thanos, S., Bonhoeffer, F., and Rutishauser, U. (1984). *Proc. Natl. Acad. Sci. U.S.A.* **81**, 1906.

Thiery, J.-P., Brackenbury, R., Ruttshauser, U., and Edelman, G. M. (1977). *J. Biol. Chem.* **252**, 6841.

Trisler, D. (1982). *Trends Neurosci.* **5**, 306.

Trisler, D. (1983). "Molecular Approaches to the Nervous System," pp. 63–72. Soc. for Neurosci. USA Short Course Syllabus.

Trisler, D., and Nirenberg, M. (1983). *Int. J. Dev. Neurosci.* **1**, 223.

Trisler, D., Schneider, M. D., and Nirenberg, M. (1981a). *Proc. Natl. Acad. Sci. U.S.A.* **78**, 2145.

Trisler, D., Schneider, M. D., Moskal, J. R., and Nirenberg, M. (1981b). *In* "Monoclonal Antibodies to Neural Antigens" (R. McKay, M. Raff, and L. Reichardt, eds.), pp. 231–245. Cold Spring Harbor Laboratory, Cold Spring Harbor, New York.

Trisler, D., Grunwald, G. B., Moskal, J., Darveniza, P., and Nirenberg, M. (1984). *In* "Neuroimmunology" (P. O. Behan and F. Spreafico, eds.), pp. 89–97. Raven, New York.

Trisler, D., Bekenstein, J., and Daniels, M. P. (1986). *Proc. Natl. Acad. Sci. U.S.A.* **83**, 4194.

Turing, A. M. (1952). *Philos. Trans. R. Soc. London Ser. B* **641**, 37.

Vanyushin, B. F., Tkacheva, S. G., and Belozersky, A. N. (1970). *Nature (London)* **225**, 948.

Vulliamy, T., Rattray, S., and Mirsky, R. (1981). *Nature (London)* **291**, 418.

Warner, A. E., Guthrie, S. C., and Gilula, N. B. (1984). *Nature (London)* **311**, 127.

Webb, M., and Woodhams, P. (1984). *J. Neuroimmunol.* **6**, 283.

Weintraub, E. (1985). *Cell* **42**, 705.

Wilcox, M., Mitchison, C. J., and Smith, R. J. (1973). *J. Cell Sci.* **12**, 707.

Wilcox, M., Brower, D. L., and Smith, R. J. (1981). *Cell* **25**, 159.

Wolpert, L. (1971). *Curr. Top. Dev. Biol.* **6**, 183–224.

Young, L., and Dowling, J. (1984). *Proc. Natl. Acad. Sci. U.S.A.* **81**, 6255.

Zipser, B., and McKay, R. (1981). *Nature (London)* **289**, 549.

CHAPTER 12

AXON–TARGET CELL INTERACTIONS IN THE DEVELOPING AUDITORY SYSTEM

Thomas N. Parks, Hunter Jackson, and John W. Conlee
DEPARTMENT OF ANATOMY
THE UNIVERSITY OF UTAH
SCHOOL OF MEDICINE
SALT LAKE CITY, UTAH 84132

I. Introduction

A principal goal of developmental neurobiology is to understand how epigenetic functional interactions among immature cells contribute to the assembly of a mature nervous system. Among the most important interactions are those between axons and their target cells, for it is the distinct features of these two elements that combine to produce the highly stereotyped pattern of connections found at each place in the nervous system. These features include the branching or divergence of individual afferent neurons, the spatial organization of axonal projections onto a group of target neurons, the structure and location of afferent synaptic contacts, the form and size of the receptive surface of the target cell, and the matching of a neurotransmitter with compatible postsynaptic receptors (Grinnell, 1977). Establishment of the appropriate variants of each of these factors during development is essential, as any alteration could be expected to lead to changes in neural function. The faithfulness with which the fundamental characteristics of relations between axons and target cells are reproduced across animals at the same locus underscores their fundamental importance.

A number of significant developmental interactions between axons and their targets have been identified. For example, the role of afferent axons in shaping dendritic trees has been most clearly revealed for the Purkinje and basket/stellate neurons of the cerebellum (Rakic, 1975; Berry *et al.*, 1980), and the influence of target cells on neuronal survival has been studied most extensively in the chick spinal cord (Oppenheim, 1981). Developmental regulation of the number of afferent axons converging on a target neuron has been most frequently examined in the peripheral nervous system (Purves and Lichtman,

309

1980; Bennett, 1983), and the contributions of sensory experience (manifested as a change in the activity of afferent pathways) to brain development have largely been revealed by work on the visual system of the cat (Sherman and Spear, 1982). As these examples reveal, the various types of interaction have generally each been studied in isolation using a particular favorable preparation. In the hope of obtaining a more comprehensive view of these processes and their interrelationships, we have been examining a variety of interactions between axons and their targets during the early life of certain auditory neurons in the avian brain stem.

The striking structural and functional specializations of brain stem auditory neurons (e.g., Lorente de Nó, 1981; Rouiller and Ryugo, 1984) make them attractive subjects for studies of cellular interactions during development of the central nervous system. The developing brain stem auditory nuclei of the chick embryo, in particular, are remarkable for the relative simplicity of their organization and their accessibility (*in ovo, in vivo,* and *in vitro*) to the experimenter. In a study on the influence of afferent synaptic connections on brain development, Levi-Montalcini (1949) introduced this preparation, the systematic investigation of which was initiated by Rubel (e.g., 1978). In this review, we focus on the results of experimental studies involving the avian auditory system but include summaries of recent work on its normal organization and development. General treatments of development in the auditory system can be found in Romand (1983a) and Rubel *et al.* (1984).

II. Organization of the Avian Brain Stem Auditory System

Hair cells in the uncoiled avian basilar papilla (cochlea) are innervated by the peripheral processes of bipolar cochlear ganglion cells. The centrally directed processes of these neurons form the the cochlear nerve (Fig. 1 and 2). As shown in Fig. 2, each of these axons branches soon after entering the brain; the lateral branch forms several types of ending in nucleus angularis (NA) and the medial branch passes to nucleus magnocellularis (NM). On the basis of several criteria, NA has been considered to comprise avian homologues of the mammalian posteroventral and dorsal cochlear nuclei (Boord, 1969; Sachs *et al.,* 1980); the neurons of NM are considered to be homologues of the "bushy cells" of the mammalian anteroventral cochlear nucleus (Boord, 1969; Jhaveri and Morest, 1982a). In NM, each cochlear nerve axon forms a characteristic large calyx-like ending on the cell body of a single NM neuron (Parks and Rubel, 1978; Jhaveri and Morest, 1982a) (see Fig. 3A,B). About two-thirds of the surface of each NM neuron is occupied

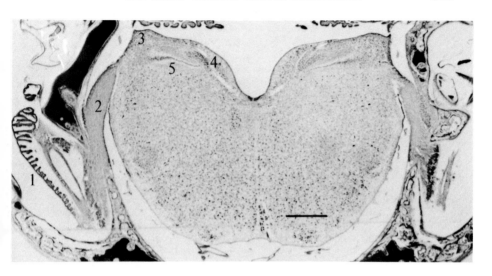

FIG. 1. Photomicrograph of a thionin-stained section through the head of a 15-day-old embryo illustrating the major central and peripheral elements of the avian auditory system discussed in this review. 1, The uncoiled basilar papilla, or cochlea; 2, cochlear nerve, arising from acoustic ganglion cells; 3, nucleus angularis; 4, nucleus magnocellularis; 5, nucleus laminaris. Bar, 0.5 mm.

by primary afferent endings (Parks, 1981a) and each neuron is innervated, on average, by just two cochlear nerve axons (Jackson and Parks, 1982).

An excitatory amino acid which acts at kainate-preferring postsynaptic receptors in NM is the probable neurotransmitter at this synapse (Nemeth *et al.*, 1983, 1985; Jackson *et al.*, 1985).[1] Stimulation of the cochlear nerve evokes an excitatory postsynaptic potential (EPSP) in NM which follows the stimulation in a one-to-one fashion; this "fast" EPSP typically has two components and is remarkable for its short latency, rapid rise time, and short duration (Hackett *et al.*, 1982). (see also Fig. 5D,E).

Nucleus magnocellularis contains a relatively homogeneous population of large spherical neurons, about 40 percent of which, in mature animals, have a single rudimentary dendrite (Conlee and Parks, 1983;

---

[1] Recently, we have isolated and characterized several compounds in the venoms of certain domestic spiders that act as extremely potent inhibitors (some, apparently irreversible) of synaptic transmission in NM (Jackson *et al.*, 1986). These compounds appear to provide useful new tools for studying the organization and development of synaptic receptors.

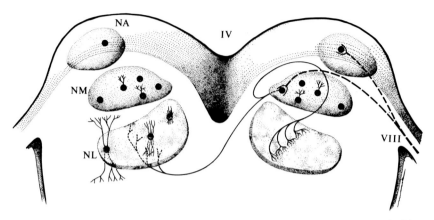

Fig. 2. Schematic representation of brain stem auditory system anatomy. Cochlear nerve (cranial nerve VIII) fibers enter the dorsolateral aspect of the brain stem, sending one branch to innervate cells of nucleus angularis (NA) and another to nucleus magnocellularis (MN). Some of the cells in NM are shown with one primary dendrite to illustrate the 40 percent of cells with such processes in adult birds. NM neurons send a bilateral projection to nucleus laminaris (NL). Axons leave NM and then bifurcate. One branch crosses the midline to innervate the ventral dendrites of contralateral NL neurons. The ipsilateral branch turns back laterally and then drops ventrally, often passing through NM, to innervate the dorsal dendrites of ipsilateral NL neurons. Terminal fields of individual NM axons in both the ipsi- and contralateral NL are oriented perpendicularly to the axis of tonotopic organization, which extends from posterolateral (low frequency) to anteromedial (high frequency). This axis also defines the gradient in NL dendritic length and number: cells possess shorter, but increasingly numerous primary dendrites as one moves from posterolateral to anteromedial.

see Fig. 3A). As shown in Fig. 2, the axon of each NM neuron forms two branches, one of which passes over and through NM to end in a fanlike spray of boutonal endings on the dorsal dendrites of neurons in the ipsilateral nucleus laminaris (NL) (Jhaveri and Morest, 1982a); NL is the avian homologue of the medial superior olivary nucleus of mammals (Boord, 1969). The second branch of each NM axon passes to the opposite side of the brain in the crossed dorsal tract (XDCT) and forms sprays of small boutonal endings on the ventral dendrites of NL neurons (Young and Rubel, 1983) (see Fig. 3D). As indicated in Fig. 2 and 3C, the neurons of NL have distinct dorsal and ventral dendritic fields, the length and complexity of which vary along the anterior-to-posterior axis of the nucleus in mature animals (Smith and Rubel, 1979). About 40 percent of the cell body surface and 65% of the dendritic surface of NL neurons are occupied by axon terminals from NM (Parks et al., 1983). Stimulation of these afferents evokes in NL neu-

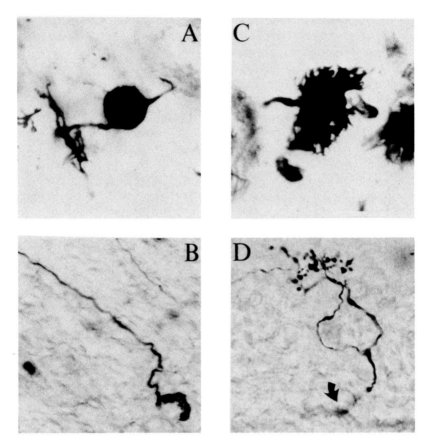

FIG. 3. Photomicrographs of NM and NL cells and their afferents. (**A**) Golgi-stained NM cell illustrating the single primary dendrite present on some mature NM neurons. The axon can be seen on the right side of the cell. (**B**) HRP-labeled cochlear nerve axon coursing through NM to end in a characteristically massive calyx-like terminal on a single NM target neuron. (**C**) Golgi-stained NL neuron illustrating the bipolar orientation of dendrites. The numerous short primary dendrites are characteristic of the anteromedial portion of NL. The axon can be seen on the left side of the cell. (**D**) A portion of the terminal arbor of a single NM axon in the ventral neuropil of the contralateral NL. This illustrates a single terminal floret arising from the parent axon (arrow) running across the bottom of the figure. Several florets, ending in small boutonal terminals, are issued at regular intervals from the parent axon as it passes from medial to lateral beneath NL. Magnifications: **A**, ×520; **B**, ×375; **C**, ×490; **D**, ×680.

rons a graded EPSP, reflecting convergence of many NM axons onto individual NL targets (Hackett *et al.*, 1982, unpublished observations).

There is an orderly representation of sound frequency in the cochlea and highly ordered topographic projections of the cochlea onto NM and NA and of NM onto NL (Parks and Rubel, 1975). These arrangements produce orderly representations of frequency (i.e., "tonotopic" maps), such that in NM and NL the highest frequencies (up to about 4,000 Hz) are represented in the most anteromedial parts of both nuclei (Rubel and Parks, 1975). Evidence is accumulating that many neurons in NA are specialized for intensity and spectral analyses of complex sounds, whereas NM and NL are adapted for analysis of the interaural time differences used to localize low-frequency sounds (Parks and Rubel, 1975; Sachs *et al.*, 1980; Konishi, 1986).

III. Early Development

The development of the auditory system in the chick embryo can be considered to begin about midway through the third day of embryonic life (E3) with formation of the otic vesicle (or otocyst) and the beginning of proliferation of the brain stem auditory nuclei (see Table I). The otocyst is the precursor of the membranous labyrinth, including the cochlea and its associated ganglion cells. At 50–60 hours of incubation, the otocyst is a prominent surface feature of the embryo and can be surgically removed to prevent formation of the inner ear and acoustico-vestibular nerve.

A. Proliferation and Migration

Most of the proliferation of NM, NL, and NA neurons in the rhombic lip germinal zone occur, respectively, during E3, E4, and E5 (Rubel *et al.*, 1976). By E4.5–5, cochlear nerve axons have entered the brain (Knowlton, 1967) and, on E5–6, the rudiments of the middle and external ears are recognizable (Jaskoll and Maderson, 1978) and accessible to experimental manipulation. Little is known about the development of NA, NM, and NL between their proliferation and the time when they can be distinguished in normal histological material (about E8–9). For lack of a reliable way to label and identify individual auditory neurons between E4 and E7, it has been impractical to study the primary migration of the neuroblasts away from the rhombic lip, the earliest stages of process growth, or early relationships between cochlear nerve fibers and central auditory neurons. Techniques that rely on retrograde transport of horseradish peroxidase, however, have shown promise for study of neuronal migration in the cochlear nucleus

TABLE I

CHRONOLOGY OF DEVELOPMENT IN THE AVIAN AUDITORY SYSTEM

Age	Event
E3	Proliferation of nucleus magnocellularis (NM); otocyst forms; otocyst removed surgically
E4	Proliferation of nucleus laminaris (NL); cochlear nerve (CN) fibers enter brain
E5	Proliferation of nucleus angularis (NA); rudiments of outer and middle ear are present
E6–9	Migration of auditory neurons; growth of anomalous NM–NM sprout after removal of otocyst; innervation of cochlea
E10	Preterminal branches of CN axons forming
E10–11	Synaptogenesis in NM and cochlea begins
E11	Acoustic stimuli evoke cochlear microphonic with a CN component; CN stimulation evokes unit responses in NM; direct stimulation of NM evokes unit responses in NL; atrophic effects of otocyst removal appear in NA and NM
E11–12	Myelination of axons in CN begins
E14	First behavioral responses to pure-tone stimuli
E14–17	Loss of preterminal branches of CN axons; loss of transient dendrites in NM; elimination of CN synapses in NM; loss of transient dendrites in NL; reorganization of dendrites in NL begins
E16	Ingrowth of non-CN afferents to NM
E17	Growth of permanent dendrites in NM begins; CN endings on NM form calyces; NL dendrites essentially mature except for spatial gradient
E18–19	Embryo breaks into airspace of egg and begins vocalization; deprivation procedures with earplugs begin
E21	Hatching
P4	Full complement of dendritic cells in NM; acoustic perception rapidly improving
P10	NM dendrites reach their mature size; acoustic deprivation from E18 produces significant retardation of dendritic growth
P25	Mature gradients of dendritic length and number are present in NL; acoustic deprivation from E19 affects dendrites in NL; tonotopic organization of the cochlea is mature
P60	Acoustic deprivation from E18 produces significant deficits in the sizes of cell bodies and dendrites in NM; function of the middle ear becomes mature.

of the opossum (Willard *et al.*, 1983) and may be useful for similar studies in the chick.

The common observation that neurons destined to be linked synaptically pass near each other during migration (e.g., Langman *et al.*, 1971) has led to the hypothesis that neuronal interactions during the migratory phase of development help determine the pattern of synaptic connections (Angevine and Sidman, 1961). As suggested by Rubel *et al.* (1976), this hypothesis might be tested directly in the chick auditory system, since proliferation of NL follows that of NM by approximately 1 day and since NL migrates ventromedially through NM. Young and Rubel (1986) make the interesting observation that growth cones from the ipsilateral branches of NM axons are in contact with the presumed precursors of NL neurons in the ventricular zone on E6; these authors go on to suggest that, as NL neurons migrate, they tow the NM axons along behind them.

Although cochlear nerve axons are in the vicinity when proliferation of NA, NL, and, perhaps, NM is still underway, these axons do not seem to have an important influence on neuronal proliferation or migration. Parks (1979) removed the otocyst unilaterally from chick embryos on E3 and compared development of NM and NA on the deafferented and normal sides of the brain from E9 through posthatching day 28 (P28). The number of neurons in NA and NM on the deafferented and normally innervated sides of the brain did not differ significantly at E9 or E11 (Fig. 4), several days after cessation of cell division in these nuclei. The large differences in neuronal number between the deafferented and normally innervated NA and NM which appear between E11 and E13 (discussed below) are the result of induced cell death rather than an effect of deafferentation on neuronal proliferation.

Parks (1979) also found that, in the absence of cochlear nerve afferents, a group of NA neurons undergoes (between E9 and E17) an abnormal "secondary" migration to an ectopic position in the brain stem. The deafferented NA neurons successfully migrate to their normal position in the brain stem by E9 but then move ventrally to an ectopic position near the descending vestibular nucleus. It is known that some neurons normally undergo secondary migrations several days after their final mitosis (e.g., Levi-Montalcini, 1949), and it is possible that many neurons retain a capacity for secondary migration that is denied expression during normal development by a stabilizing influence of afferent connections.

B. GROWTH OF PROCESSES

By E6–7, the undifferentiated sensory epithelium of the cochlea is penetrated by the peripheral processes of cochlear ganglion cells (Knowlton, 1967); by E9, the cochlear hair cells are acquiring their characteristic form and are in close apposition to nerve cell processes (Hirokawa, 1978).

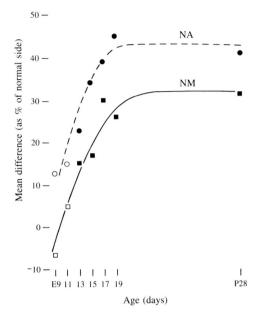

Fɪɢ. 4. The effect of unilateral removal of the otocyst on the number of neurons in NA and NM between embryonic day 9 (E9) and 28 days after hatching (P28). Each symbol represents the mean of the percentage differences obtained by comparing neuronal number on the deafferented and normal sides of the brain in each of three animals at that age. Statistically significant decreases in neuronal number in both nuclei (● and ■) are evident only after the time (E11) when the cochlear nerve normally forms functional synapses there.

By E6, the contralateral branches of some NM axons have reached the opposite side of the brain, and the ipsilateral axonal branch ramifies in the dorsolateral area of the medulla from which newly born NL neurons are migrating ventrally and medially (Young and Rubel, 1986; unpublished observations). Unilateral removal of the otocyst on E3 results in the formation, by E7, of anomalous functional collaterals from the normally innervated NM to the NM deprived of cochlear nerve afferents (Jackson and Parks, 1987); the nature and significance of this projection is discussed in Section IV,C below.

At E7–8, NM neurons, as seen in Golgi preparations, have multipolar perikarya from which extend several long branched processes. These processes, whose length may be up to four times the diameter of the cell body, often bear on their distal tips a nodular swelling that is sometimes found at the ventricular surface (Jhaveri and Morest, 1982b). The axons of NM neurons, as described above, have already

formed ipsilateral and contralateral branches at this stage; these axonal arbors undergo some structural modifications between E8 and
E14 (Young and Rubel, 1986; unpublished observations). At E8–9, NM
neurons have more complex processes than at E7 and short spines
studding the cell body; each neuron may span up to one-third of the
mediolateral extent of NM. The distal tips of the long somatic processes (or dendrites) frequently end in dendritic growth cones complete
with filopodia. The appearance of increased numbers of progressively
more complex processes characterizes the development of NM from E7
through E11 (Jhaveri and Morest, 1982a; Parks and Jackson, 1984;
unpublished observations).

Smith (1981) has studied the early development of NL neurons in
Golgi preparations and, for the period between E8 and E10, found that
these cells have about 6–8 primary dendrites distributed in characteristic bipolar fashion on the dorsal and ventral sides of the neuron.
These dendrites reveal their immaturity by the presence of filopodia
along the dendritic shaft and occasional growth cones with filopodia at
the distal tips. No spatial gradient in the number of primary dendrites
or the total length of dendrites is apparent within NL during this
period (Smith, 1981).

IV. Formation and Refinement of Synaptic Connections

A. SYNAPTOGENESIS AND ONSET OF FUNCTION

Data from two studies indicate that synapses between cochlear
nerve axons and NM targets are first formed at about E10–11. Jackson
et al. (1982) were first able to elicit unit responses in NM by direct
electrical stimulation of the cochlear nerve at E11. Some time passes,
of course, between the initial formation of contacts and their functional maturation to the point where they are capable of evoking action potentials in the target cells (cf. Dennis and Ort, 1976). Although
that interval cannot be precisely specified, the morphological description provided by Jhaveri and Morest (1982b) suggests that synapses
are probably not formed much before E10. Those authors reported
that, in Golgi-stained material from E10 animals, many cochlear
nerve axons appear to end in growth cones, although some end in
swellings without filopodia. Jhaveri and Morest suggest that these
swellings may be retraction clubs formed as axons readjust their direction of growth. Alternatively, they may represent early synaptic terminals. In any case, these findings indicate that most axons at E10 are
probably still growing to their targets or just beginning to form terminal specializations. Unfortunately, electron microscopic data that
might further delineate the initial stages of synaptogenesis in NM are

not yet available. Although there are no morphological data on synapse formation in NL, Jackson *et al.* (1982) have found that direct electrical stimulation of NM can evoke unit responses in NL no earlier than E11, suggesting that NM and NL become responsive to synaptically mediated stimuli at about the same time in development.

Given that some synapses in NM are mature enough by E11 to transmit impulse activity from the cochlear nerve, is such impulse activity normally present at this stage? Hirokawa (1978), in an electron microscopic study of the developing chick cochlea, found the first recognizable synapses between hair cells and ganglion cells at E11. In agreement with this morphological result, Saunders *et al.* (1973) recorded a cochlear microphonic with a cochlear nerve component at E11 whereas only the receptor potential could be evoked at E10. These data indicate a capacity for the system to respond to acoustic stimuli at E11. The intensity of stimuli required to evoke a cochlear microphonic (125–135 dB sound pressure level; Saunders *et al.*, 1973), however, suggests that environmental acoustic stimuli do not normally drive the system until later stages of development (see below). Spontaneous activity may be present in these young embryos, although in light of data from mammals (Romand, 1983b) one might expect the level of spontaneous activity to be quite low during early stages of functional development.

B. REFINEMENT OF SYNAPTIC CONNECTIONS

Increases in the responsiveness of NM to direct cochlear nerve stimulation (Jackson *et al.*, 1982) and the proliferation of preterminal branches on cochlear nerve axons (Jhaveri and Morest, 1982b) indicate that synapse formation continues in NM until about E13. At that point begins a period when these connections are apparently "refined." NM neurons have a number of somatic processes at E13, and virtually all cochlear nerve synapses are located on those processes (Jhaveri and Morest, 1982c). Individual NM neurons receive innervation from an average of four cochlear nerve afferents at this stage. Over the next 4 days, afferent convergence is cut in half; by E17–18 an average of only 2.2 axons innervate individual target cells in NM (Jackson and Parks, 1982) (see Fig. 5D,E). This convergence ratio remains the same at least through 4 days after hatching (see also Hackett *et al.*, 1982).

The reduction in functional convergence is accompanied by a parallel simplification of both pre- and postsynaptic structures. NM neurons begin to lose their dendrites at about E13–14 (Jhaveri and Morest, 1982b; Parks and Jackson, 1984); by E16–17, the cells are largely adendritic (Figs. 6 and 7). As NM cells are losing their dendrites,

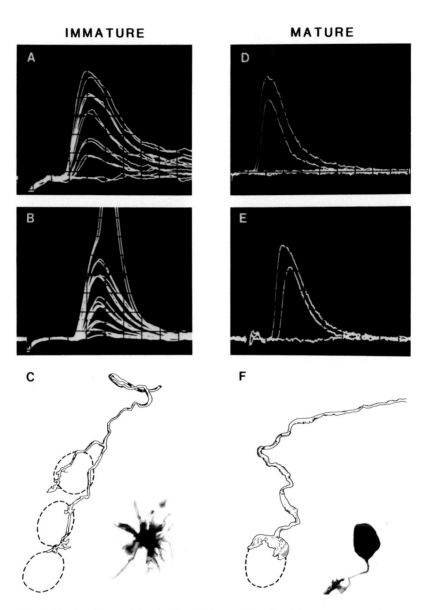

IMMATURE   MATURE

FIG. 5. The developmental reduction in the number of cochlear nerve axons innervating individual NM neurons and the corresponding reduction in terminal branching of cochlear nerve axons. (**A**) Intracellular record from an NM neuron at E13. Gradual variation of cochlear nerve stimulus intensity revealed five components in this excitatory postsynaptic potential (EPSP). Several traces are superimposed to show the unitary nature and reproducibility of the components upon repeated sweeps of stimulus inten-

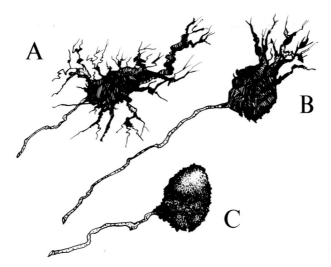

FIG. 6. Stylized drawings of HRP-stained NM neurons depicting the temporal and spatial gradients of structural transformation that occur during normal development. At E11–12, neurons throughout NM have numerous dendrites (**A**); these are rapidly lost between E13 and E16 along an anteromedialpto-posterolateral gradient within NM. At E14, neurons about midway along this axis have only a few dendrites (**B**), while more anterior cells (**C**) have lost all of their dendrites. By E16–E17, all neurons in NM resemble that illustrated in (**C**); about 40 percent of the neurons in NM grow "permanent" dendrites between E17 and P4 (cf. Fig. 3A). Removal of the otocyst on E3 has no effect on these structural transformations in NM.

---

sity. (**B**) Record from another NM neuron at E13 taken from a different preparation than that shown in A. At least six EPSP components and an action potential can be seen in this record. Traces generated during repeated sweeps of stimulus intensity are shown. (**C**) Camera lucida drawing of an HRP-labeled cochlear nerve axon at E14. Note the terminal branching of the axon within NM. The inset photomicrograph shows an E13 NM neuron stained by the Golgi–Cox method; note the numerous long somatic processes (cf. Fig. 6A). (**D**) Record from an NM neuron at P4, showing only two EPSP components. (**E**) Record from an NM neuron at E18, again indicating innervation by only two cochlear nerve axons. (**F**) Drawing of a labeled cochlear nerve fiber at P4. No terminal branches were seen at this age; individual axons terminated in the characteristic calyx-like ending apposed to a single NM neuron (cf. Fig. 3B). The inset photomicrograph shows an E17 NM neuron stained by the Golgi–Rio Hortega method; note the complete absence of somatic processes (cf. Fig. 6C). Calibration for electrophysiological records, 2 mV, 2 msec in **A**; 5 mV, 2 msec in **B**, **D**, and **E**. (From Jackson and Parks, 1982.)

321

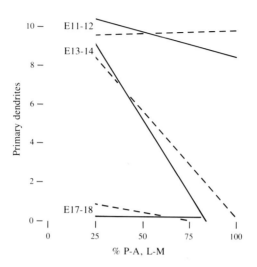

Fɪɢ. 7. Six multiple linear regression lines showing the spatiotemporal gradient of dendritic loss in NM on each side of the brain in animals from which the right otocyst had been removed on E3. Neurons were stained (1) with HRP by backfilling them through their axons after an HRP injection at the midline, or (2) by a Golgi method. From camera lucida drawings, the number of primary dendrites was counted on each of 114 NM neurons from 3 animals at E11–12, 134 neurons from 3 animals at E13–14, and 168 neurons from 4 animals at E17–18. The percentile coordinate position of each neuron along the posterior-to-anterior (%P-A) and lateral-to-medial (%L-M) axes of NM was also determined. Multiple linear regressions of dendritic number on position were then performed separately for the normal (solid lines) and deafferented (dashed lines) sides of the brain at each age to yield the curves shown here; increasing values of %P-A, L-M on the abscissa represent increasingly anteromedial positions with NM. There is no reliable relationship between the number of dendrites and position at E11–12 on either side of the brain; regardless of their position, cells have an average of 9–10 dendrites. At E13–14, there is a very significant relationship between position and the number of dendrites on both sides of the brain, with the most anteromedial NM neurons having lost most or all of their dendrites. By E17–E18, cells have lost all of their transient dendrites and there is no longer a relationship between position and dendritic number on either side of the brain. Afferents from the cochlear nerve are not required for the structural transformation of NM cells to proceed normally.

cochlear nerve axons are losing their preterminal branches (cf. Fig. 5C and 5F). Axonal branching peaks at about E13 and then declines until, by E17, these fibers are unbranched and end on a single NM target cell. During this period there seems to be little death in the acoustic ganglion cell population (Ard and Morest, 1984; unpublished observations), indicating that the elimination of functional synapses in NM can be explained almost entirely by the loss of preterminal branches from afferent axons (Jackson and Parks, 1982).

Establishment of synaptic connections in this system thus seems to involve two distinct phases. The first is one of proliferation, during which both pre- and postsynaptic elements form an apparent overabundance of processes. Following the formation of synaptic contacts, the second phase, one of consolidation, begins. This consolidation involves the elimination of some synapses and concomitant simplification of pre- and postsynaptic structures. It seems likely that this general sequence of events is repeated at most sites within the developing CNS. For example, several types of neurons have been shown to go through a phase of exuberant growth of dendrites followed by a period of remodeling (Ramón y Cajal, 1929; Morest, 1969a,b; Landmesser and Pilar, 1972; Smith, 1981). Certainly the elimination of functional synapses has been well documented at a number of sites in the peripheral nervous system (Purves and Lichtman, 1980) and evidence is accumulating for the general occurrence of this phenomenon in the CNS (e.g., Crepel et al., 1976; Ivy et al., 1979; Shatz and Kirkwood, 1984).

If it is accepted that this is a typical developmental sequence leading to the establishment of mature patterns of innervation, it becomes a matter of some importance to determine the factors that regulate the process. A striking feature of afferent—target cell development in NM is the parallel nature of the changes in the two elements. This correlation between structural changes in afferent axons and their target cells suggests that these transformations may be controlled and synchronized by either the pre- or postsynaptic element. Descriptive studies of normal development have tended to implicate the afferent axon as the element that regulates several aspects of development during this period of synapse formation and remodeling. For example, many developmental processes in the avian auditory system proceed over time along an anterior-to-posterior gradient. These include proliferation and elimination of dendrites in NL (Smith, 1981) and NM (Parks and Jackson, 1984), cell death in NL and increases in the volume of NM and NL (Rubel et al., 1976), and the onset of postsynaptic responsiveness to afferent stimulation (Jackson et al., 1982). As discussed earlier, the cochlea projects topographically to NM, and this topography is maintained in NM's projection onto NL. In terms of this projection, the anterior-to-posterior gradients in NM and NL correspond to a basal-to-apical gradient in the cochlea, and it is along this basal-to-apical gradient that the cochlea matures (Rubel et al., 1984). In light of this, and the fact that the developmental gradients in the brain stem auditory nuclei become evident just as cochlear nerve fibers form synapses with NM targets, it is tempting to suggest that the gradients in NM and NL, which mirror the gradient in the cochlea, are imposed on the brain by the developing cochlear nerve afferents. This, of course, implies that the afferents, through contact and/or activity, provide signals that regulate a variety of developmental changes in their targets. This sort of afferent control of target-cell development has been suggested, for example, by Rubel (1978) to explain the morphogenetic gradient in NM and by Smith (1981) to explain the development of gradients of dendritic length in NL.

There is also reason to suspect, however, that the target cells might regulate various aspects of their afferent input. For example, it is easy to imagine that the growth of NM dendrites prior to synapse formation serves to increase both the chance of eventual contact between afferents and targets (cf. Landmesser and Pilar, 1972) and the number of afferents that can form synapses with a given cell. Subsequent loss of these processes might force converging afferents to compete for a diminishing target, resulting in the elimination of some afferents and the ultimate establishment of the mature convergence ratio. In this way the growth of dendrites could regulate initial synapse formation, and their loss could determine the final level of innervation. Certainly the temporal correlation between the loss of NM dendrites and the elimination of preterminal axon branches and functional synapses supports this scheme (Jackson and Parks, 1982). Further support is offered by the work of Purves and colleagues (e.g, Purves and Hume, 1981; Forehand and Purves, 1984; Purves, 1983), who have extensively studied the relationship between dendrites and converging afferents in autonomic ganglia and have found a strong relationship between dendritic number and the number of axons innervating a given target cell.

C. Experimental Manipulations of Development

One seemingly straightforward approach to evaluating the roles of afferents and targets in regulating development in this system is to prevent formation of the cochlear nerve axons. This can be achieved by early (E3) removal of the otocyst. As discussed above, the otocyst gives rise to the inner ear, including the cochlear ganglion cells, so that removal of the otocyst prevents formation of the cochlear nerve. This manipulation severely affects the development of NM and NA, as documented by Levi-Montalcini (1949) and Parks (1979). As discussed in Section III,A, a group of NA neurons migrates to an ectopic position between E9 and E17. The normal increase in the volume of NM and NA is progressively retarded, as is the size of neurons in these nuclei. Perhaps most dramatic is the loss of 40 percent of NA neurons and 30 percent of NM neurons (Fig. 4). Unilateral removal of the otocyst also affects the growth of NL dendrites (Parks, 1981a); the affected dendrites are much shorter than normal but show an essentially normal spatial gradient of dendritic length. The results of this experiment are discussed in greater detail in Section V,B,2 below. Finally, we have recently examined the development of the transient NM dendrites following removal of the otocyst (Parks and Jackson, 1984). In the absence of cochlear nerve axons, these dendrites seem to be issued and lost in an apparently normal fashion (Fig. 7).

The results of these experiments leave little doubt that cochlear nerve afferents exert a major developmental influence on their targets. However, the nature of that influence remains difficult to specify for two reasons. First, although certain characteristics of brain stem auditory neurons were severely affected by otocyst removal, others were unchanged. It seems as though cochlear nerve afferents provide their targets with general trophic support but do not induce or regulate the expression of distinctive phenotypic characteristics during this period. Thus, lack of normal cochlear nerve afferents may result in target cell atrophy, including retarded growth and even cell death, but specific features of these neurons, such as the growth and loss of transient dendrites in NM and the gradient of dendritic length across NL seem unaffected.

It would appear, then, that certain aspects of morphogenesis in these nuclei are under the control of the neurons themselves. However-er, the recent finding that otocyst removal induces the formation of an anomalous NM–NM collateral projection renders this conclusion premature (Jackson and Parks, 1987). Neurons in NM are not deafferented by otocyst removal as was previously thought. Rather, they are innervated by a novel class of afferents. This information requires that previous conclusions drawn from the experiments using otocyst removal be reevaluated. Certainly it cannot now be assumed that most cells in NM do not require innervation for their survival; the anomalous NM–NM projection may well be what saves the 70 percent of NM neurons that survive after otocyst removal (Parks and Jackson, 1984). These sprouted collaterals may also serve to regulate such factors as dendritic growth and loss, although if they do, it would indicate that such regulation is not a property specific to the normal afferents contacting these cells. Although this anomalous projection complicates the interpretation of the otocyst removal experiments, it is in itself an interesting case of collateral sprouting and offers unusual opportunities for experimental analysis of developmental interactions between afferents and their targets. This projection and its usefulness are discussed in the following section.

D. THE INDUCED NM–NM PROJECTION

Our preliminary analysis has been directed toward answering several basic questions about the anomalous NM–NM projection.

1. Is this a *de novo* projection or does it represent the maintenance and expansion of a small, normally occurring pathway? We have never seen any indication of such a projection in normal animals from E11 through several days after hatching, so if such a projection is normally occurring, it must be a transient one that disappears by the time of cochlear nerve synapse formation in NM. This, too, seems unlikely, since we have thus far examined labeled NM axons in embryos as young as E7 and have found no collaterals from the contralateral NM on the unoperated side of the brain. This certainly does not preclude the possibility that in normal animals an occasional stray axon from NM may enter the contralateral NM early in development (Young and Rubel, 1986). Our analysis, however, suggests that there is normally no consistent growth of NM axons into the contralateral NM and, therefore, that the collateral projection found following otocyst removal represents an essentially novel innervation of those NM targets.

If this projection does in fact arise *de novo,* as now appears likely, it is quite unusual when compared to other examples of sprouted connections, which typically represent only a quantitative increase and/or rearrangement of normal connections with a given target (Cotman *et al.,* 1981). This may be related to the fact that, whereas most other cases of collateral sprouting are induced by removal of normal afferents, removal of the otocyst prevents those afferents from ever forming. Perhaps neurons are more receptive to different classes of afferents at the time of initial innervation than they are after one group of afferents has established itself and then been experimentally removed.

2. At what stage does the anomalous projection arise? As shown in Figs. 8 and 9, the anomalous NM–NM projection is formed by vertically directed offshoots from the main branches of NM axons passing laterally along the ventral surface of the contralateral NL. Such collaterals have been seen in E7 animals, the youngest we have thus far studied (Fig. 8). The appearance of the anomalous collateral at E7 precedes by several days the normal onset of synaptogenesis between cochlear nerve axons and NM cells and the atrophic changes induced in NM by removal of the otocyst. It thus seems that the signal initiating ingrowth of the anomalous collaterals is not related to the failure of normal synapses to form in NM. It is plausible that ingrowth of the

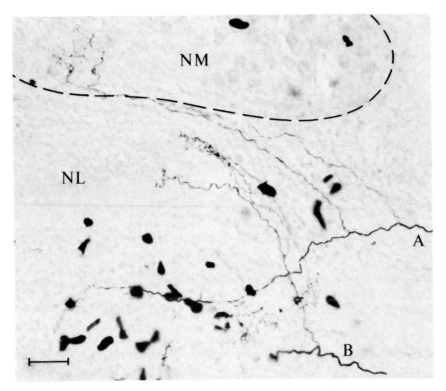

FIG. 8. Photomicrograph from an E16 animal illustrating anomalous collaterals formed after removal of the otocyst. A small HRP injection was made at the midline to label NM axons. Two axons (**A** and **B**) can be seen coursing from the midline (right) toward their normal targets in NL. The main branches of both axons pass below NL. A collateral from axon **B** ascends toward the ventral neuropil of NL and there forms a normal spray of terminal boutons. Other collaterals bend around the medial edge of NL and terminate in the dorsal neuropil. Axons from the contralateral NM typically end in the ventral neuropil of NL, so these terminals from axon **B** suggest some sprouting to the dorsal neuropil may occur after removal of the otocyst. The collaterals seen here arising from axon **A** are clearly anomalous as they ascend to end in sizable arborizations in the lateral portion of NM. Bar, 50 μm.

anomalous axons might be stimulated by a diffusable factor produced by NM target cells; in normal animals, such a factor would serve to attract cochlear nerve axons (Berg, 1984). This factor would normally be "neutralized" by cochlear nerve axons (cf. Diamond, 1982), which reach NM by E6. In the absence of these fibers, the diffusable signal would remain active and able to stimulate collateral growth in fibers from the contralateral NM.

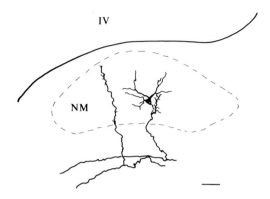

FIG. 9. Camera lucida drawing from an E7 subject from which the otocyst was removed unilaterally. HRP was injected at the midline (to the left of the figure) to label NM axons. Two axons are seen here coursing toward their normal targets in posterolateral NL (not seen in this section). As they pass NM, each axon gives off an anomalous collateral which travels dorsally to penetrate deep into NM. These collaterals are immature, showing no gross terminal specializations or arborizations. An NM neuron backfilled by the HRP injection is also shown. IV, fourth ventricle. Bar, 40 μm.

As attractive as this hypothesis is, it raises some troublesome questions regarding the specificity of the signal. For example, the signal would have to be nonspecific enough that it could attract axons from both the cochlear nerve and NM. On the other hand, if the vestibular nuclei, which have also been deprived of primary afferents by removal of the otocyst, also secrete a "sprouting factor," it appears to be sufficiently specific that it does not attract NM axons since none are apparent in the vestibular nuclei. Perhaps most difficult is that NM neurons could produce a signal that attracts collaterals from axons arising in the contralateral NM without inducing significant growth of recurrent collaterals from their own axons.

3. Is intact cochlear nerve innervation necessary to support the growth or maintenance of anomalous collaterals by NM neurons? This question relates to the possible role of stimulation in sprouting (e.g., Diamond, 1982). To answer this question, we removed the otocyst bilaterally and examined several subjects as old as E17. In each case, we found reciprocal NM–NM projections linking the two nuclei (Jackson and Parks, 1987). This indicates that acoustically evoked activity is not required for the survival of the majority of NM neurons or their normal and anomalous axonal projections.

4. Are the anomalous terminals functional? Electrophysiological experiments indicate that the synapses are functional and capable of driving their NM target cells. Direct electrical stimulation of either

the intact contralateral cochlear nerve or contralateral NM evokes unit responses in NM on the side of otocyst removal. A negative field potential, which apparently represents postsynaptic currents in NM on the operated side, can also be recorded in response to stimulation of the contralateral nerve or NM. These findings suggest that NM neurons innervated by these collaterals are driven by acoustic stimuli *in vivo*. Interestingly, the anomalous terminals formed subsequent to bilateral otocyst removal also seem to be capable of function, as indicated by unit responses recorded in NM after direct electrical stimulation of the contralateral NM (unpublished observations). Whether these synapses are ever active *in vivo* is open to question. Not only is there no peripheral input, there may also be no spontaneous activity. In birds, as in mammals (Koerber *et al.*, 1966), the spontaneous activity recorded in the cochlear nuclei seems to be generated in the periphery. After destruction of the cochlea in the chicken, no activity is recorded in NM (E. W. Rubel, personal communication). This raises the intriguing possibility that bilateral removal of the otocyst may result in a central auditory system built on a foundation of reciprocal silent synapses.

The NM–NM synapses appear to offer an unusual experimental opportunity to examine the relative importance of afferents and their targets in determining the characteristics of mature connections. As described above, the normal innervation of NM by cochlear nerve afferents is distinctly different, both structurally and functionally, from the normal innervation of NL by NM axons. These differences allow one to ask testable questions regarding the characteristics of the novel NM–NM synapses. As a simple illustration, the gross morphology of normal cochlear nerve terminals on NM neurons is quite distinctive and clearly different from that of terminals normally formed by NM axons in NL (cf. Fig. 3B and 3D). If NM neurons normally play a role in determining the morphology of those normal cochlear nerve endings, one would expect the anomalous terminals formed by NM axons on NM target cells to develop some of the characteristics of the cochlear nerve calyx-type terminals. We expect that this approach can be used to analyze a number of synaptic features and, ultimately, to provide an overview of the interactions between afferents and their targets that serve to establish the highly stereotyped characteristics of innervation at a given site.

V. Development after Hatching

In birds, the process of hatching involves an ensemble of anatomical, physiological, and behavioral changes that, in the domestic fowl, extend over the last several days of embryonic life (Oppenheim, 1973).

Midway through E18, the embryo pushes its beak into the airspace and begins pulmonary respiration. The outer ear is free of mesenchyme at this time and the auditory pathway is functional. Embryos begin vocalizing at this stage and can be influenced by sounds, notably their own vocalizations and those of other birds (Vince, 1973; Gottlieb, 1981). By E19, the embryo has broken through the eggshell with its beak to begin the laborious final phase of hatching, which is not complete until E21. Thus, from E19 onwards, the animal is exposed directly to the external acoustic environment. Because of the great changes in the embryo's exposure to sound that occur after E17, it is useful to define the posthatching period of development for the auditory system as beginning on E18.

In addition to changes in the exposure of the animal to sound, the weeks between E17 and about P60 bring gradual maturation of the sensitivity of the ear (Saunders *et al.,* 1983; Lippe and Rubel, 1983) and rapid experience-dependent improvement in the ability of the bird to discriminate the characteristic vocalizations of its species (e.g., Gottlieb, 1981) and changes in the intensity and frequency of tones (e.g., Gray and Rubel, 1981).

By E17 the numbers of neurons in NA, NM, and NL have reached levels that are maintained into maturity (see Fig. 10), although there is considerable increase in the size of NM and NA for at least 2 months after hatching (Parks, 1979; Conlee and Parks, 1981) (see Fig. 11). By E17, the cochlear nerve's innervation of NM and NM's innervation of

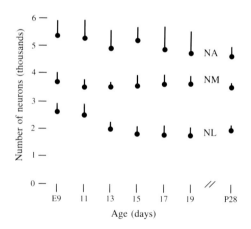

FIG. 10. Neuronal number in the cochlear nuclei and nucleus laminaris during normal development. Each symbol represents the mean and standard error of cell counts from three animals (two at P28). Only NL shows significant normal cell death.

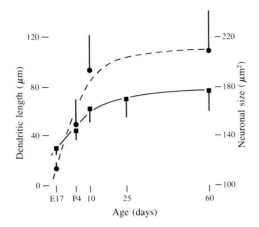

FIG. 11. Normal development of neuronal cross-sectional area (size) and total dendritic length per cell in NM. Mean neuronal size is plotted in relation to the right ordinate in μm² *while dendritic length* is plotted in relation to the left ordinate in μm; the standard error of each mean is shown. Cell size (■) increases steadily between E17–P10 and begins to plateau after P25. Dendritic length (●) grows rapidly between E17 and P10 and virtually ceases thereafter.

NL are, in most important respects, mature. Studies on the auditory nuclei after hatching have therefore focused on the development of dendrites in NM and NL and the role of afferent synaptic connections and acoustic experience in fostering the continued growth and maintenance of these neurons.

A. NUCLEUS MAGNOCELLULARIS

1. Normal Development

When studied in Golgi preparations, 14 percent of the neurons in NM are found to have a single primitive dendrite at E17. Only 8 days later (P4), nearly 40 percent of NM's neurons have dendrites. This division into approximately 40% "dendritic" and 60% "adendritic" neurons is maintained for at least 2 months after hatching (Conlee and Parks, 1983).

The debut of permanent dendrites in NM at E17 is thus quickly followed by sprouting of dendrites from all those neurons destined to have them. E17 is extraordinarily late in development for the initiation of dendrites; most, if not all, other large neurons in this region of the brain stem have well-established dendrites by this time. The probable explanation for this unusually late appearance of dendrites is that, unlike many other neurons, primitive cell processes present early in development are not maintained and transformed into permanent dendrites but are lost between E13 and E16, as described above. The late dendrogenesis in NM may be timed to coincide with the arrival at E16–17 of noncochlear afferents to NM (Parks, 1981b; Jhaveri and Morest, 1982a,b). Study of the synaptic relations of NM dendrites during development will be needed to resolve this question.

The majority of dendritic cells in NM bear a single dendrite which, during the perihatching period, has a relatively uncomplicated arborization with few branches (Jhaveri and Morest, 1982b; Conlee and Parks, 1983) (see Fig. 3A). By P10, the average total length of each cell's dendrite has risen from a mean of about 15 μm at E17 to approximately 100 μm, a value that does not change significantly through at least P60 (see Fig. 11). Dendrites from older animals generally exhibit longer primary dendritic shafts and more extensive higher-order branching than are found in younger animals. As late as P60, however, there is remarkable variation in the size and complexity of dendrites, with dendritic lengths ranging from less than 10 μm to more than 400 μm. The developmental factors controlling which NM neurons will bear dendrites and the ultimate length of each particular dendrite are poorly understood, although it is clear that acoustic stimulation is essential for the normal increase in average dendritic length between E17 and P60 (Conlee and Parks, 1983; discussed below).

2. Effects of Manipulating Afferent Connections or Activity

The severity of neuronal loss and shrinkage in NM after destruction of the inner ear at E3 (Parks, 1979), P4 (Jackson and Rubel, 1976), P14, P42, and 66 weeks after hatching (Rubel et al., 1984) has been studied. Deafferentation at the earliest four ages produces cell loss averaging about 25–30 percent and shrinkage of 15–30 percent over quite short survival periods; degeneration of some NM cells and their axons in NL is apparent within 5 days after deafferentation at P4–P10 (Jackson and Rubel, 1976; Parks and Rubel, 1978). By 66 weeks after hatching (middle age for a fowl), destruction of the cochlea produces less than 10% loss of neurons and little if any shrinkage (Rubel et al., 1984). The dependence of NM neurons on cochlear nerve afferents seems to decline gradually with age rather than changing abruptly during some "critical period" in development, such as the onset of auditory function, hatching, etc. The physiological mechanisms of this increasing autonomy from the effects of deafferentation are unknown, although it is interesting to note that inhibition of the incorporation of amino acids into deafferented NM neurons also declines with the age at which the ear is destroyed (Rubel et al., 1984).

Deafferentation may be considered to represent the most severe of a continuum of possible modifications of afferent input since it deprives the target neurons of the total range of influences produced by the presynaptic axons. Variation in the activity of intact afferents provides a more physiological and subtle influence on developing neurons and is undoubtedly the mechanism by which experience influences brain development (Globus, 1975). The electrical activity of neurons in NM and NL is almost entirely dependent upon synaptic stimulation from the cochlea via the cochlear nerve. Thus, the afferent activity impinging on NM and NL can be manipulated by changing the activity of the cochlear nerve through acoustic deprivation or selective stimulation.

A severe but reversible conductive hearing loss can be produced in chick embryos on E18 by gently injecting a liquid silicone-based hearing aid sealant into the external auditory canal. When this earplug hardens, it has been shown to produce a 40 dB attenuation across the audible range of the animal (Kerr et al., 1979). The earplugs can be removed at any time and replaced. Although the effects of these earplugs on electrical activity in the cochlear nerve have not yet been studied thoroughly, it appears that they produce a reduction in the electrical activity generated by the cochlea (Tucci and Rubel, 1985).

The effects of a unilateral earplug on the subsequent growth of NM cells have been investigated. Conlee and Parks (1981) inserted earplugs in E18 embryos and measured the cross-sectional areas of neuronal cell bodies on the deprived and nondeprived sides of the brain (and in control animals) at P4, P10, P25, and P60. Measurements were taken from three sectors along the anterior-to-posterior axis of NM to determine if the effects of earplugging were dependent upon position within the nucleus. Significant retardation in the growth of neuronal size was not evident anywhere in NM until P25; at this age, effects of deprivation were apparent only in regions of NM that represent mid- to high-frequency tones. By P60, all regions of NM showed an effect of earplugging on neuronal size, which was reduced by an average of 12 percent on the plugged side (Fig. 12).

No effect of earplugging on neuronal number in NM and no compensatory increase in the size of NM neurons on the unmanipulated side of the brain were found (Conlee and Parks, 1981). This last finding suggests that the retarded growth of affected NM neurons arises not from their failure to compete with unaffected NM neurons for synaptic space in NL but from a "pure" effect of hearing loss (cf. Hickey et al., 1977). The finding that earplugging first affects mid- to anterior portions of NM may reflect the earlier functional develop-

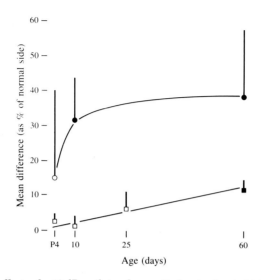

FIG. 12. The effects of a 40 dB unilateral acoustic deprivation initiated at E18 on the development of neuronal cross-sectional area (squares) and total dendritic length per cell (circles) in NM. Each symbol represents the mean and standard error of the deprived vs normal percentage differences calculated for each of 4–9 individual animals. Acoustic deprivation retards the growth of NM neurons, producing statistically significant differences (● and ■) in dendritic length by P10 and in cell body size by P60.

ment of the basal portion of the cochlea, which projects to the anterior portion of NM. The basal cochlea is the first region to begin transducing acoustic stimuli and it is fully functional well before more apical regions (Rubel *et al.*, 1984). Thus, it is probable that the *effective* period of deprivation for the mid-to-anterior region of NM was, at any given age, of longer duration than at more posterior locations.

The fact that permanent dendrites in NM do not begin to grow until about E17–E18 provided a unique opportunity to study the effects of sensory deprivation on the entire sequence of dendritic development in a vertebrate sensory neuron. Conlee and Parks (1983) used earplugs to produce unilateral acoustic deprivation in E18 chick embryos. The proportion of NM neurons with dendrites and the mean length of these dendrites at P4, P10, and P60 were compared on the deprived and nondeprived sides of brains stained by a Golgi method. Prolonged earplugging had no effect on the proportion of "dendritic" neurons in NM but did produce a highly significant 38 percent retardation in dendritic length by P60 (Fig. 12). As in the study of cell body size, there was no compensatory hypertrophy of dendrites in the unplugged NM when these were compared with dendrites from sham-operated control animals.

No regional differences within NM in the effects of earplugging on dendritic growth were found. Interestingly, the length but not the diameter of dendrites was affected by this manipulation. Parks (1981a), in studying the effects of otocyst removal on dendrites in NL at E17, also found large changes in the length of dendrites without any accompanying effects on their diameter. The length and diameter of dendrites appear to be regulated by different mechanisms. Conlee and Parks (1981, 1983) concluded that normal acoustic stimulation is necessary for growth of NM neurons in the posthatching period but that earplugging has no effect on the proportion of neurons which bear dendrites. Studies on acoustic deprivation in mammals are generally consistent with these conclusions (Brugge, 1983).

B. NUCLEUS LAMINARIS

1. Normal Development

As described above, NL neurons pass through a period of exuberant growth of dendrites between E10–E13. Many processes are then eliminated between E14 and E16; both proliferation and elimination of dendrites progress along an anteromedial-to-posterolateral gradient within the nucleus (Smith, 1981). Because the wave of dendritic proliferation begins in the anterior part of NL and lasts longer there than in more posterior regions, neurons in the anterior NL have a larger number of primary dendrites from about E10 onwards. Because dendrites in the posterior NL continue growing while the wave of dendritic elimination sweeps toward them, they have a greater total dendritic length than more anterior NL neurons by about E15 (Smith, 1981). Thus, by about E15 the foundations have already been established for the striking gradients in dendritic number and length which characterize these neurons in mature animals (Smith and Rubel, 1979). The ratio between the total dendritic lengths per neuron at the extreme caudolateral and anteromedial poles of NL rises from about 1.8 at E15 to 2.5 at E17 (Parks, 1981a) to 5 at E19 and 13 at P25 (Smith, 1981). These findings reveal that dendrites in the caudal NL (which represent low sound frequencies) grow more rapidly from about E14 through P25 than more anterior dendrites (which represent higher sound frequencies). Because low frequency sounds were present at substantially higher intensities than higher frequency sounds in the environment where his animals were kept, Smith (1981) concluded that the differential growth rates of NL dendrites were due to differences in acoustically evoked afferent synaptic stimulation.

2. Experimental Manipulations

The structural integrity of dendrites in NL is profoundly dependent upon afferent synaptic input from NM axons. Benes *et al.* (1977) deafferented the ventral dendrites of NL in P5–P10 animals by sectioning the crossed dorsal cochlear tract (XDCT) carrying afferent axons from NM. By 96 hours after this lesion, morphometric analyses of electron micrographs revealed that the deafferented dendrites had shrunk by about 80 percent, without any significant change in the size of the normally innervated dendrites on the opposite side of the same cells. Deitch and Rubel (1984) report that NL dendrites have shrunken by 16% only 2 hours after sectioning of the XDCT. These results provide a dramatic example of the power and specificity that afferent synaptic connections can exert on the structural integrity of target neurons.

The role of acoustic experience in shaping dendritic structure in NL has been studied by Smith *et al.* (1983), who inserted unilateral earplugs of the sort described above into chick embryos at E19. From hatching through P25, the animals were raised in a controlled acoustic environment where all sound frequencies audible to the animal were present at equivalent intensities. These investigators found significant differences between the lengths of deprived and nondeprived NL dendrites, supporting the contention that acoustically evoked afferent activity in NM axons regulates dendritic growth in NL.

Parks (1981a) attempted to evaluate the hypothesis that acoustically evoked input controls the rate of growth of NL dendrites and thus the development of spatial gradients of dendritic length. The otocyst was removed unilaterally and the dendrites studied in Golgi material at E17. Although the NL dendrites connected to the deafferented NM were 44% shorter than "unmanipulated" dendrites on the opposite side of the same neurons (Fig. 13B), the two sets of dendrites did not differ significantly in the strengths of their gradients of dendritic length (Fig. 13A). Because of the likelihood that sprouting of NM axons across the cell body lamina in NL occurs after destruction of the otocyst (Jackson and Parks, 1987) or cochlea (Rubel *et al.*, 1981), it has been necessary to examine dendritic gradients in NL after bilateral destruction of the otocyst. Significant gradients of dendritic length are present at E17 in the NL of animals lacking both otocysts (Parks *et al.*, 1987), demonstrating that neither cochlear nerve axons nor the acoustically evoked activity they transmit to the brain are necessary for the establishment of these gradients.

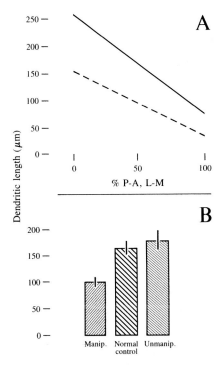

FIG. 13. The effects of unilateral removal of the right otocyst on total dendritic length per neuron in NL at E17. Camera lucida drawings of 50 Golgi-stained neurons from six animals were made. The total dendritic length of the dorsal and ventral dendritic fields of each neuron were separately measured and the neuron's percentile position along the posterior-to-anterior (%P-A) and lateral-to-medial (%L-M) axes of the nucleus determined. Removal of the right otocyst manipulates the right NM's afferents to the dorsal dendrites of the ipsilateral NL and the ventral dendrites of the contralateral NL; these are termed the "manipulated" dendrites. Dendrites on the opposite side of each neuron are thus the "unmanipulated" dendrites. (A) Two multiple regression lines show the relationship between dendritic length and position within NL for the unmanipulated (solid line) and manipulated (dashed line) dendrites; increasing values of %P-A, L-M on the abscissa represent increasingly anteromedial positions within NL. Although removal of the otocyst reduces the length of manipulated dendrites, it does not greatly affect the spatial gradient of total length along the posterolateral-to-anteromedial axis of NL. (B) Mean total dendritic lengths (and standard errors) for the manipulated (left) and unmanipulated (right) dendrites are shown along with comparable data from normal control animals (center). The lengths of unmanipulated and normal control dendrites are not reliably different, but both are significantly greater than manipulated dendrites.

The facilitatory effects of afferent stimulation on developing neurons are well documented, particularly in the auditory and visual systems (e.g., Harris, 1981; Brugge, 1983; Sherman and Spear, 1982). Continued progess in this field would seem to require studies that (1) continue to investigate which aspects of neuronal development are experience-sensitive and which are not (e.g., Gottlieb, 1976), (2) accurately correlate changes in the target neuron with changes in presynaptic activity during development (e.g., Archer *et al.*, 1982), and (3) investigate the cellular mechanisms by which a change in afferent stimulation leads to altered structure or function (e.g., Zigmond and Bowers, 1981; Black, 1982). The avian auditory nuclei appear to provide an advantageous system for pursuing these questions.

ACKNOWLEDGMENTS

Our research has been supported by grants from the U. S. Public Health Service (NS 15132 and NS 17257), the March of Dimes Birth Defects Foundation (5-246), and the Deafness Research Foundation, and greatly expedited by the skilled technical help of Patricia Collins, Danna Gray, Cardy Romero, Dwan Taylor, and Sarvjit Gill.

REFERENCES

Angevine, J. B., Jr., and Sidman, R. L. (1961). *Nature (London)* **192,** 766–768.
Archer, S. M., Dubin, M. W., and Stark, L. A. (1983). *Science* **217,** 743–745.
Ard, M. D., and Morest, D. K. (1984). *Int. J. Dev. Neurosci.* **2,** 535–547.
Benes, F. M., Parks, T. N., and Rubel, E. W. (1977). *Brain Res.* **122,** 1–13.
Bennett, M. R. (1983). *Physiol. Rev.* **63,** 915–1048.
Berg, D. K. (1984). *Annu. Rev. Neurosci.* **7,** 149–170.
Berry, M., McConnell, P., and Sievers, J. (1980). *Curr. Top. Dev. Biol.* **15,** 67–101.
Black, I. B. (1982). *Science* **215,** 170–176.
Boord, R. L. (1969). *Ann. N.Y. Acad. Sci.* **167,** 186–198.
Brugge, J. F. (1983). *In* "Development of Auditory and Vestibular Systems" (R. Romand, ed.), pp. 89–120. Academic Press, New York.
Conlee, J. W., and Parks, T. N. (1981). *J. Comp. Neurol.* **202,** 373–384.
Conlee, J. W., and Parks, T. N. (1983). *J. Comp. Neurol.* **217,** 216–226.
Cotman, C. W., Nieto-Sampedro, M., and Harris, E. W. (1981). *Physiol. Rev.* **61,** 684–784.
Crepel, F., Mariani, J., and Delhaye-Bouchaud, N. (1976). *J. Neurobiol.* **7,** 567–578.
Deitch, J. S., and Rubel, E. W. (1984). *J. Comp. Neurol.* **229,** 66–79.
Dennis, M. J., and Ort, C. A. (1976). *Cold Spring Harbor Symp. Quant. Biol.* **40,** 435–442.
Diamond, J. (1982). *Curr. Top. Dev. Biol.* **17,** 147–205.
Forehand, C. J., and Purves, D. (1984). *J. Neurosci.* **4,** 1–12.
Globus, A. (1975). *In* "The Developmental Neuropsychology of Sensory Deprivation" (A. H. Riesen, ed.), pp. 9–91. Academic Press, New York.
Gottlieb, G. (1976). *In* "Behavioral Embryology" (G. Gottlieb, ed.), Vol. 3, pp. 237–280. Academic Press, New York.
Gottlieb, G. (1981). *In* "Development of Perception, Vol. 1" (R. N. Aslin *et al.*, eds.), pp. 5–44. Academic Press, New York.

Gray, L., and Rubel, E. W. (1981). *J. Comp. Physiol. Psychol.* **95**, 188–198.

Grinnell, A. D. (1977). *In* "Handbook of Physiology, Sect. 1. The Nervous System." (E. R. Kandel, ed.), Vol. 1, Part 2, pp. 803–853. American Physiological Society, Bethesda, MD.

Hackett, J. T., Jackson, H., and Rubel, E. W. (1982). *Neuroscience* **7**, 1465–1469.

Harris, W. A. (1981). *Annu. Rev. Physiol.* **43**, 689–710.

Hickey, T. L., Spear, P. D., and Kratz, K. E. (1977). *J. Comp. Neurol.* **172**, 265–282.

Hirokawa, N. (1978). *J. Neurocytol.* **7**, 283–300.

Ivy, G. O., Akers, R. M., and Killackey, H. P. (1979). *Brain Res.* **173**, 532–537.

Jackson, H., and Parks, T. N. (1982). *J. Neurosci.* **2**, 1736–1742.

Jackson, H., and Parks, T. N. (1987). Submitted for publication.

Jackson, H., and Rubel, E. W. (1976). *Anat. Rec.* **184**, 434–435.

Jackson, H., and Rubel, E. W. (1978). *J. Comp. Physiol. Psychol.* **92**, 682–696.

Jackson, H., Hackett, J. T., and Rubel, E. W. (1982). *J. Comp. Neurol.* **210**, 80–86.

Jackson, H., Nemeth, E. F., and Parks, T. N. (1985). *Neuroscience*, **16**, 171–179.

Jackson, H., Urnes, M., Gray, W. R., and Parks, T. N. (1986). *In* "Excitatory Amino Acid Transmission" (T. P. Hicks *et al.*, eds.), pp. 47–51. Liss, New York.

Jaskoll, T. F., and Maderson, P. F. A. (1978). *Anat. Rec.* **190**, 177–200.

Jhaveri, S. R., and Morest, D. K. (1982a). *Neuroscience* **7**, 809–835.

Jhaveri, S. R., and Morest, D. K. (1982b). *Neuroscience* **7**, 837–853.

Jhaveri, S. R., and Morest, D. K. (1982c). *Neuroscience* **7**, 855–870.

Kerr, L. M., Ostapoff, E. M., and Rubel, E. W. (1979). *J. Exp. Psychol.* **5**, 97–115.

Knowlton, V. Y. (1967). *J. Morphol.* **21**, 179–208.

Koerber, K. C., Pfeiffer, W. B., and Kiang, N. Y.-S. (1966). *Exp. Neurol.* **16**, 119–130.

Konishi, M. (1986). *Trends Neurosci.* **9**, 163–168.

Landmesser, L. T., and Pilar, G. (1972). *J. Physiol. (London)* **222**, 691–713.

Langman, J., Shimada, M., and Haden, C. (1971). *In* "Cellular Aspects of Neural Growth and Differentiation" (D. C. Pease, ed.), pp. 33–59. Univ. of Calif. Press, Los Angeles.

Levi-Montalcini, R. (1949). *J. Comp. Neurol.* **91**, 209–241.

Lippe, W., and Rubel, E. W. (1983). *Science* **219**, 514–516.

Lorente de Nó, R. (1981). "The Primary Acoustic Nuclei." Raven, New York.

Morest, D. K. (1969a). *Z. Anat. Entwicklungsgesch.* **128**, 271–289.

Morest, D. K. (1969b). *Z. Anat. Entwicklungsgesch.* **128**, 290–317.

Nemeth, E. F., Jackson, H., and Parks, T. N. (1983). *Neurosci. Lett.* **40**, 39–44.

Nemeth, E. F., Jackson, H., and Parks, T. N. (1985). *Neurosci. Lett.*, **59**, 297–301.

Oppenheim, R. W. (1973). *In* "Behavioral Embryology" (G. Gottlieb, ed.), Vol. 1, pp. 163–244. Academic Press, New York.

Oppenheim, R. W. (1981). *In* "Studies in Developmental Neurobiology" (W. M. Cowan, ed.), pp. 74–133. Oxford Univ. Press, New York.

Parks, T. N. (1979). *J. Comp. Neurol.* **183**, 665–678.

Parks, T. N. (1981a). *J. Comp. Neurol.* **202**, 47–57.

Parks, T. N. (1981b). *J. Comp. Neurol.* **203**, 425–440.

Parks, T. N., and Jackson, H. (1984). *J. Comp. Neurol.* **227**, 459–466.

Parks, T. N., Gill, S., and Jackson, H. (1987). *J. Comp. Neurol.*, in press.

Parks, T. N., and Rubel, E. W. (1975). *J. Comp. Neurol.* **164**, 435–448.

Parks, T. N., and Rubel, E. W. (1978). *J. Comp. Neurol.* **180**, 439–448.

Parks, T. N., Collins, P., and Conlee, J. W. (1983). *J. Comp. Neurol.* **214**, 32–42.

Purves, D. (1983). *Trends Neurosci.* **6**, 10–16.

Purves, D., and Hume, R. I. (1981). *J. Neurosci.* **1**, 441–452.

Purves, D., and Lichtman, J. W. (1980). *Science* **210**, 153–157.

Rakic, P. (1975). *Adv. Neurol.* **12**, 117–134.

Ramón y Cajal, S. (1929). "Studies on Vertebrate Neurogenesis" (L. Guth, trans.). Thomas, Springfield, Illinois (1960 reissue).

Romand, R. (1983a). "Development of Auditory and Vestibular Systems." Academic Press, New York.

Romand, R. (1983b). *In* "Development of Auditory and Vestibular Systems" (R. Romand, ed.), pp. 47–88. Academic Press, New York.

Rouiller, E. M., and Ryugo, D. K. (1984). *J. Comp. Neurol.* **225**, 167–186.

Rubel, E. W. (1978). *In* "Handbook of Sensory Physiology" (M. Jacobson, ed.), Vol. IX, pp. 135–237. Springer, New York.

Rubel, E. W., and Parks. T. N. (1975). *J. Comp. Neurol.* **164**, 411–434.

Rubel, E. W., Smith, D. J., and Miller, L. C. (1976). *J. Comp. Neurol.* **166**, 469–490.

Rubel, E. W., Smith, D. J., and Steward, O. (1981). *J. Comp. Neurol.* **202**, 397–414.

Rubel, E. W., Born, D. E., Deitch, J. S., and Durham, D. (1984). *In* "Recent Developments in Hearing Science" (C. Berlin, ed.), pp. 109–157. College Hill Press, San Diego, California.

Sachs, M. B., Woolf, N. B., and Sinnott, J. M. (1980). *In* "Comparative Studies of Hearing in Vertebrates" (A. N. Popper and R. R. Fay, eds.), pp. 323–353. Springer, New York.

Saunders, J. C., Coles, R. B., and Gates, G. R. (1973). *Brain Res.* **63**, 59–74.

Saunders, J. C., Kaltenbach, J. A., and Relkin, E. M. (1983). *In* "Development of Auditory and Vestibular Systems" (R. Romand, ed.), pp. 3–25. Academic Press, New York.

Shatz, C. J., and Kirkwood, P. A. (1984). *J. Neurosci.* **4**, 1378–1397.

Sherman,S. M., and Spear, P. D. (1982). *Physiol. Rev.* **62**, 738–855.

Smith, D. J., and Rubel, E. W. (1979). *J. Comp. Neurol.* **186**, 213–240.

Smith, Z. D. J. (1981). *J. Comp. Neurol.* **203**, 309–333.

Smith, Z. D. J., Gray, L., and Rubel, E. W. (1983). *J. Comp. Neurol.* **220**, 199–205.

Sullivan, W. E., and M. Konishi (1984). *J. Neurosci.* **4**, 1787–1799.

Takahashi, T., Moiseff, A., and Konishi, M. (1984). *J. Neurosci.* **4**, 1781–1786.

Tucci, D. L., and Rubel, E. W. (1985). *J. Comp. Neurol.* **238**, 371–381.

Vince, M. A. (1973). *In* "Behavioral Embryology" (G. Gottlieb, ed.), Vol. 1, pp. 285–323. Academic Press, New York.

Willard, F. H., Thompson, D. M., and Martin, G. F. (1983). *Soc. Neurosci. Abstr.* **9**, 377.

Young, S. R., and Rubel, E. W. (1983). *J. Neurosci.* **3**, 1373–1378.

Young, S. R., and Rubel, E. W. (1986). *J. Comp. Neur.,* in press.

CHAPTER 13

NEURAL REORGANIZATION AND ITS ENDOCRINE CONTROL DURING INSECT METAMORPHOSIS

Richard B. Levine

ARIZONA RESEARCH LABORATORIES
DIVISION OF NEUROBIOLOGY
THE UNIVERSITY OF ARIZONA
TUCSON, ARIZONA 85721

I. Introduction

Insect metamorphosis is a particularly dramatic example of the extent to which organisms can continue to develop postembryonically. In a volume devoted to recent advances in the study of neural development, however, it is worth considering whether the process of metamorphosis merely represents an interesting specialization or actually assumes more general relevance by offering conveniently blatent examples of mechanisms common to all developing organisms. A close look at the cellular reorganization of the insect nervous system during metamorphosis suggests that the latter view has a great deal of merit. During this process a functioning larva with a characteristic set of behavioral programs is first transformed into a relatively quiescent but still behaving pupa, and then into an adult which possesses new sensory structures, new appendages, and new behavior. Although in many cases larval portions of the nervous system are scrapped to make way for new adult neurons, an equally important aspect of neural reorganization is the recycling of the larval interneurons, motorneurons, and even sensory neurons to participate in different behavior at different stages. This recycling can, in many respects, be viewed as a partial redifferentiation. Basic characteristics of several differentiated neurons, such as their dendritic morphology and synaptic interactions, are altered. These metamorphic changes may be considered simply delayed aspects of the normal differentiative process, with neurons delaying completion of their development temporarily until the proper signal enables them to finish. The larval, pupal, and adult stages of life may also be viewed as alternative, rather than sequential forms, with neurons expressing a different set of characteristics depending upon a variety of possible external influences (Wigglesworth, 1954).

CURRENT TOPICS IN
DEVELOPMENTAL BIOLOGY, VOL. 21

It is this type of plasticity, coupled with the unusual ease with which it can be studied at the single-cell level, which makes insect metamorphosis so relevant to current research in developmental neurobiology. Similar postembryonic changes have recently been described in the vertebrate autonomic nervous system (Purves and Hadley, 1985) and also in the central nervous system (CNS) where they have been implicated in learning (Nottebohm, 1981) and in the acquisition of sexually dimorphic behavior (DeVoogd and Nottebohm, 1981a,b; Gurney, 1981). Due to the punctuated nature of metamorphic development, the holometabolous insects offer a particularly convenient system in which to investigate the induction and regulation of such phenomena and their relationship to the original embryonic development of nerve cells.

The neural reorganization that accompanies insect metamorphosis requires two types of control mechanisms. First, individual cells must be induced to express their internal developmental programs at the proper time. Second, on a more rapid time scale, new circuits must become functional at the appropriate time in the animal's life. The latter is especially important because the new circuits are incorporated into a functioning nervous system and must not disrupt ongoing behavior. Understanding these control mechanisms requires that the cellular changes associated with metamorphosis be described in detail. The first part of this review is devoted to summarizing this work. Later sections describe recent evidence that endocrine signals are important for inducing and regulating metamorphic changes.

Most of the work to be described was performed on the tobacco hawkmoth, *Manduca sexta*. Its life cycle will serve as a representative example, although the details vary in different holometabolous insects. The larvae spend a great deal of time feeding or seeking a new food source and, as they grow, must periodically form a new cuticular coat underneath the old one. This molting process culminates in a series of patterned muscular contractions, termed ecdysis, during which the larva leaves its old skin. Although most larval behavior is no longer displayed after entry into the pupal stage, there are reflex movements which are unique to this transitory period and the necessity of controlling them places obvious constraints upon the developing adult nervous system. Adult behavior, such as walking, flight, mating, and egg laying, is markedly different from that of the previous stages and requires new, or more sophisticated, sensory systems as well as motor control over newly developed appendages.

Insect metamorphosis is controlled by two hormones: the steroid 20-hydroxyecdysone (20-HE) and the sesquiterpenoid juvenile hormone

(JH). The most detailed investigations of their actions have been performed on the epidermal cells which are responsible for synthesizing and secreting the proteins comprising the cuticular exoskeleton (see Riddiford, 1985, for review). During each of four larval molts the epidermal cells secrete a new larval cuticle, but at the end of the fifth a more rigid pupal cuticle is secreted. The adult forms inside within approximately 18 days, during which time the epidermal cells secrete yet a third cuticular type. Epidermal cells are induced to synthesize and secrete new cuticular proteins by the direct action of 20-HE, but the type of proteins secreted varies for larval, pupal, and adult cuticles and is regulated by JH. In the presence of JH, larval cuticle is secreted, while in its absence pupal, and then adult, cuticle forms. Thus, as demonstrated *in vitro*, epidermal cells are polymorphic, with JH and 20-HE acting on each cell to influence its pattern of mRNA and protein synthesis. During the normal postembryonic development of *Manduca*, JH is released from the corpora allata and titers are high until midway through the last larval instar, whereas 20-HE is released from the prothoracic glands prior to each molt. At the end of the last larval instar there are two peaks of 20-HE in the blood; the first is important for epidermal cell commitment to a pupal pattern of mRNA synthesis, while the second induces the secretion of new cuticular proteins. Following pupation, 20-HE titers again rise to stimulate adult development.

II. Postembryonic Neurogenesis

A. DEVELOPMENT OF THE ADULT BRAIN

The fate of the neurons which comprise the larval nervous system and the origin of the adult nervous system varies depending upon the species and the region of the nervous system. The adult eyes and antennae are derived from imaginal discs and send thousands of afferent axons into the brain. Within the optic and antennal lobes new adult interneurons are generated from retained neuroblasts and either differentiate alongside larval interneurons or replace them (Edwards, 1969). The incorporation of tritiated thymidine into the DNA of dividing cells has been used to establish the birthdays of neurons in the monarch butterfly, *Danaus* (Nordlander and Edwards, 1969a,b, 1970), and in *Drosophila* (White and Kankel, 1978). In both, neuroblasts begin dividing and generating interneurons in the early larval instars and continue through the early pupal stage.

Within the developing optic lobes, newly generated cells surround the rudimentary larval center. Massive death of larval cells was not observed in either *Danaus* (Nordlander and Edwards, 1969a) or *Drosophila* (White and Kankel, 1978), although Nordlander and Edwards (1969b) reported the degeneration of some differentiated neurons during the midpupal stage and attributed it to the elimination of redundant adult neurons. Furthermore, in *Drosophila,* unlabeled cells were present even in larvae exposed continuously to tritiated thymidine. Both studies concluded, therefore, that some larval neurons were incorporated into the adult optic lobes. The elaboration of the highly structured optic neuropiles has been detailed in several excellent reviews (Edwards, 1969; Meinertzhagen, 1973; Bate, 1978; Kankel *et al.,* 1980).

In contrast to the optic lobes, degenerating neurons were observed in the developing antennal lobes at the time of pupation, suggesting the elimination of at least some larval neurons (Nordlander and Edwards, 1970). The characteristic glomerular structure of the neuropil is lost at the end of the larval stage, only to reappear during the pupal stage when adult interneurons extend their dendrites and antennal afferents enter (Nordlander and Edwards, 1970). The development of the adult antennal lobe has been examined in detail in *Manduca* (see Hildebrand 1985, for review). Antennal sensory neurons are born on days 1 and 2, and their axons enter the developing antennal lobe between days 3 and 10 of the pupal stage (roughly 10–50 percent of adult development). Meanwhile, the interneurons begin to elaborate dendrites and continue through day 12. Ultrastructural as well as physiological evidence indicates that most synapses between antennal afferents and the interneurons mature between days 9 and 13 (Tolbert *et al.,* 1983).

The imaginal discs that produce the adult antenna can be extirpated and transplanted before differentiation of the sensory neurons begins. Antennal lobe interneurons survive early removal of an antennal disc, and although the glomeruli are reduced to small "protoglomeruli," the neuropil is still segregated into synaptic and nonsynaptic areas (Tolbert *et al.,* 1983). An inductive relationship between the antennal afferents and the central interneurons is clearly demonstrated when antennal discs are transplanted between males and females (Schneiderman *et al.,* 1982). Normally the male is distinguished by its large macroglomerular complex (MGC) which receives the axons of pheromone receptors on the antenna. Genetic females which receive a male disc develop an MGC and interneurons which respond to the pheromone, while males receiving a female disc lack the MGC as adults. Similarly, the temporal relationship between the entry of retinular axons and the formation of the optic neuropils suggests an important developmental interaction (Meinertzhagen,

1973) which has been confirmed by genetic as well as surgical perturbations (Meyerowitz and Kankel, 1978; Nassel and Geiger, 1983; Fischbach, 1983).

In summary, the sensory processing areas are enlarged significantly during metamorphosis by the differentiation of postembryonically generated interneurons. The role of embryonically generated neurons in the adult brain remains unclear, although the absence of massive cell death in many areas suggests that some are retained. In addition, since neuroblasts begin generating neurons early in larval life, the possibility remains that some of these new neurons participate in larval behavior. Interneurons clearly do play both larval and adult roles in other parts of the brain. Kenyon cells in the *Drosophila* corpora pendunculata lose processes at the end of the larval stage and extend new ones during metamorphosis (Technau and Heisenberg, 1982).

B. NEUROGENESIS IN ABDOMINAL AND THORACIC GANGLIA

Although postembryonic neurogenesis is most common within the sensory processing areas of the brain, new neurons are also produced in thoracic as well as abdominal ganglia (Booker and Truman, 1986). One group of newly generated cells in the abdominal ganglia become neurosecretory cells and extend axons out of the CNS (Tublitz and Truman, 1985). It is clear, however, that most neurons produced postembryonically must become local or interganglionic interneurons, since all adult motor neurons are already present and functional in the larva (Levine and Truman, 1985). In this respect the neuroblasts which divide postembryonically may be similar to those which complete their proliferation during the embryonic period. One carefully studied neuroblast in grasshopper embryos, for example, first produces motor neurons, followed by interganglionic, then intraganglionic interneurons (Goodman *et al.*, 1980). Therefore, in holometabolous insects some neuroblasts may simply delay the production of their final offspring. Of particular interest is the identity of those neuroblasts that are destined to display this delayed pattern and the signal responsible for renewing the proliferative process.

III. Motor Neuron Recycling

A. ABDOMINAL MUSCULATURE

While the olfactory and visual areas of the insect brain provide dramatic examples of metamorphic change, the large number of neurons involved has made it difficult to analyze the process at the level of individual cells. Substantial progress in this regard has been achieved in recent analyses of the abdominal motor system of *Manduca*. Although the behavior of the abdomen changes during metamorphosis, the number of neurons involved is sufficiently small that individuals may be identified and followed through the animal's life.

Movements of the abdomen are controlled by a network of body-wall muscles which is different in larva and adult (Levine and Truman, 1985). Larval movements are dominated by the large intersegmental muscles (ISM). These persist through the pupal stage and participate in adult emergence only to die soon thereafter (Schwartz and Truman, 1982). A second group of smaller external muscles lies immediately adjacent to the body wall and in the larval stage controls subtle movements of each abdominal segment. All of these muscles die at the end of the larval stage and are replaced during metamorphosis by newly generated adult muscles. Due to the loss of the larger ISM, the external muscles of the adult have a greater behavioral role than that of their larval counterparts (Truman and Levine, 1983).

B. The ISM Motor Neurons

Unlike the interneurons that comprise the sensory processing areas of the insect brain, all abdominal motor neurons of the adult were functional larval motor neurons (Taylor and Truman, 1974; Levine and Truman, 1985). The large ISM of the larval abdomen are each innervated by a single excitatory motor neuron. The dendrites of these motor neurons display an interesting relationship to the location of their target muscles (Fig. 1). Those motor neurons innervating midline targets have bilateral dendritic trees, while the dendrites of motor neurons innervating lateral targets are confined to one side of the ganglion. This relationship may arise because midline muscles contract in synergy with their contralateral homologues during most larval behavior and must therefore receive similar synaptic inputs, whereas lateral motor neurons normally act as antagonists to their contralateral homologues. The functional rationale for this somatotopic relationship, however, has yet to be fully resolved. In the pupal stage the ISM and their motor neurons are retained and used for characteristic rotary movements, as well as the pupal-specific gin-trap reflex (Section V,B). After playing a key role in adult emergence, the ISM motor neurons die along with their targets (Truman, 1983; see review by Truman, this volume). Although differences in the behavior in which these motor neurons are involved suggest that there must be changes in their synaptic inputs during metamorphosis, this is not reflected in the dendritic morphology of the motor neurons (Fig. 1). There is an increase in the overall extent of fine processes, but den-

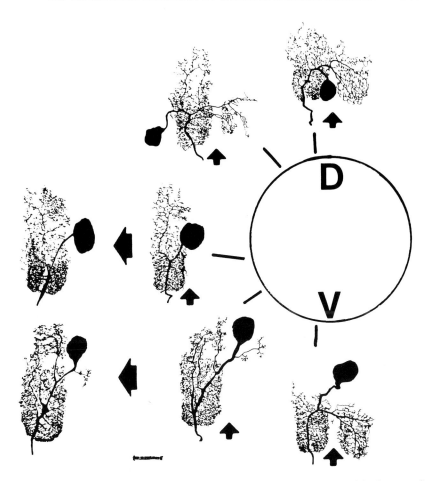

FIG. 1. Motor neurons innervating the intersegmental muscles (ISM) of the larva and newly emerged adult. The drawing shows a cross section of a larval abdominal segment, with dorsal (D) at the top. Around the segment five ISM motor neurons are shown with the position of their target muscles indicated. The small arrows point to the midline of the nervous system in each drawing. Note that motor neurons innervating targets near the dorsal or ventral midlines have bilateral dendritic processes, while those with more lateral targets tend to have dendrites restricted to one half of the neuropil. At the extreme left, two of the motor neurons are shown as they appear in the newly emerged adult. Motor neurons were stained by intracellular cobalt injection. Bar, 50 μm.

drites do not enter new areas of neuropil. Therefore, the changes in circuitry observed must reflect alterations in sensory neurons or interneurons (see below).

C. RESPECIFIED MOTOR NEURONS

The remaining motor neurons lose their target muscles at the close of the larval stage. Although several die following entry into the pupal stage (Weeks and Truman, 1985; Truman, this volume), the majority survive to innervate newly generated adult muscles. At the time of pupation these "respecified" motor neurons begin a slight reduction in their dendritic profiles (Truman and Reiss, 1976; Levine and Truman, 1982, 1985; Weeks and Truman, 1984). Meanwhile, their peripheral axons retract from the degenerating larval muscles and begin to invade the developing adult muscles (Stocker and Nuesch, 1975). Once adult development has started, the respecified motor neurons begin growing new dendritic processes which expand over the next 2 weeks to reach the adult form shortly before emergence (Levine and Truman, 1982, 1985). The abdominal motor neurons are not unique in their persistence through metamorphosis. Thoracic motor neurons are also respecified to innervate new adult targets. For example, the motor neurons controlling wing movements in adult *Manduca* are retained from the larval stage and extend new dendrites during metamorphosis (Casaday and Camhi, 1976).

Just as the dendritic morphology of each identified neuron is a characteristic of that cell and intimately related to its function, so the new dendritic growth occurring during metamorphosis is specific for each respecified motor neuron (Fig. 2). Several motor neurons extend dendrites across midline commissures to become bilateral, some invade more anterior regions of the neuropil, while still others enter posterior regions. Therefore there is no indication that the signal for dendritic growth is random or that all motor neurons send new processes into one or a few particular regions.

At this point it is worth considering the nature of the signal for dendritic growth. Many neurons lose their larval targets, but only some survive this loss and acquire new dendrites. Therefore, while it is unlikely that target loss is the only signal involved, the myoblasts near the axon terminals of respecified motor neurons could serve an inductive role. The acquisition of new synaptic inputs (to be described below) may also be important. Finally, the changing hormonal environment may act directly upon specific motor neurons to induce the growth of new dendrites. Hormonal cues have been clearly implicated in the dendritic reduction and motor neuron death that occur at the

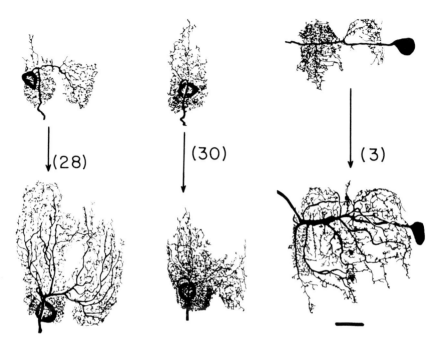

Fig. 2. The larval (top) and adult (bottom) morphologies of three respecified motor neurons. Each loses its larval target and innervates a newly generated muscle in the adult. The first motor neuron, MN-28 (28), sends new dendritic processes into anterior portions of the ganglion; MN-30 (30) grows new dendrites contralateral to its axon, while MN-3 (3) expands into posterior regions of the neuropil. Bar, 50 μm.

close of the larval stage (Weeks and Truman, 1985). These possibilities will be explored further in a later section.

IV. Relationship between Dendritic Growth and New Circuitry

A. CHANGES IN THE STRETCH REFLEX

It might be expected that the dendritic reorganization of a re-specified motor neuron would reflect the acquisition of new synaptic inputs required for its new behavioral role. In order to verify this correlation, neural circuits underlying prominent components of larval behavior were characterized and followed as the animal acquired first pupal and then adult behavior. One easily studied synaptic input to most motor neurons is derived from a pair of stretch receptors located in each abdominal segment of the larva. These sensory structures each contain a single sensory neuron that is activated when the segment is stretched (Weevers, 1966). The structures are present throughout the life of the animal and, while the central branching pattern of the sensory neuron is not altered appreciably during metamorphosis (Levine and Truman, 1982), its effect on various motor neurons does change in a manner that is consistent with their dendritic reorganization.

In one specific example, motor neuron MN-1 receives direct excitatory synaptic input from the stretch receptor ipsilateral to its target muscle. The larval motor neuron is also inhibited by the contralateral stretch receptor through a polysynaptic pathway (Fig. 3). At this stage MN-1 has a unilateral dendritic field which overlaps with processes of the ipsilateral, but not contralateral, stretch receptor. By contrast, in the adult the ipsilateral stretch receptor still excites MN-1, but instead of inhibiting the motor neuron the contralateral stretch receptor now excites it through a short-latency pathway that is probably monosynaptic (Fig. 3). Furthermore, during adult development MN-1 has acquired new dendrites which overlap with processes of the contralateral stretch receptor. Although the direct nature of the new connection has not been confirmed with the electron microscope, physiological evidence suggests that the pathway is direct (Levine and Truman, 1982). In behavioral terms, the larval circuit represents a simple stretch reflex which counteracts lateral flexion of the abdomen. The change observed during metamorphosis is in line with an observed shift from right/left antagonism in the larva to dorsal/ventral antagonism in the adult. The larval targets of MN-1, which are small external muscles near the dorsal midline, act as antagonists, and this is reinforced by the stretch receptors. In the adult, however, the new MN-1 targets serve as synergists, as suggested by the presence of numerous common inputs, including those from the two stretch receptors. Therefore, the new dendritic growth that accompanies metamorphosis is correlated with behavioral, and consequently synaptic, changes.

B. RAPID LOSS OF THE INHIBITORY PATHWAY AT ADULT EMERGENCE

In following the development of the new stretch receptor pathway, an additional feature of the metamorphic process was revealed. The new excitatory connection appeared gradually during the final week of adult development, but the larval inhibitory connection persisted. Just prior to adult emergence, stimulation of the contralateral stretch receptor evoked a biphasic response in MN-1, with the larval inhibitory potential being preceeded by a new short-latency excitatory potential (Fig. 3). During trains of stimuli which mimicked the natural activity of the sensory neuron, the inhibition predominated so that the net

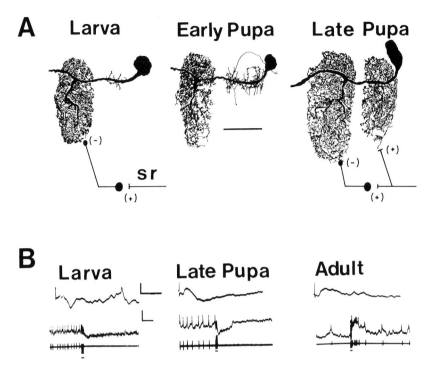

FIG. 3. Relationship between new dendritic growth and altered neural circuitry. (A) A respecified motor neuron (MN-1) grows a new dendritic tree contralateral to its axon (and target muscle) during metamorphosis. (B) In the *larva,* MN-1 is inhibited indirectly by a stretch receptor contralateral to its target. (Top), Single stimuli to the stretch receptor cause a long-latency inhibitory postsynaptic potential (IPSP), while a burst of stimuli causes prolonged inhibition (bottom). In the *adult,* however, the same stretch receptor excites the motor neuron through monosynaptic connections that form upon the newly acquired dendrites. In the *late pupal stage,* single stretch receptor action potentials cause a biphasic response in MN-1, with a new excitatory postsynaptic potential (EPSP) preceding the inhibitory response. However, a burst of stimuli (which is how the sensory neuron normally fires) still inhibits the motorneuron (bottom). The inhibitory component disappears within 30 minutes of adult emergence so that, in the *adult,* stretch receptor activity excites MN-1. The change is shown schematically in A. SR, Stretch receptor. Bar 100 μm; Calibrations in B: top, 5 mV, 100 mseconds; bottom, 10 mV, 500 mseconds.

result was inhibition of MN-1. Within 30 minutes of adult emergence, however, the inhibitory pathway was no longer effective and the adult excitatory response was observed (Levine and Truman, 1982).

Thus, the inhibitory pathway was maintained during adult development but deactivated rapidly at adult emergence. This sequence of events prevents the premature loss of the stretch reflex, which is an important component of pupal behavior, and allows the expression of adult behavior at the appropriate time. Whether the deactivation involves the death of an inhibitory interneuron, or merely a rapid block of its function, is unclear. Although several interneurons die at this time (Truman, 1983), cell death seems too slow to account for the rapid switch observed. Similar rapid changes associated with pupal or adult emergence have been attributed to the action of the peptide eclosion hormone (see below). Perhaps the two mechanisms act in concert, with the peptide-mediated deactivation of an inhibitory interneuron being followed by its death.

V. Recycling of Mechanosensory Neurons

A. TOPOGRAPHIC PROJECTION PATTERNS OF THE LARVAL MECHANOSENSORY NEURONS

Another example of the neuronal reorganization that accompanies metamorphosis comes from the study of a second sensory system. The surface of *Manduca* larvae is covered with mechanoreceptive sensory hairs, and at the base of each is a single sensory cell which sends its axons into the CNS. Since in each abdominal segment there are over 700 of these sensory neurons sending processes to the CNS, they constitute an important portion of the neural circuitry in operation at this time. Analysis of the central branching patterns of these afferents has provided new information regarding the organization of the abdominal ganglia during the larval stage.

Just as there is a somatotopic map of the mammalian body surface in the cerebral cortex, the larva carries a map of its surface in the CNS. Thus, sensory neurons innervating hairs that are anterior in the segment terminate in anterior regions of the neuropil, while lateral neurons branch laterally (Fig. 4) (Levine *et al.*, 1985). This quite literal topographic mapping is common in other insects (Murphey, 1981) and means that each position on the body surface is represented by a distinct region of the CNS. Upon stimulation of these receptors, the value of this precise pattern of termination becomes clear. The animal's motor response depends upon the area of the surface being stimulated. Larvae withdraw locally, or bend towards the side being stimulated, with a slow movement that involves the contraction of the ISM in each segment along one side of the body. Thus, ipsilateral motor neurons in several segments are excited, while those innervating targets on the opposite side of the body are inhibited (Levine *et al.*, 1985). One consequence of this response pattern is that motor neurons innervating medial targets must be excited by sensilla on either side of the body

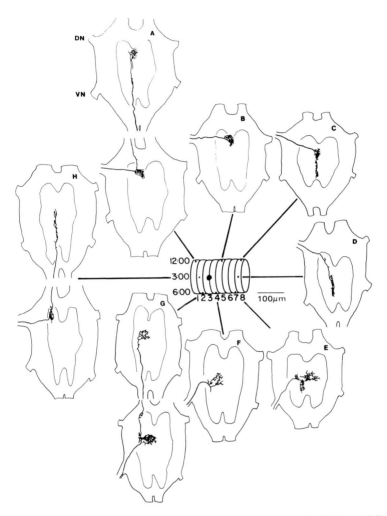

Fig. 4. (A–H) Branching patterns of sensory neurons innervating hairs on different parts of the second abdominal segment of a fifth instar larva. Insert shows the positions of the hairs on a schematic abdominal segment with anterior to the left, dorsal to the top. The large spot near the center of the segment denotes the position of the spiracle. The numbers at the bottom refer to annulus number, while those to the left refer to positions on an imaginary clock face. Note that anterior sensory neurons extend into the next anterior ganglion before terminating, while posterior ones turn in the opposite direction. Midline sensory neurons terminate near the midline of the ganglion. Dorsal and lateral sensory neurons enter the dorsal nerve (DN), while ventral afferents enter the ventral segment nerve (VN). Bar, 100 μm.

midline and, indeed, midline motor neurons and sensilla have bilateral processes. Cross sections of stained motor and sensory neurons suggest, however, that connections between the two are not direct (Levine *et al.*, 1985). Recently, interneurons have been identified which receive direct input from sensory hairs on specific regions of the body surface and drive body-wall motor neurons (R. B. Levine, unpublished). The dendritic profiles of these interneurons lie in neuropilar regions that are consistent with their receptive fields. Thus, as in other insect sensory systems (Bacon and Murphey, 1984; Murphey, 1985), the somatotopic termination pattern of afferent axons allows interneurons to send dendrites to specific regions of the neuropil and thereby read information from a particular region of sensory space.

Even within the larval stage, the motor effects of these sensory neurons are not constant. During the early larval instars, tactile stimulation of individual or small groups of sensilla evokes a clear behavioral response consisting of lateral flexion or crawling. In the last larval stage, however, the same stimuli evoke extremely weak behavioral responses even though there are more hairs to activate. Animals transiently become more excitable on the day prior to pupation, but sensitivity again declines in the hours preceding pupal ecdysis. The significance of these changes is unclear.

B. Persistence of Some Sensory Neurons into the Pupal Stage

During the final 4 days of their larval stage, animals begin the transition into pupal life. Epidermal cells secrete a new cuticle, imaginal discs evert, and cuticular pits form bilaterally at the anterior lateral margins of abdominal segments 5, 6, and 7. Each of these "gin traps" contain roughly 20 mechanoreceptive hairs which evoke a powerful contraction of the ipsilateral ISM muscles of the next anterior segment. As a result two segments are drawn together rapidly to crush objects within the trap. This defensive "gin-trap reflex" is characteristic of the pupal stage and is functional throughout this period (Bate, 1973a).

Several lines of evidence indicate that the sensory neurons innervating the gin-trap sensilla are retained from larval life. Cross sections of the developing gin trap reveal dendrites of sensory neurons simultaneously attached to both larval and pupal hairs just prior to the larval–pupal ecdysis (Bate, 1972). Furthermore, the afferent axons and central ramifications of the sensory neurons can be stained with cobalt at any time during the larval–pupal transition (R. B. Levine and D. Linn, unpublished). Finally, the central processes of sensory neurons stained during larval life by exposing the dendrite to HRP are still present and can be visualized in the pupal stage after processing the nervous system to reveal reaction products of the enzyme (R. B. Levine, M. Jacobs, and F. Rothenberg, unpublished).

Several motor neurons also persist into the pupal stage, and their response to stimulation of the retained mechanosensory neurons is quite different than in the larva. While motorneurons in several segments are excited weakly by single sensilla in the larval stage, stimulation of a single pupal gin-trap hair evokes a large depolarization and a high-frequency burst of action potentials in a restricted set of motor neurons which innervate the ipsilateral ISM of the next anterior segment (Bate, 1973b; Levine and Truman, 1983). Contralateral motor neurons, and those innervating other segments, are inhibited during the reflex. In addition, the larval response is gated by information from abdominal stretch receptors, but the pupal reflex is unaffected by the animal's posture (Bate, 1972, 1973c).

This physiological change is accompanied by a reorganization of the central arborizations of the sensory neurons. The total length and number of branch points of their terminal processes increase at least 3-fold during the final 4 days of the larval stage (Levine et al., 1985) (Fig. 5). The relationship between the new afferent arborizations and the altered behavioral response is not entirely clear since the pathway involved in the reflex is polysynaptic and the interposed interneurons have not been identified, but evidence detailed below suggests that the growth is crucial to the acquisition of the new behavior. Thus, like the motor neurons, sensory neurons can be used for different purposes at different stages of life. Although the sensory neuron reorganization discussed here occurs at an earlier stage than that of the motor neurons, the inductive mechanisms involved are likely to be similar.

C. Possible Inductive Mechanisms

As with the motor neurons, there are several alternative inductive mechanisms that can be invoked to explain the onset of sensory neuron terminal growth. In his original description of the system, Bate (1973a) noted that only those sensilla from a discrete anterior/lateral region of certain larval segments would come to participate in the gintrap reflex. He suggested that the peripheral position of the sensory cell body determined whether it would form central connections with the gin-trap circuitry. The fact that even in the early larva the central arborizations of these anterior/lateral sensory neurons differ from those of sensilla with other peripheral positions (Fig. 4) lends support to Bate's hypothesis. Recent experiments in other insect systems further support the idea that information available to peripheral cell bodies influences the central connections that sensory neurons make

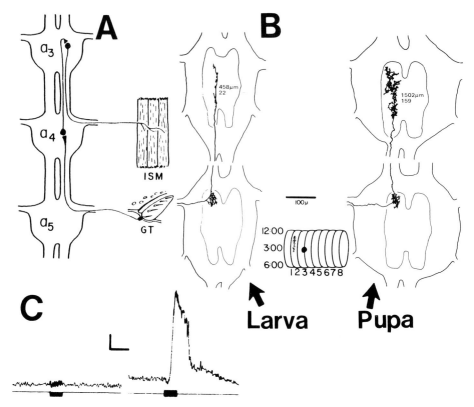

FIG. 5. Development of the gin-trap reflex. (A) Schematic representation of the neural circuit responsible for the pupal reflex. Three abdominal ganglia are required (a_3–a_5). Afferents from the gin trap (GT) enter a_5 and terminate in a_4 upon an unidentified interneuron, which contacts ISM motorneurons in a_3. (B) Growth of sensory neurons during the larval–pupal transition. Left, A sensory neuron from the presumptive gin-trap region of an early fifth instar larva. Right, A neuron innervating a hair in the pupal gin trap. Numbers refer to the total length, and total number of branch points in the anterior ganglion. Insert shows an abdominal segment with the gin-trap region shaded. (C) Effect of eclosion hormone, response of an ISM motor neuron to electrical stimulation of the gin-trap afferents (see marker in lower trace) in an animal dissected 3 hours prior to the normal time of pupal ecdysis. Before the addition of eclosion hormone (EH) to the bath there was no response (left). About 1 hour after addition of EH (right), however, the motor neuron responded to the same stimulis with a large depolarization and a burst of action potentials. The same activation occurred in different preparations when only a_3 was exposed to EH, while exposure of a_4 alone was without effect (calibrations, 5 mV, 400 mseconds).

(Walthall and Murphey, 1984). This hypothesis suggests that a peripheral gradient of a morphogenetic substance induces those sensory cells in a specific location to grow new central processes. An additional peripheral cue that could act separately, or in concert with such a gradient, is the hormonal environment of the sensory cell bodies. Upon sensing the appropriate combination of JH and 20-HE, prespecified sensory neurons may be induced to form new central connections. A second class of mechanisms does not involve the peripheral environment but relies instead upon local cues available to the terminal processes of the sensory neurons in the CNS. These include the hormonal environment of the terminals, the presence of new target cells, or the partial deafferentation that results when several sensory neurons that surround the gin-trap region die at pupation (Bate, 1972).

VI. Rapid Activation of the Gin-Trap Reflex

A. THE ROLE OF ECLOSION HORMONE

Just as the reorganized stretch reflex did not immediately assume its adult role in behavior as it was formed, the gin-trap reflex is not functional even minutes prior to pupal ecdysis. Three to five hours prior to pupal ecydsis, the gin trap is fully tanned, and by removing the overlying larval cuticle it is possible to show that its sensilla respond to tactile stimuli (Levine and Truman, 1983) but fail to evoke the gin-trap reflex response. Responsiveness appears abruptly during the brief ecdysis behavior as the pupa emerges from the old skin. This lack of function is not due to a general inhibition of the entire nervous system, since other motor neuron inputs, such as those from the stretch receptors, are functional during the preecdysial period. Therefore, some component of the polysynaptic circuit must be nonfunctional, or specifically inhibited, prior to ecdysis.

Pupal ecydsis is triggered by the peptide eclosion hormone (EH) which is released from neurosecretory cells into the hemolymph of the animal shortly before each ecdysis (Truman et al., 1981). Injection of EH into prepupae several hours before the normal time of its release not only results in premature ecdysis (Truman et al., 1980) but also activates the gin-trap reflex (Levine and Truman, 1983). In addition, EH activated the reflex circuit in isolated prepupal nerve cords, suggesting a direct action on the nervous system. The site of EH action could be further distinguished by taking advantage of the distributed nature of the circuit. Individual ganglia of semiintact prepupal preparations were exposed to the peptide in a small chamber without disrupting the integrity of the interganglionic connections involved in the reflex (Fig. 5). EH presented only to the ganglion containing the sensory neuron terminals was without effect. In contrast, EH presented to the ganglion containing the motorneuron dendrites activated the entire reflex, suggesting that the peptide influences (directly or indirectly) the interneuron–motor neuron connection. In neither case was the ecdysis motor program generated, suggesting that there are multiple sites of EH action within the CNS. Thus, the transition from larval to pupal behavior involves two steps in this case. First, the sensory neurons, and perhaps interneurons, involved in the reflex must grow new processes within the CNS. This occurs gradually over the 3–4 days just prior to ecdysis. Second, the circuit must be activated by EH. This step occurs rapidly and is tied temporally to pupal ecdysis.

B. RAPID ACTIVATION OF ADULT BEHAVIOR

The rapid validation of circuits required for new stages of life appears to be a general rule. In addition to the gin-trap circuit which is activated at pupal ecdysis, several components of adult behavior are inhibited on the day before emergence (Blest, 1960; Truman, 1976; Kammer and Kinnamon, 1977), perhaps to prevent rupturing of soft new structures or disruption of pupal behavior. At emergence, as the animal is shedding one skin and officially entering the next phase of its life, the behaviors appropriate for that stage become functional. The new excitatory connection between MN-1 and the contralateral abdominal stretch receptor is, as described previously, not expressed prior to adult emergence because an inhibitory pathway is coactivated. This example may not be completely analogous to the behavioral repression just described, since the inhibition is retained from an earlier stage rather than being imposed near the end of development. Nevertheless, the stretch receptor system could serve as a useful model for the release of inhibition at the time of adult emergence. The role of EH has not been established in this case, but the rapid deactivation of the inhibitory path is coincident with the appearance of EH in the hemolymph. In addition to these rapid changes in the CNS, synapses between motor neurons and newly formed adult muscles are enhanced markedly on the day of emergence by octopamine (Klaassen and Kammer, 1985).

VII. Endocrine Control of Neuron Form and Function

A. POSSIBLE ROLE OF JUVENILE HORMONE AND 20-HYDROXYECDYSONE

In the previous sections I have described two sets of neurons that are used for different purposes at different stages of life. Both the motor neurons and the mechanosensory neurons reshape their morphology and reorganize their patterns of synaptic interactions during metamorphosis. These differentiated neurons must begin the second stage in their development at the proper time and must know which stage they are about to enter. As described earlier, the epidermal cells of *Manduca* and other holometabolous insects display a similar type of polymorphism during metamorphosis by secreting different cuticular proteins for each of the three stages of life in response to specific hormonal cues (Riddiford, 1985). It seems quite likely that the hormones 20-HE and JH also determine the developmental fates of individual neurons. Recent evidence suggests strongly that programmed cell death is induced, at least in part, by changes in the 20-HE titer (Truman and Schwartz, 1984; Weeks and Truman, 1985; Truman, this volume).

The possibility of endocrine control is exciting because it suggests a direct parallel between the insect nervous system and sexually dimorphic regions of the vertebrate CNS. Although artificial manipulations of the steroid environment alter the morphology of neurons within these regions (DeVoogd and Nottebohm, 1981b; Gurney, 1981), it is difficult to prove that the hormones act directly. The primary targets could be other neuronal or nonneuronal cells that subsequently trigger changes in the sexually dimorphic neurons by direct contact or release of diffusable substances (Arnold and Gorski, 1984). In attempting to investigate the hormonal induction of dendritic growth in *Manduca* motorneurons, the same problem arises. Exposure of the entire CNS, or even a single ganglion, to an altered endocrine environment may induce metamorphic changes, but it is impossible to know whether the hormones are acting directly upon the motor neurons because their processes are immersed within a dense neuropil. Even autoradiographic demonstration of steroid binding is not sufficient because steroids may bind to neurons for reasons unrelated to the structural changes under consideration.

B. Endocrine Manipulation of Sensory Neuron Development

With these limitations in mind, the mechanoreceptive neurons discussed above offer several distinct advantages. Their somata are spaced far apart in the epidermis, and their hormonal environment may be manipulated without exposing the entire nervous system, including the terminal processes of the sensory neurons themselves, to exogenous hormones. The following set of experiments was made possible by the fact that JH, when applied topically, will only affect those cells underlying the site of application. A JH analogue (methoprene) was applied to the presumptive gin-trap region of last instar larvae during the period previously shown to be crucial for pupal commitment of the epidermal cells (Truman et al., 1974). Thus, the sensory neurons as well as the cells which generate the hair and socket of the sensilla failed to experience the drop in JH which is necessary for pupal commitment.

At the time of pupation, treated animals appeared normal except for the presence of a small larval patch at the site of methoprene application (Levine et al., 1986). The hairs within the larval region retained a larval appearance in that their shafts were longer and not set upon a pedestal as is characteristic of pupal gin-trap hairs (Levine et al., 1985). While tactile stimulation of these hairs caused the sensory neurons to respond, the gin-trap reflex was not evoked. Other gin traps, including the one contralateral to the treated area, were normal in appearance and evoked the typical pupal reflex behavior.

Intracellular recordings from the motor neurons confirmed these behavioral observations. Mechanical or electrical stimulation of treated sensilla evoked no response in those motor neurons that would normally be driven strongly in the pupal stage. This lack of response is reminiscent of the last larval instar when single sensilla rarely evoke a motor neuron response, and suggests that the treated sensory neurons failed to make connections necessary for the new reflex. The untreated contralateral sensory neurons, however, were able to evoke a typical gin-trap response in the appropriate motor neurons (Fig. 6) (Levine et al., 1986).

This apparent retension of the larval behavioral pattern was reflected in the morphology of the sensory neurons. The central arborizations of the treated sensory neurons remained larval in appearance, with the total length and number of branch points of treated afferents being similar to those of normal larvae (Levine et al., 1986) (Fig. 6). Untreated contralateral sensory neurons, however, grew to their normal pupal extent, suggesting that the CNS had not been directly affected by the JH treatment. Further evidence that the CNS was normal came from experiments in which small doses of the JH analogue induced the formation of small larval patches within an otherwise normal pupal gin trap. Sensory neurons within the larval areas did not evoke the gin-trap reflex while sensilla in the pupal area did, suggesting that target interneurons were present and able to accept synaptic inputs from the normal sensory neurons as well as evoke the gin-trap reflex. The sensory neurons within the larval patches retained a larval branching pattern in the same neuropil area where their normal peripheral neighbors had pupal-like arbors.

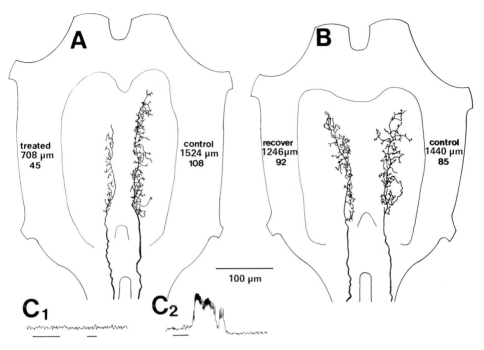

FIG. 6. Hormonal control of developmental changes in the sensory neurons. (A) Axonal arborizations of a sensory neuron treated with a JH analogue and unable to evoke the gin-trap reflex (left) and a control neuron from the normal contralateral gin trap of the same animal (right). The cell bodies of these sensory neurons send their axons into the fifth abdominal ganglion where they branch minimally before proceeding to the fourth segmental ganglion to terminate (shown). (B) Same as A, but the treated sensilla were allowed to recover the ability to evoke the gin-trap reflex during the first 4 days of the pupal stage before staining. C_1, Tactile stimulation of a treated sensillum (solid lines underlying record) evokes no response in a motorneuron innervating the ipsilateral side of the next anterior segment. C_2, Stimulation of a normal gin-trap hair from the control side of the same animal results in a high-frequency burst of action potentials, in an ipsilateral motor neuron, that is terminated by inhibition. Calibrations, 5 mV, 200 msec.

These results suggest that the exogenous JH treatment resulted in the formation of heterochronic mosaics, with larval sensory neurons entering a pupal CNS. Three important conclusions can be drawn from this experiment. First, the expansion of sensory neuron terminal arborizations is necessary for their participation in the gin-trap reflex, since treated sensilla that failed to attain the pupal growth pattern failed to evoke the reflex. Second, the change requires that the sensory neurons experience a specific hormonal environment: a decline in JH accompanied by a pulse of 20-HE. Whether the hormones induce the production of new stage-specific mRNA by the sensory neurons, as they do in the epidermal cells (Riddiford, 1985), is unclear. Finally the hormones do not induce terminal arbor growth solely by modifying local cues available to the growing afferent terminals in the CNS. Even though the terminal arbors of the sensory neurons experience a pupal endocrine environment within the CNS as well as the presence of their normal pupal target cells, neither is sufficient to induce growth. Therefore, the hormonal environment of the peripheral sensory cell bodies either induces the growth of terminal arbors directly or instructs the terminals to respond to local cues in the CNS.

One possibility that is not ruled out by the experiments described above is that the peripheral endocrine environment is important for the formation of a peripheral gradient which directs certain afferents to connect with the gin-trap circuitry (Bate, 1973a). This hypothesis is unable to explain the observation that, with appropriately timed applications of the JH analogue, gin traps of normal pupal morphology appeared in the appropriate place but the sensilla within them could not evoke the gin-trap behavior. Furthermore, the terminal arbors of the sensory neurons within these nonfunctional traps were of the larval morphology. Earlier work with topically applied JH had demonstrated clearly that different cell types become committed at different times during the larval–pupal transition (Truman et al., 1974). This suggests that the sensory neurons may have a slightly different time of commitment than that of the surrounding cell types, making it possible to prevent the pupal commitment of the sensory cells without influencing the surrounding epidermal cells or even the cells that produce the hair and socket. Because the gin-trap pit forms in the normal place and sensilla appear in the correct location in these animals, it seems unlikely that the formation of a morphogenetic gradient has been disrupted. It is, however, possible that by blocking pupal commitment the endocrine treatment prevented the sensory neurons from responding appropriately to such a gradient.

Following entry into the pupal stage, there is another surge in the 20-HE titer that triggers adult development (Riddiford, 1985). This is the first time that the treated sensory neurons have experienced 20-HE in the absence of JH and, accordingly, they begin their pupal differentiation. Larval patches lose their green color, as the epidermis normally does near the end of larval life, and the epidermal cells secrete a new pupal cuticle underneath the patch. Within 4 days of pupation the sensory neurons within treated areas have gained the ability to evoke the gin-trap reflex and acquired the pupal morphology within the CNS (Levine et al., 1986) (Fig. 6).

VIII. Future Directions

The process of metamorphosis provides a convenient system in which to explore the significance of postembryonic neural reorganization. One clear goal will be to understand alterations in the morphology and synaptic interactions of neurons in the context of the obvious behavioral transformations that accompany postembryonic development. Although two examples have been discussed in detail in this review, further advances will require more detailed knowledge of the neural circuits underlying various components of behavior including, in particular, the identity of the interneurons involved. Since interneurons are often more specific than motor neurons or sensory neurons in terms of their behavioral roles, we can expect that those retained from one stage to the next will display a significant degree of plasticity.

Beyond the relationship between neural reorganization and behavior, future work must be directed at understanding the induction and regulation of neural transformations. While *Manduca* have been useful because of the relative ease with which conventional intracellular recording methods can be applied and our detailed understanding of their endocrinology, we can expect significant advances through use of the powerful genetic and molecular techniques which may be employed in *Drosophila*. Particularly useful has been the recent development of monoclonal antibodies that are stage specific (White *et al.*, 1983) or recognize specific neuropilar regions (Teugels and Ghysen, 1983, 1985). One such antibody recognizes the leg neuromeres of the adult central nervous system so that, in combination with other antibodies, the development of these areas can be followed from the early larval stages (Ghysen *et al.*, 1985). Furthermore, it was possible to show that segmental specialization within the nervous system is controlled by the bithorax gene complex and that the development of the leg neuromeres during metamorphosis (at least in terms of their gross features) does not depend upon the presence of the corresponding legs.

Work in *Manduca* has suggested that two types of endocrine control are exerted upon the nervous system during metamorphosis. JH and 20-HE together regulate long-term developmental changes in neuronal morphology and synaptic interactions, while the peptide EH activates new circuits at the appropriate time. This knowledge is useful because hormones represent readily obtainable trophic factors that can serve as models for other types of inductive signals, and they provide a convenient method of precisely manipulating the developmental process in an effort to understand the importance of various cellular interactions. In addition, the relative simplicity of the insect CNS suggests that it will serve as a convenient model in the effort to understand the cellular and molecular mechanisms by which these hormones act upon the nervous system.

ACKNOWLEDGMENTS

Some of the work described in this review was performed while the author was a postdoctoral fellow in the laboratory of James W. Truman. Recent work from the author's laboratory was supported by a grant from the NSF.

REFERENCES

Arnold, A. P., and Gorski, R. A. (1984). *Annu. Rev. Neurosci.* **7**, 413–442.
Bacon, J. P., and Murphy, R. K. (1984). *J. Physiol. (London)* **352**, 601–623.
Bate, C. M. (1972). Ph.D. thesis, Cambridge University.
Bate, C. M. (1973a). *J. Exp. Biol.* **59**, 94–107.
Bate, C. M. (1973b). *J. Exp. Biol.* **59**, 109–119.
Bate, C. M. (1973c). *J. Exp. Biol.* **59**, 121–135.
Bate, C. M. (1978). *In* "Handbook of Sensory Physiology, Vol. IX. Development of Sensory Systems" (M. Jacobsen, ed.), pp. 1–53. Springer-Verlag, Berlin and New York.
Blest, A. D. (1960). *Behaviour* **16**, 188–253.
Booker, R., and Truman, J. W. (1986). *J. Comp. Neurol.,* in press.
Casaday, G. B., and Camhi, J. M. (1976). *J. Comp. Physiol.* **112**, 143–158.
DeVoogd, T. J., and Nottebohm, F. (1981a). *J. Comp. Neurol.* **196**, 309–316.
DeVoogd, T. J., and Nottebohm, F. (1981b). *Science* **214**, 202–204.
Edwards, J. S. (1969). *Adv. Insect Physiol.* **6**, 79–137.
Fischbach, K. F. (1983). *Dev. Biol.* **95**, 1–18.
Ghysen, A., Jan, L. Y., and Jan, Y. N. (1985). *Cell* **40**, 943–948.
Goodman, C. S., Pearson, K. G., and Spitzer, N. C. (1980). *Proc. Natl. Acad. Sci. U.S.A.* **77**, 1676–1680.
Gurney, M. E. (1981). *J. Neurosci.* **1**, 658–673.
Hildebrand, J. G. (1985). *In* "Model Neuronal Networks and Behavior" (A. I. Selverston, ed.), pp. 129–148. Plenum, New York.
Kammer, A. E., and Kinnamon, S. C. (1977). *J. Comp. Physiol.* **114**, 313–326.
Kankel, D. R., Ferrus, A., Gaven, S. H., Harte, P. J., and Lewis, P. E. (1980). *In* "The Genetics and Biology of Drosophila" (M. Asburner and T. R. F. Wright, eds.), pp. 295–368. Academic Press, New York.
Klaasen, L. W., and Kammer, A. E. (1985). *J. Neurobiol.* **16**, 227–243.
Levine, R. B., and Truman, J. W. (1982). *Nature (London)* **299**, 250–252.
Levine, R. B., and Truman, J. W. (1983). *Brain Res.* **279**, 335–338.
Levine, R. B., and Truman, J. W. (1985). *J. Neurosci,* **5**, 2424–2431.
Levine, R. B., Pak, C., and Linn, D. (1985). *J. Comp. Physiol.* **157**, 1–13.
Levine, R. B., Truman, J. W., Linn, D., and Bate, C. M. (1986). *J. Neurosci.* **6**, 293–299.
Meinertzhagen, I. A. (1973). *In* "Developmental Neurobiology of Arthropods" (D. Young, ed.), pp. 51–104. Cambridge Univ. Press, London.
Meyerowitz, E. M., and Kankel, D. R. (1978). *Dev. Biol.* **62**, 112–142.
Murphey, R. K. (1981). *Dev. Biol.* **88**, 236–246.
Murphey, R. K. (1985). *J. Comp. Physiol.* **156**, 357–367.

Nassel, D. R., and Geiger, G. (1983). *J. Comp. Neurol.* **217,** 86–102.
Nordlander, R. H., and Edwards, J. S. (1969a). *Wilhelm Roux Arch.* **162,** 197–217.
Nordlander, R. H., and Edwards, J. S. (1969b). *Wilhelm Roux Arch.* **163,** 197–220.
Nordlander, R. H., and Edwards, J. S. (1970). *Wilhelm Roux Arch.* **164,** 247–260.
Nottebohm, F. (1981). *Science* **214,** 1368–1370.
Purves, D., and Hadley, R. D. (1985). *Nature (London)* **315,** 404–406.
Riddiford, L. M. (1985). *In* "Comprehensive Insect Physiology, Biochemistry, and Pharmacology" (G. A. Kerkert and L. I. Gilbert, eds.), Plenum, New York.
Schneiderman, A. M., Matsumoto, S. G., and Hildebrand, J. G. (1982). *Nature (London)* **298,** 844–846.
Schwartz, L. M., and Truman, J. W. (1982). *Science* **215,** 1420–1421.
Stocker, R. F., and Nuesch, H. (1975). *Cell Tissue Res.* **159,** 245–266.
Taylor, H. M., and Truman, J. W. (1974). *J. Comp. Physiol.* **90,** 367–388.
Technau, G., and Heisenberg, M. (1982). *Nature (London)* **295,** 405–407.
Teugels, E., and Ghysen, A. (1983). *Nature (London)* **304,** 440–442.
Teugels, E., and Ghysen, A. (1985). *Nature (London)* **314,** 558–561.
Tolbert, L. P., Matsumoto, S. G., and Hildebrand, J. G. (1983). *J. Neurosci,* **3,** 1158–1175.
Truman, J. W. (1976). *J. Comp. Physiol.* **107,** 39–48.
Truman, J. W. (1983). *J. Comp. Neurol.* **216,** 445–452.
Truman, J. W., and Levine, R. B. (1983). *Curr. Methods Cell. Neurobiol.* pp. 16–48. Wiley, New York.
Truman, J. W., and Reiss, S. E. (1976). *Science* **192,** 477–479.
Truman, J. W., and Schwartz, L. M. (1984). *J. Neurosci.* **4,** 274–280.
Truman, J. W., Riddiford, L. M., and Safranek, L. (1974). *Dev. Biol.* **39,** 247–262.
Truman, J. W., Taghert, P. H., and Reynolds, S. E. (1980). *J. Exp. Biol.* **88,** 327–337.
Truman, J. W., Taghert, P. H., Copenhaver, P. F., Tublitz, N. T., and Schwartz, L. M. (1981). *Nature (London)* **291,** 70–71.
Tublitz, N. J., and Truman, J. W. (1985). *Science* **228,** 1013–1015.
Walthall, W. W., and Murphey, R. K. (1984). *Nature (London)* **311,** 57–59.
Weeks, J. C., and Truman, J. W. (1984). *J. Comp. Physiol.* **155,** 423–433.
Weeks, J. C., and Truman, J. W. (1985). *J. Neurosci.* **5,** 2290–2300.
Weevers, R. De G. (1966). *J. Exp. Biol.* **44,** 177–194.
White, K., and Kankel, D. R. (1978). *Dev. Biol.* **65,** 296–321.
White, K., Periera, A., and Cannon, L. E. (1983). *Dev. Biol.* **98,** 239–244.
Wigglesworth, V. B. (1954). "The Physiology of Insect Metamorphosis." Cambridge Univ. Press, London.

CHAPTER 14

NOREPINEPHRINE HYPOTHESIS FOR VISUAL CORTICAL PLASTICITY: THESIS, ANTITHESIS, AND RECENT DEVELOPMENT

Takuji Kasamatsu

THE SMITH-KETTLEWELL EYE RESEARCH FOUNDATION
AT PACIFIC PRESBYTERIAN MEDICAL CENTER
SAN FRANCISCO, CALIFORNIA 94115

I. Introduction

Sometime ago, we proposed that the norepinephrine (NE)-containing system in the brain is necessary to maintain and enhance neuronal plasticity which is typically found in the immature visual cortex of young kittens during the postnatal susceptible period (Kasamatsu and Pettigrew, 1976, 1979; Pettigrew and Kasamatsu, 1978; Kasamatsu *et al.*, 1979, 1981b). The background, as well as the later development, of this NE hypothesis for visual cortical plasticity has been reviewed in detail (Kasamatsu, 1983). Recently, two groups of investigators in other laboratories confirmed a part of our early findings. By directly perfusing the kitten visual cortex with a catecholamine (CA)-related neurotoxin, 6-hydroxydopamine (6-OHDA), they noted a concurrent loss of the plasticity in the cortex due, presumably, to morphological destruction specific to NE terminal fields within the kitten visual cortex (Daw *et al.*, 1983; Bear *et al.*, 1983; Paradiso *et al.*, 1983).

This interpretation, however, has been challenged lately by these investigators and others who have reported a set of "negative" results with 6-OHDA used in paradigms other than that of direct, cortical perfusion. Despite a substantial reduction in the NE content in the visual cortex, as a result of global changes in the central NE projection, visual cortical plasticity appeared to be sustained in the cortex (see below). Unfortunately, these seemingly straightforward results have confused rather than helped us in the quest for cellular mechanisms underlying visual cortical plasticity.

The current review is composed of two parts. First, a critical evaluation of negative results will be presented against a background of the general scope of the NE hypothesis which is illustrated in Fig. 1.

367

*CURRENT TOPICS IN
DEVELOPMENTAL BIOLOGY, VOL. 21*

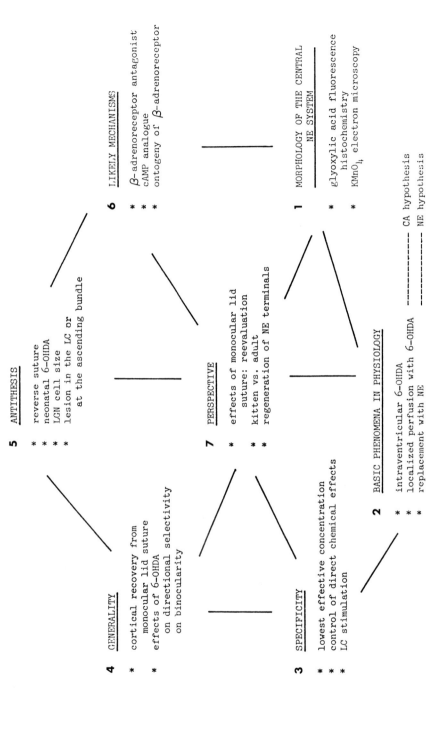

Second, I present some recent developments that we have made in five related areas surrounding our main theme, the regulation of visual cortical plasticity by the central NE system. Extensive reviews are available for visual cortical plasticity (Movshon and Van Sluyters, 1981; Sherman and Spear, 1982; Frégnac and Imbert, 1984).

II. Antithesis

Some topics in this category have been recently reviewed by other authors (Frégnac and Imbert, 1984; Daw et al., 1985c).

A. NEONATAL, SYSTEMIC INJECTIONS OF 6-OHDA

It is well known that one can modify, biochemically and morphologically, postnatal ontogeny of the central CA system by systemically injecting 6-OHDA into neonatal rat pups. Since 6-OHDA usually does not cross the blood–brain barrier, this modification is obviously dependent on the postnatal maturation of the barrier. 6-OHDA, when systemically injected within a few days after birth, reaches the brain and eventually leads to long-lasting selective morphological destruction of CA terminal fields in the neocortex, which is accompanied by their hypertrophy in the brainstem near the locus coeruleus (LC) and the cerebellum (e.g., Sachs and Jonsson, 1973, 1975; Schmidt and Bhatnagar, 1979; Onténiente et al., 1980). Jonsson and associates explained this regional difference in the effects of 6-OHDA as due to "pruning" effects, a concept originally proposed by Schneider (reviews, 1973, 1981).

Bear and associates (1983) systemically injected neonatal kittens with 6-OHDA hoping to detect no sign of neuronal plasticity in the visual cortex. Contrary to their expectations, they found a substantial loss of binocular cells and a preponderance of monocular cells which exclusively responded to stimulation of the nondeprived eye during the height of cortical susceptibility to monocular lid suture. In the same study they also reported a substantial reduction of endogenous CAs (NE and dopamine) in such cortex.

A physiological assay with microelectrodes of the ocular dominance distribution in a given visual cortex is subject, more or less, to sam-

---

FIG. 1. Summary and scope of the norepinephrine (NE) hypothesis of visual cortical plasticity. Some aspects such as morphology of the central NE system (1) and generality (4) were not included in the present review but mentioned elsewhere by the same author (Kasamatsu, 1983; Kasamatsu et al., 1984). KMnO$_4$, Potassium permanganate; 6-OHDA, 6-hydroxydopamine; CA, catecholamine; LC, locus coeruleus; LGN, lateral geniculate nucleus; cAMP adenosine cyclic monophosphate.

pling bias of neurons which belong to either the left or right eye column. Ideally, it is possible to reconstruct a given electrode track in an autoradiogram which gives us an indication of ocular dominance boundaries (e.g., Kennedy *et al.*, 1976; Shatz and Stryker, 1978; Hubel *et al.*, 1977). At the moment, however, this approach is not readily available in everyday practice. The effect of sampling bias is usually minimized by characterizing several factors involved during recording sessions, such as types of microelectrodes used, the distance between successively recorded neurons, the length of an electrode track, the laminar location of recorded neurons, and others. Thus, it should be remembered that any interpretation of changes in ocular dominance is valid only under a given set of known biases.

Assuming the width of the ocular dominance hypercolumn is about 1 mm in the kitten visual cortex, as shown for the monkey counterpart (Hubel *et al.*, 1978), a near vertical electrode track which is less than 1 mm in length is obviously too short, even under an ideal condition, to allow us to sample isolated single neurons evenly from the left and right eye column. This is exactly the reason why we collected 30 visually responsive neurons along a single track longer than 3 mm in almost all of our recordings; our microelectrode, angled 5–10° from the vertical in the parasagittal plane and 5° in the frontal plane, is usually allowed to traverse the ocular dominance boundary more than a few times, registering, on average, one neuron every 100 μm. Since we have recorded so far several thousand neurons from the visual cortex of normal and drug-treated kittens and cats, I believe we have a data base with a known sampling bias.

Let us assume, for the sake of the following discussion, that the shortest acceptable length of an electrode track is 2 mm. Since Bear and Daniels (1983) also collected single units approximately every 100 μm, the minimal acceptable number of units per penetration may be 20 in their studies. Then, two (K127, K131) of the three animals in the *neonatal 6-OHDA group* were suitable for the final analysis of correlation between physiology and biochemistry, since both physiological and biochemical assays were made in the same cortices of these animals. The average NE content was <18 ng/g and binocularity (proportion of binocularly driven neurons to the total number of visually active neurons) was 0.44. These are directly compared with the average values (NE, 97.6 ng/g; binocularity, 0.41) calculated for the five controls (K116, K122, K129, K130, K132) which met the same criteria. Likewise, in the parallel study of the effects of *local 6-OHDA perfusion,* the average binocularity was 0.73, and the NE content was 88.2

ng/g, in five experimental animals (K135, K136, K137, K138L, K138R). These values are compared with 0.50 and 647.0 ng/g, respectively, calculated for four control animals (K135, K136, K140, K143).

Apparently, the sample size in the studies by Bear and associates was too small to reach any sensible conclusions from comparison of these numbers, especially for the neonatal 6-OHDA study. Nevertheless, a few remarks may be appropriate. First, it is a great surprise that the most quoted conclusion of this study, i.e., the persistence of plasticity in the kitten visual cortex without much endogenous NE, was in fact based on the results from two animals. To make the results more difficult to evaluate, one (K127) of the two had been exposed to monocular deprivation for 47 days (all others for 10 days) since postnatal day 7. It is quite possible that if this kitten had not been injected with 6-OHDA neonatally, it would have shown the totally shifted ocular dominance distribution with nearly zero binocularity under the comparable condition. In an age-matched control (K129), binocularity was 0.30, which is at least smaller than the value of 0.38 shown by K127. Second, the variability in control measures was very large in both the biochemical and binocularity assays (see below). Third, binocularity in the two controls, 0.41 for the neonatal 6-OHDA study and 0.50 for the local 6-OHDA study, was larger than the usual value (~0.2) obtained under brief (~7 days) monocular deprivation at the corresponding age (5–8 weeks old) (e.g., Fig. 5A in Kasamatsu et al., 1981b; see also Hubel and Wiesel, 1970; Olson and Freeman, 1975). It should be pointed out that a control kitten (K132) had an unusually high number of binocular cells, despite monocular deprivation, resulting in binocularity of 0.68 (Fig. 4 of their publication). Fourth, the level of endogenous CAs, as measured by high-pressure liquid chromatography (HPLC), in the control was also unusually high for the age of animals included (6–8 weeks old) (cf. Fig. 1 in Jonsson and Kasamatsu, 1983). A few more general comments will be given below.

Since the visual cortex is not the only region of the brain that is affected by 6-OHDA injected systemically, the rationale underlying seemingly straightforward and simple application of Jonsson's paradigm to studying its effect on visual cortical plasticity is questionable. Needless to say, large doses of systemically injected 6-OHDA would severely interfere with the peripheral NE system, resulting in an alarming high rate of mortality of kittens (Bear et al., 1983). The long time interval between the date of 6-OHDA injections and physiological recordings is another concern. In order to interpret findings by Bear and associates properly, one should know at least the level of intracortical NE and its distribution within the visual cortex, especially around the beginning of the postnatal susceptibility period (~4 weeks old). Furthermore, in the light of a recent finding by Jonsson and Kasamatsu (1983) which showed a broad peak of β-adrenergic receptor binding during the postnatal susceptibility period (see below), it may be appropriate to correlate the plasticity with changes in the number of β-adrenergic receptors rather than the NE content *per se,* which, more or less, continuously increases toward adulthood (Jonsson and Kasamatsu, 1983). Thus, "negative" findings by Bear and associates might be explained, at least in part, by the long-lasting activation of β-adrenergic receptors preceded by an abrupt destruction of NE terminal fields in the brain due to neonatal 6-OHDA injections. Jonsson and associates (1979), in fact, found a few lines of evidence in support of the above possibility in adult rats which had been neonatally injected with 6-OHDA. When biochemical assays were carried out in adulthood (2–3 months old), they noted (1) a major NE metabolite, 3-methyoxy-4-hydroxy-phenylglycol, was reduced by 70 percent in the neocortex while endogenous NE was depleted by 92 percent; (2) NE turnover in remaining NE terminals in the neocortex, however, was increased; (3) a 20–50% increase in α- and β-adrenoreceptor binding sites was obtained in the neocortex of 6-OHDA-treated animals; and (4) NE-induced formation of adenosine cyclic monophosphate (cAMP) *in vitro* doubled.

The conclusion of Bear and associates may in fact be valid, since some technical problems mentioned above have been overcome in recent follow up studies by the same group and another (M. Bear and R. Baughman, personal communications). It is likely that unspecified compensation processes, independent of the NE system, may be involved in their study (Bear and Daniels, 1983). Nevertheless, their study still faces the same fundamental problems in its experimental design. Their findings, as they stand, should not be interpreted as being inconsistent with the NE hypothesis for visual cortical plasticity.

B. Local Injections of 6-OHDA into the LC Area

Adrien *et al.* (1985) used the following two methods to deplete the NE content in the kitten visual cortex. In the first group, a total of 12 mg 6-OHDA was injected over 11 days into the lateral ventricle, and in the second, 8 μg 6-OHDA was bilaterally injected into each of three sites (total, 48 μg) in the LC area of young kittens. Enzymatic assay for CAs carried out after the end of physiological recordings showed a substantial reduction (<40 ng/g) of endogenous NE in the visual cortex. Four of the eight LC-injected kittens (M.d.-4, -5, -8, and -11), for example, had a significantly ($p < 0.05$) greater number of binocular cells (binocularity: 0.47, 0.42, 0.54, and 0.43) than the control following monocular lid suture which started 6–13 days after the LC lesion and lasted for 7–16 days. However, two of the remaining four kittens (M.d.-6 and -7) had been monocularly deprived two times longer (14–19 days) than others 1 week after the LC lesion, showing binocularity of 0.31 and 0.14. The other two kittens (M.d.-9 and -10) were also subjected to monocular deprivation for 16–19 days, 2 weeks after the LC lesion, resulting in binocularity of 0.23 and 0.24. The binocularity values from the last four kittens were not significantly different from the control. Since they failed to correlate these changes in binocularity values and others with decreases in endogenous NE in the visual cortex of the LC-injected kittens, they concluded that noradrenergic projections are not essential in the expression of epigenetic modifications of visual cortical receptive fields. Nevertheless, when they quantitatively assessed the same set of data using "iso-duration time," they, in fact, noted that almost all experimental animals showed higher binocularity and lower open-eye dependency (shift) following monocular deprivation than would usually be expected in a control, although the difference was small. These results seem to suggest that LC lesion somehow intervened in the expected shift in ocular dominance following monocular deprivation.

An obvious shortcoming in their paradigm seems to be the timing between denervation of NE terminals in the brain and the start of monocular lid suture. Thus, the above results may be explained, at least partly, by either the fact that supersensitivity of NE-related receptors may have reached its peak during the 1–2 week period of monocular lid suture or the presence of regenerative NE terminals, especially after the intraventricular 6-OHDA injections. Alternatively, the "negative" results by Adrien and associates may be just secondary to some unspecified changes which occurred in the vast NE terminal fields in brain areas other than the visual cortex. These two possibilities, which are not mutually exclusive, cannot be separately analyzed in their paradigm.

C. Massive Lesion at the Ascending CA Bundles

Daw and associates unilaterally placed an electrolytic lesion at the dorsolateral hypothalamus in order to deplete endogenous NE in the visual cortex of several-week-old kittens (1984). Despite a large decrease in endogenous NE as measured later by an HPLC assay (down to 10–50% of control), this treatment apparently failed to bring about any changes in neuronal plasticity; the shift in ocular dominance following brief monocular lid suture was observed as usual. They recently added another finding to their list of "negative" results by studying the visual cortex of monocularly deprived kittens which had been intraperitoneally injected with DSP-4, a potential CA neurotoxin which has not been fully characterized (Daw *et al.*, 1985a).

Although interpreted by some critics as being inconsistent with the NE hypothesis, the study by Daw and associates also shares the same shortcomings as the studies by Bear *et al.* and Adrian *et al.* A common feature of these studies is that they all lack consideration of *specificity* of cause–effect relations; it appears as if they assumed that the visual cortex was the only site in which endogenous NE was substantially reduced by various means used in their studies. Obviously, this assumption is not valid, since the visual cortex is located at the far end of innervation by ascending NE fibers originating from the LC. More seriously, those studies did not pay attention to the state of NE-related receptors through which NE molecules exert their effect on visual cortical plasticity. For example, in the above-mentioned study with DSP-4, Daw and associates (1985a) found no changes in β adrenoreceptor binding in the kitten visual cortex, although endogenous NE was depleted significantly. No explanation was offered to these unusual findings which are inconsistent with the pharmacological common knowledge (i.e., an increase in receptor binding after deafferentation).

To make the situation more complicated, it is not immediately clear how large the depletion of an endogenous chemical in the brain should be before definite changes in physiology and then behavior, both of which are eventually assigned to a chemical in question, become obvious. In this context, only a positive correlation may be useful in searching for further linkage between brain chemicals and behavior. For example, in the kitten visual cortex which had been locally perfused with 4 mM 6-OHDA for a week, we still detected ~10% (~20 ng/g) of endogenous NE, although the neuronal plasticity was absent in such cortex (Kasamatsu *et al.*, 1981a). We would like to know the cellular and subcellular location of this minor but 6-OHDA-resistant pool of endogenous NE in the neocortex. Apparently, biochemical assays for remaining NE in the 6-OHDA-affected cortical area are necessary but not sufficient to tell us how remaining NE is distributed and behaves, which is crucial for interpreting results in physiology.

We should also point out another dimension of complexity: the plausible reversal trend of the reduced neuronal plasticity which may be set in motion due to the remarkably quick reappearance of NE terminals within the cortical area denervated by direct perfusion of 6-OHDA (Nakai *et al.*, 1981, 1987). Since the amount of initial depletion of endogenous NE was variable and incomplete, though substantial, the regeneration-induced restoration of neuronal plasticity may also account in part for some of the reported "negative" results.

D. OTHER PARADIGMS

Two more experimental conditions were reported previously in which no role of NE for neuronal plasticity was implied. Since these have been already discussed elsewhere (Kasamatsu, 1983), I am going to mention them only briefly. First, the function of the central NE system seems *not* to be involved in a type of neuronal plasticity underlying a reverse suture paradigm. We noted the usual shift back of ocular dominance when the side of monocular lid suture was reversed from one eye to the other in the visual cortex which had been treated with 6-OHDA either intraventricularly or locally (Ary *et al.*, 1979). This result may be related to the fact that no binocular interaction is possible in the reverse suture paradigm.

Second, using the lateral geniculate nuclei derived from the several kittens which had been physiologically recorded by us previously (Kasamatsu *et al.*, 1979), Hitchcock and Hickey (1982) measured geniculate cell size as an assay of the effects of monocular deprivation. They concluded that 6-OHDA injected intraventricularly did not prevent deprivation-induced changes in geniculate cell size despite its blocking effect on ocular dominance shift in the cortical physiology. In addition to a technical argument against their handling of a set of numbers, which was presented elsewhere (Kasamatsu, 1983), one of our recent results raised the question of the adequacy of this usual method for quantifying the effects of monocular lid suture on the geniculate cells. We obtained a significant change in genicular cell size following a very brief blockade (1 week) of total retinal output from one eye by twice injecting with 10 μg tetrodotoxin (Kuppermann and Kasamatsu, 1983). This change was obtained not only in the binocular segment but also in the monocular segment of the lamina A in which binocular competition is unlikely to occur. Thus, a major determinant of changes in geniculate cell size may not be binocular competition as elaborated previously (Guillery, 1972), but tonic retinal (dark) discharges which continually exert the powerful metabolic influence on geniculate cells and which cannot be suppressed by lid suture.

E. INTRAVENTRICULAR 6-OHDA:REEXAMINATION

Faced with the "negative" results that visual cortical plasticity persists despite a substantial reduction of endogenous NE in the kitten visual cortex, as briefly reviewed above, several investigators reexamined the effects of 6-OHDA which was repeatedly injected into the lateral ventricle, the very first paradigm used by us 10 years ago. Endogenous monoamines were measured by either a radioenzymatic method or an HPLC method.

In the study already mentioned above, Adrien and associates (1985) concluded that intraventricular injections of 6-OHDA failed to protect visual cortex from the loss of binocular interaction produced by monocular occlusion. There is a problem, however, with the timing of physiological assays. For two (K.P.-1 and -2) of the five kittens used here, the low binocularity (0.25 and 0.14) may be related to the fact that monocular deprivation outlasted 6-OHDA injection by 2 weeks. The prototype of these two kittens was TJ13 in our early study (Kasamatsu and Pettigrew, 1979). That kitten had binocularity of 0.35 when recorded a second time 19 days after the end of 6-OHDA injections. It was 0.58 when first recorded 10 days earlier. Another problem in this study comes from biochemistry. The depletion of dopamine was said to be complete in K.P.-3 (2 ng/g, Table 3 in Adrien *et al.*, 1985), but the dopamine content obtained from the remaining two kittens, K.P.-4 and -5, (20–22 ng/g) was about two times higher than that of age-matched control kittens (Fig. 3 in Adrien *et al.*, 1985). In these two kittens, however, the NE level was lowered to 38 and 10 percent of control, respectively. These biochemical findings are not self-consistent. Daw and associates (1985b) also concluded that intraventricular 6-OHDA did not decrease plasticity, although some of their kittens had more binocular cells following monocular deprivation than they were supposed to have.

Trombley and associates (1986) obtained basically similar results to those of Daw and associates but reached a slightly different conclusion. They concluded that intraventricular 6-OHDA can alter cortical plasticity, but the decrease in plasticity is not related to the depletion of endogenous NE in the visual cortex. They obtained the plasticity-suppressing effect of 6-OHDA only at a total dose of 4.8 mg or more, while 0.4–4.0 mg of 6-OHDA caused the maximal depletion of NE (nearly maximal depletion at 0.2 mg 6-OHDA).

Obvious difficulties in this study are derived from biochemical findings. First, it is left unexplained why the "low dose" of 6-OHDA (0.2 mg or less) was more effective than the "high dose" (4.8 mg or more) in depleting endogenous dopamine which was assayed concurrently with NE using the same cortical tissues. The maximal depletion (62% decrease) of dopamine was obtained at the dose of 0.1–6.25 μg of 6-OHDA which reduced NE by 53%. This is inconsistent with the generally known fact that the depletion effect of 6-OHDA is far stronger on NE than dopamine in a wide range of doses. Second, the more puzzling result was obtained for the effect of 6-OHDA on endogenous serotonin. There was an *inversely* dose-dependent depletion of serotonin by 6-OHDA; the maximal depletion (56% decrease) of endogenous serotonin was obtained in the visual cortex at the lowest dose of 6-OHDA in the visual cortex (0.01 μg) injected into the lateral ventricle. Again, no explanation was offered for these unusual findings in biochemistry which are little short of suggesting that, in their study, the primary effect of 6-OHDA was to deplete serotonin and dopamine but not NE in the visual cortex.

In summary, two of three reexaminations had uneasy problems in their biochemistry of 6-OHDA's effects on monoamines. In these studies, the suppressive effect of 6-OHDA on the shift in ocular dominance was smaller than that we observed earlier. The discrepancy remains, at best, unexplained at the moment.

III. Recent Development in the NE Hypothesis

As discussed above most of the recent "negative" results are derived from the studies which suffered from innate problems in their design, execution, and interpretation. Therefore, they would be by no means inconsistent with, as they stand, the NE hypothesis. Instead, they seem in effect to confuse the issue rather than redirect our search for cellular mechanisms and chemical factors underlying the visual cortical plasticity. Comparing all findings reported to date in this matter, the balance of evidence seems to favor, qualitatively as well as quantitatively, the NE hypothesis over the antithesis. The role of the central NE system in the regulation of visual cortical plasticity should be further investigated rather than simply dismissed on the basis of our current ignorance of underlying mechanisms for the link. It may turn out to be a sterile effort to explain how cortical 6-OHDA perfusion produces its effect that is independent of the central NE sytem, the study which was recently suggested by an anonymous critic.

Rather, what is most needed at the moment is to pursue cascades of neuronal events in the visual cortex following activation of NE-related receptors as a likely cellular basis of enhancing the plasticity. Indeed, one of our recent results (see Section III,D) in which we restore plasticity into the mature cortex by LC stimulation requires an explanation in its own right since no exogenous chemicals were used and the visual cortex was left totally intact. At the same time, knowing the multiplicity of the postnatal susceptibility period for experience-dependent plasticity in the visual cortex (e.g., Daw and Wyatt, 1976; Le Vay *et al.*, 1980), one should be open to any factors involved in the regulation of the plasticity other than the NE system.

A. POSTNATAL ONTOGENY OF β-ADRENERGIC RECEPTORS

Our preliminary results suggested the involvement of β-adrenergic receptors in the regulation of visual cortical plasticity (Kasamatsu, 1979, 1980; see also Section III,E). Therefore, we studied postnatal changes in endogenous NE as well as β-adrenergic receptor binding in the visual cortex of normally developing kittens (Jonsson and Kasamatsu, 1983). We used an HPLC method in combination with electrochemical detection developed by Keller and associates (1976) to measure endogenous CAs. We also used [^{3}H]dihydroalprenolol as a radioligand to label β-adrenergic receptors and quantify the number of their binding sites. The maturation curves of the two measures were quite different from each other. Endogenous NE increased, more or less, continuously toward adulthood with a small peak at the fourth and eleventh week. However, the total number of specific β-adrenergic receptor binding sites increased rapidly with age, showing a broad peak larger than the adult value between the fifth and thirteenth week after birth (maximally, 150% of the adult value) (Fig. 2). However, during the same period, the binding affinity remained constant. Thus, the postnatal maturation of β-adrenergic receptors, rather than the increase of NE content itself, may provide us with a likely biochemical basis for the sensitivity of cortical cells to visual experience in the early postnatal life. Further studies are needed to determine which of the two subtypes of β adrenoreceptors, β1 and β2 (Minneman *et al.*, 1981), contribute more to the peak of β adrenoreceptor binding than the other.

B. EXOGENOUS NE IN THE NORMAL VISUAL CORTEX

There may be arguments that the plasticity-enhancing effect of exogenous NE is due primarily to its direct chemical interference with

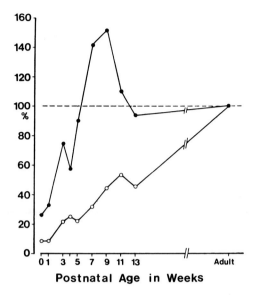

FIG. 2. Postnatal changes in β adrenoreceptor binding (●) and endogenous NE (○) in the occipital cortex of cats. (●) The tissue homogenate was incubated in 1 nM [³H]dihydroalprenolol ([³H]DHA, 45 Ci/mmol, New England Nuclear) at 37°C for 15 minutes. Nonspecific binding was determined by incubating homogenates in a mixture of 1 nM [³H]DHA and 1 μM (−)alprenolol. Specific binding of [³H]DHA was defined as the difference between the total and nonspecific bindings. The IC$_{50}$ value for (−)alprenolol was 1–2 nM. Hill coefficients were close to unity. (○) Cortical tissue was homogenized by ultrasonication in 310 μl 0.1 M perchloric acid containing an appropriate amount of internal standard (10–30 pmol α-methyl-dopamine). The extracts were purified by an Al₂O₃-absorption–desorption step and 10–20 μl of the extract was injected into the column for high pressure liquid chromatography with electrochemical detection (Keller et $al.$, 1976). The values were corrected for 50–70 percent recovery (reproduced from Jonsson and Kasamatsu, 1983).

the neuronal circuit responsible for ocular dominance rather than an increase in sensitivity of modifying such circuit. We have, however, ruled out such a possibility as follows: by perfusing the normal visual cortex with exogenous NE without altering the visual environment or visual input, we first obtained an unexpected bias in ocular dominance toward the contralateral eye, resulting in a very skewed distribution in favor of the cells receiving dominant input from the contralateral eye (Kasamatsu et $al.$, 1979; Kuppermann and Kasamatsu, 1984). Interestingly, however, the proportion of binocular cells remained close to normal ($\sim$70%). Next, we found that about 1 week was needed before such change became obvious. Finally, if the kittens were placed in darkness, depriving them of any visual experience, during the period of NE perfusion, no change whatsoever in ocular dominance was induced (Kuppermann and Kasamatsu, 1984). We interpreted the initial finding of the skewed ocular dominance as a result of an exaggerated influence of ongoing binocular interaction in the NE-pefused, otherwise normal, visual cortex in such a way that the innate dominance of the contralateral eye over the ipsilateral eye (Hubel and Wiesel, 1962; Blakemore and Pettigrew, 1970) is strengthened. Thus, we concluded that these results are consistent with and supportive of the NE hypothesis for visual cortical plasticity.

It may be added that the skewed distribution of ocular dominance seen in the NE-perfused cortex predicts a net result of ongoing binocular interaction in the normal visual cortex. The present result strongly suggests that we need to modify our customary explanation for the shift of ocular dominance in monocularly deprived animals as being due to binocular competition. The shift in ocular dominance may be a consequence of the lack rather than the enhancement of binocular competition (see Kasamatsu *et al.*, 1984, for further discussion).

C. Changes in Ocular Dominance in Acutely Anesthetized and Paralyzed Preparations

In 1976, I reported that ocular dominance in the visual cortex of acutely anesthetized and paralyzed young kittens could be changed quickly, in the order of a few minutes, by total but reversible blockade of one optic nerve before the optic chiasm with direct application of a small amount of local anesthetic (Kasamatsu, 1976). The rationale behind this study was that a brief binocular imbalance acutely created in total output from the retina may lead to an alteration of ocular dominance of given cortical cells if the intensity of the imbalance is strong enough to destabilize the existing synaptic connections in the visual cortex. In 60 percent of cortical units in that early study, I obtained enhancement of visually evoked unit responses from one eye at the expense of those from the other eye ipsilateral to the blocked optic nerve. A similar treatment of the adult visual cortex did not produce any notable changes. This preliminary finding, which was based on a relatively small number of cells ($N = 40$), has not yet been independently confirmed. Later, however, a related finding was reported based on the changes in the amplitude of field potentials which were evoked in the kitten visual cortex by stimulating the optic nerve. The amplitude of cortical evoked potentials was greatly augmented for several hours by preceding tetanic stimulation of one optic nerve. At the same time, similar potentials in response to stimulation of the other optic nerve were suppressed (Tsumoto and Suda, 1979).

On the other hand, a few groups of investigators have suggested that a high level of vigilance is necessary to alter ocular dominance of cortical cells in response to changes in visual experience. These conditions may be satisfied when the animal moves its eyeballs either passively under paralysis (Freeman and Bonds, 1979; Frégnac and Bienenstock, 1981) or actively (Tsumoto and Freeman, 1981), and possibly when the animal is visually attending at a target in visual space (e.g., Singer, 1982).

In the light of these contradictory suggestions, it is important to clarify whether ocular dominance can be changed in acutely anesthetized and totally paralyzed preparations. An unequivocal answer to this question should lead us to insight into cellular mechanisms underlying visual cortical plasticity, which would be independent of behavioral contexts of experimental manipulations and thus minimize the number of factors involved. When 0.5 M NE was iontophoresed, under general anesthesia and paralysis, into the visual cortex of a normal animal or animals pretreated with 6-OHDA, we noted a low incidence of recording binocular cells (Heggelund and Kasamatsu, 1981; Kasamatsu and Heggelund, 1982). Likewise, a substantial reduction of binocular cells (binocularity, 20–40%) was noted in the normal visual cortex perfused directly with 0.5 mM exogenous NE, which is 10-fold higher than the threshold concentration needed to restore plasticity to the 6-OHDA-pretreated cortex (Heggelund and Kasamatsu, 1981). It should be emphasized that this concentration refers to the NE solution stored in the continuous microperfusion system with an Alzet osmotic minipump (model 2001) and that the actual concentration of NE at the recording site is presumably at least 100-fold lower (Kasamatsu et al., 1981a). In the comparable recording condition, without exogenous NE, we usually obtain many binocularly driven cells (binocularity, 0.75 ± 0.10) (Kasamatsu et al., 1985) since the animals are raised in the normal visual environment throughout. What causes this difference in binocularity in the two experimental protocols? We wanted to interpret the significant reduction in binocularity in the NE-perfused, otherwise normal visual cortex as caused by an enhanced "squint" effect (Hubel and Wiesel, 1965), which is set in action due to breaking the binocular convergence during paralysis (acute paralysis squint). Apparently, the disturbing effects of the acute paralysis squint, which usually reaches several prism diopters, are tolerated by most binocular cells in the normal visual cortex with absence of exogenous NE.

A crucial test of the above interpretation gave us a satisfactory answer: in the NE-perfused, otherwise normal cortex, we obtained the normal distribution of ocular dominance only when a prism with an appropriate power was placed in front of one eye in order to superimpose approximately corresponding receptive fields for the two eyes (Heggelund *et al.*, in preparation). This finding, however, is subject to various alternative explanations since it depends on the comparison of two distribution patterns of ocular dominance obtained from two populations of cells recorded several hours apart from one visual cortex. We do not yet know whether one can truly change ocular dominance of a given cell at will during its recording.

D. RESTORATION OF PLASTICITY TO ADULT CORTEX BY ELECTRICAL STIMULATION OF THE LC

In our previous studies, we used exogenous NE to restore neuronal plasticity to the *aplastic* visual cortex by either pretreatment with 6-OHDA (kittens) or outgrowing of the usual duration of the postnatal susceptibility period (adult cats) (Pettigrew and Kasamatsu, 1978; Kasamatsu *et al.*, 1979). More recently, we wanted to restore plasticity to the matured visual cortex by increasing secretion of endogenous NE in response to electrical stimulation given to NE cells in the LC, thus leaving the visual cortex itself totally intact.

The animals were monocularly exposed to usual laboratory surroundings for 2 hours daily and otherwise kept in the dark for 6 consecutive days. During this brief period of monocular exposure, the animals received electrical stimulation with a train of short pulses delivered through a pair of chronically implanted bipolar electrodes in the LC. When we recorded from these animals on the seventh day, we found a significant loss of binocular cells (binocularity, ~41%) from the visual cortex, while no such change was noted in the cortex of control animals treated similarly without accompanying LC stimulation (Kasamatsu *et al.*, 1985). The changes in ocular dominance were not obtained in the visual cortex of the LC-stimulated cats if the cortex had been pretreated with the direct perfusion of 6-OHDA.

The animals were revived from the recordings and thereafter subjected to one of the following two treatments. (1) Daily brief monocular exposure was repeated without accompanying LC stimulation; in this case, we found that the plasticity thus restored in the mature cortex could be sustained, although with a declining trend, for at least the next three weeks, which was the longest term tested in this study. (2) The animals were subjected to the usual form of monocular lid suture and kept caged in a cat colony (light/dark = 16/8); in this case no such declining trend in the restored plasticity was noted, but the reduced state of binocularity either continued or even decreased further. Thus, we concluded that enhanced release of endogenous NE restores plasticity to the mature cortex and that such change can be maintained for at least 3 weeks thereafter (Kasamatsu *et al.,* 1985).

E. CONCENTRATION-DEPENDENT SUPPRESSION BY β-ADRENERGIC ANTAGONISTS OF VISUAL CORTICAL PLASTICITY

Previously, I reported a preliminary finding that the plasticity-suppressing effect of the 6-OHDA perfusion may be replicated by the local and continuous perfusion of the kitten visual cortex with D,L-propranolol, a common antagonist of β-adrenergic receptors (Kasamatsu, 1979). We followed up this series of studies by extending our examination to include dose–effect relationships as well as application of sotalol, a "cleaner" β blocker which does not have a "local anesthetic effect." The concentration of β blockers was changed widely from 10^{-6} to $10^{-2}\,M$ in an Alzet osmotic minipump. The expected shift in ocular dominance following monocular lid suture was suppressed in a concentration-dependent manner. The half-maximum effect was obtained at about $10^{-4}\,M$ propranolol and $10^{-5}\,M$ sotalol in an osmotic minipump (Fig. 3) (see also Shirokawa and Kasamatsu, 1986). The concentration of propranolol at the recording site, which was usually ~2 mm from the perfusion center, was calculated to be as $5.8 \times 10^{-7}\,M$. It should be pointed out that even with 10 mM D,L-propranolol, the highest concentration used in the present study, we were able to record many visually active cells with normal receptive field properties. Taken together with other preliminary findings that locally perfused dibutyryl cAMP, a chemical analogue of cAMP, can enhance the plasticity (Kasamatsu, 1980), the present results strongly suggest that β-adrenergic receptors are involved in the execution of NE-modulated neuronal plasticity in the visual cortex (Kasamatsu and Shirokawa, 1985; Shirokawa and Kasamatsu, 1986). Our recent study on the ontogeny of β-adrenergic receptors in the visual cortex (Jonsson and Kasamatsu, 1983) is certainly in harmony with this conclusion.

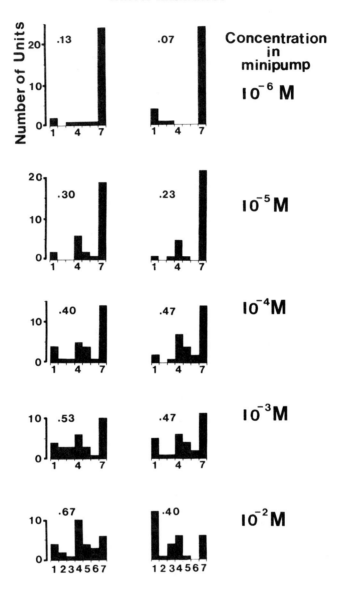

FIG. 3. Concentration-dependent blockade by β-adrenergic antagonists of the shift in ocular dominance following monocular lid suture. The visual cortex was locally perfused with various concentration (10^{-6} to 10^{-2} M in an Alzet minipump 2001) of either propranolol (left) or sotalol (right) during the week of monocular lid suture at the contralateral eye. Each ocular dominance histogram was derived from 30 visually active cells recorded along a single electrode track in the medial bank of area 17 of the kitten

IV. Conclusion and Perspectives

In the above discussion, I reviewed several reports which have been interpreted by some critics as being inconsistent with the NE hypothesis for visual cortical plasticity. The current evaluation was based primarily on the soundness of techniques used in such studies as well as the merits of their findings without directly referring to the NE hypothesis. It appears as if no consideration was given to the *specificity* of cause–effect relations in designing these experiments. For example, in their paradigms the visual cortex was not a unique region of the brain which had lost endogenous NE substantially (a question of the site of action). No information was available about how the remaining NE is distributed in the visual cortex of animals treated with 6-OHDA. No particular attention was paid to the condition of NE-related receptors through which NE should exert its actions. It should be stressed again that the absolute level of endogenous NE itself is not sufficient to indicate the strength of plasticity operating in a given visual cortex. Our recent study with β-adrenergic antagonists clearly showed that the disappearance of cortical plasticity could occur even in the presence of endogenous NE if β adrenoreceptors are functionally blocked within the visual cortex. Another 6-OHDA-free paradigm was successfully tested using clonidine, an α2-adrenergic agonist which is thought to decrease the release of NE from terminals besides other actions (Nelson *et al.,* 1985).

I strongly believe that what is needed most at the moment is not to explain how the local 6-OHDA perfusion produces its effect without reference to the central NE system but to pursue cascades of cellular events in the cortex following activation of β-adrenergic receptors by endogenous NE. Various results briefly summarized in Section III illustrate the current state of our quest for cellular mechanisms underlying the visual cortical plasticity. As a logical extension of the original NE hypothesis, we now propose the involvement of β-adrenergic

---

visual cortex. Ocular dominance groupings are based on Hubel and Wiesel's 7-group scheme (1962). The number given for each histogram refers to binocularity (number of group 2–6 cells divided by the total number of visually active cells). When the concentration of β blockers increased, not only did the binocularity increase, but the proportion of group 7 cells decreased, suggesting a decrease in neuronal plasticity. Note that the increase in binocularity became saturated at a lower concentration of sotalol than that of propranolol, due likely to an increase in the number of monocular cells which exclusively responded to stimulation of the deprived, contralateral eye (reproduced with minimal modification from Shirokawa and Kasamatsu, 1986).

receptors and cyclic nucleotides in visual cortical plasticity. Thus, we suggest that endogenous NE released from the NE terminals in the visual cortex works through the β-adrenoreceptor-cyclic nucleotide system and imprints the summary of visual experience, resulting in alterations in synaptic connectivity in the neocortex. It is worth emphasizing again that all the results that we have managed to obtain so far are consistent and can be uniquely explained in terms of the function of the central NE system.

To avoid unnecessary confusion, the following two points should be mentioned. First, under normal conditions, the plasticity-enhancing capability of the NE system is *not* saturated, since it is further augmented in response to an increase in availablility of NE to β adrenoreceptors. Second, current discussions do *not* necessarily exclude possible contributions of various factors other than the NE-β adrenoreceptor system in this matter. One such possibility was recently provided by studying the cholinergic projection system in the kitten brain. Bear and Singer (1986) reported that ocular dominance plasticity disappeared only when both NE and cholinergic inputs to the visual cortex had been reduced.

Recently, it was reported that the level of *in vitro* phosphorylation of microtubule-associated protein 2 (MAP-2) was low in the visual cortex of dark-reared cats but increased quickly to normal values upon brief exposure to light (Aoki and Siekevitz, 1985). This finding may be related to earlier results from the rat visual cortex which showed that a similar increase in tubulin biosynthesis occurred by briefly exposing dark-reared rats to light (Rose *et al.*, 1976; Perry and Cronly-Dillon, 1978). It is quite intriguing that the metabolism of cytoskeletal proteins in the visual cortex may be altered by manipulation of an animal's visual experience. In this connection, the following two findings may deserve particular attention, since both show the differential distribution of a few species of structural proteins within the same neuron. First, in the rat brain, MAP-2 appears to exist preferentially in the dendrites of Purkinje cells in the cerebellar cortex and pyramidal cells in the neocortex, whereas another species of MAP, termed MAP-1, can be immunocytochemically detected diffusely in various segments of the same neurons (Bloom *et al.*, 1984; Huber and Matus, 1984; De Camilli *et al.*, 1984). A similar observation with MAP-2 was made in the hippocampus and the cerebellum of rat brain (Caceres *et al.*, 1984). In addition, the association of MAP-2 immunoreactivity with postsynaptic densities was shown under electron microscopy (Caceres *et al.*, 1984). Second, in both the chicken cerebellar cortex

(Lazarides and Nelson, 1983) and the retina (Lazarides *et al.*, 1984), the subcellular distribution of erythroid (α,β',β)-spectrin appeared to be restricted to the cytoplasmic surface of the plasma membrane of neuronal somata and dendrites and to be absent from the axonal membrane. This contrasted with the distribution of brain (α,γ)-spectrin or fodrin which is present throughout in the cell body and axon of the neuron (Levine and Willard, 1981). In addition to possible changes in presynaptic terminals in the visual system as generally assumed, experience-dependent changes in structural proteins at the *postsynaptic* membrane may be another important factor involved in alteration of synaptic connectivity. It is obvious, however, that much must be learned before one can make a factual link between the experience-dependent modification of neuronal connectivity and changes in accumulation of special species of structural proteins within neurons in the involved brain area. This line of research may eventually enable us to devise a direct test of Hebb's postulate (1949), for example, by following a model proposed by Stent (1973) some years ago.

ACKNOWLEDGMENTS

I am grateful to W. J. Nelson for improving English in the text and M. King for typing the manuscript. Supported by USPHS Grant EY-05549.

REFERENCES

Adrien, J., Blanc, G., Buisseret, P., Frégnac, Y., Gary-Bobo, E., Imbert, M., Tassin, J.-P., and Trotter, Y. (1985). *J. Physiol. (London)* **367**, 73–98.
Aoki, C., and Siekevitz, P. (1985). *J. Neurosci.* **5**, 2465–2483.
Ary, M., Pettigrew, J. D., and Kasamatsu, T. (1979). *ARVO Abstr. Suppl. Invest. Ophthalmol. Vis. Sci.* **18**, 136.
Bear, M. F., and Daniels, J. D. (1983). *J. Neurosci,* **3**, 407–416.
Bear, M., and Singer, W. (1986). *Nature (London)* **320**, 172–177.
Bear, M. F., Paradiso, M. A., Schwartz, M., Nelson, S. B., Carnes, K. M., and Daniels, J. D. (1983). *Nature (London)* **302**, 245–247.
Blakemore, C., and Pettigrew, J. D. (1970). *Nature (London)* **225**, 426–429.
Bloom, G., Schoenfeld, T. A., and Vallee, R. B. (1984). *J. Cell Biol.* **98**, 320–330.
Caceres, A., Binder, L. I., Payne, M. R., Bender, P., Rebhun, L., and Steward, O. (1984). *J. Neurosci.* **4**, 394–410.
Daw, N. W., and Wyatt, H. J. (1976). *J. Physiol. (London)* **257**, 155–170.
Daw, N. W., Rader, R. K., Robertson, T. W., and Ariel, M. (1983). *J. Neurosci.* **3**, 907–914.
Daw, N. W., Robertson, T. W., Rader, R. K., Videen, T. O., and Coscia, C. J. (1984). *J. Neurosci.* **4**, 1354–1360.
Daw, N. W., Videen, T. O., Parkinson, D., and Rader, R. K. (1985a). *J. Neurosci.* **5**, 1925–1933.

Daw, N. W., Videen, T. O., Radar, R. K., Robertson, T. W., and Coscia, C. J. (1985b). *Exp. Brain Res.* **59**, 30–35.

Daw, N. W., Videen, T. O., Robertson, T. W., and Rader, R. K. (1985c). *In* "The Visual System" (A. Fine, ed.), pp. 133–144. Liss, New York.

De Camilli, P., Milles, P. E., Navone, F., Theurkauf, W. E., and Vallee, R. B. (1984). *Neuroscience* **11**, 819–846.

Freeman, R. D., and Bonds, A. B. (1979). *Science* **206**, 1093–1095.

Frégnac, Y., and Bienenstock, E. (1981). *Doc. Opthalmol. Proc. Ser.* **30**, 100–108.

Frégnac, Y., and Imbert, M. (1984). *Physiol Rev.* **64**, 325–434.

Guillery, R. W. (1972). *J. Comp. Neurol.* **144**, 117–130.

Hebb, D. O. (1949). "Organization of Behavior." Wiley, New York.

Heggelund, P., and Kasamatsu, T. (1981). *Adv. Physiol. Sci.* **36**, 233–242.

Heggelund, P., Imamura, K., and Kasamatsu, T. (1987). In preparation.

Hitchcock, P. F., and Hickey, T. L. (1982). *J. Neurosci.* **2**, 681–686.

Hubel, D. H., and Wiesel, T. N. (1962). *J. Physiol. (London)* **160**, 106–154.

Hubel, D. H., and Wiesel, T. N. (1965). *J. Neurophysiol.* **28**, 1041–1059.

Hubel, D. H., and Wiesel, T. N. (1970). *J. Physiol. (London)* **206**, 419–436.

Hubel, D. H., Wiesel, T. N., and Le Vay, S. (1977). *Philos. Trans. R. Soc. London Ser.* B **278**, 377–409.

Hubel, D. H., Wiesel, T. N., and Stryker, M. P. (1978). *J. Comp. Neurol.* **177**, 361–379.

Huber, G., and Matus, A. (1984). *J. Neurosci.* **4**, 151–160.

Jonsson, G., and Kasamatsu, T. (1983). *Exp. Brain Res.* **50**, 449–458.

Jonsson, G., Wiesel, F.-A., and Hallman, H. (1979). *J. Neurobiol.* **10**, 337–353.

Kasamatsu, T. (1976). *Exp. Brain Res.* **26**, 487–494.

Kasamatsu, T. (1979). *ARVO Abstr. Suppl. Invest. Ophthalmol. Vis. Sci.* **18**, 135.

Kasamatsu, T. (1980). *Abstr. Soc. Neurosci.* **6**, 494.

Kasamatsu, T. (1983). *Prog. Psychobiol. Physiol. Psychol.* **10**, 1–112.

Kasamatsu, T. (1986). *ARVO Abstr. Suppl. Invest. Ophthalmol. Vis. Sci.* **27**, 153.

Kasamatsu, T., and Heggelund, P. (1982). *Exp. Brain Res.* **45**, 317–327.

Kasamatsu, T., and Pettigrew, J. D. (1976). *Science* **194**, 206–209.

Kasamatsu, T., and Pettigrew, J. D. (1979). *J. Comp. Neurol.* **185**, 139–162.

Kasamatsu, T., and Shirokawa, T. (1985). *Exp. Brain Res.* **59**, 507–514.

Kasamatsu, T., Pettigrew, J. D., and Ary, M. (1979). *J. Comp. Neurol.* **185**, 163–182.

Kasamtsu, T., Itakura, T., and Jonsson, G. (1981a). *J. Pharmacol. Exp. Ther.* **217**, 841–850.

Kasamatsu, T., Pettigrew, J. D., and Ary, M. (1981b). *J. Neurophysiol.* **45**, 254–266.

Kasamatsu, T., Watanabe, K., Heggelund, P., and Schöller E. (1985). *Neurosci. Res.* **2**, 365–386.

Kasamatsu, T., Itakura, T., Jonsson, G., Heggelund, P., Pettigrew, J. D., Nakai, K., Watabe, K., Kuppermann, B. D., and Ary, M. (1984). *In* "Monoamine Innervation of Cerebral Cortex" (L. Descarries, T. A. Reader, and H. H. Jasper, eds.), pp. 301–319. Liss, New York.

Keller, R., Oke, A., Mefford, I., and Adams, R. N. (1976). *Life Sci.* **19**, 995–1004.

Kennedy, C., Des Rosiers, M. H., Sakurada, O., Shinohara, M., Reivich, M., Jehle, J. W., and Sokoloff, L. (1976). *Proc. Natl. Acad. Sci. U.S.A.* **73**, 4230–4234.

Kuppermann, B. D., and Kasamatsu, T. (1983). *Nature (London)* **306**, 465–468.

Kuppermann, D. B., and Kasamatsu, T. (1984). *Brain Res.* **302**, 91–99.

Lazarides, E., and Nelson, W. J. (1983). *Science* **220**, 1295–1296.

Lazarides, E., Nelson, W. J., and Kasamatsu, T. (1984). *Cell* **36**, 269–278.

Le Vay, S., Wiesel, T. N., and Hubel, D. H. (1980). *J. Comp. Neurol.* **191**, 1–51.

Levine, J., and Willard, M. (1981). *J. Cell Biol.* **90,** 631–643.

Minneman, K. P., Pittman, R. N., and Molinoff, P. B. (1981). *Annu. Rev. Neurosci.* **4,** 419–461.

Movshon, J. A., and Van Sluyters, R. C. (1981). *Annu. Rev. Psychol.* **32,** 477–522.

Nakai, K., Jonsson, G., and Kasamatsu, T. (1981). *Abstr. Soc. Neurosci.* **7,** 675.

Nakai, K., Jonsson, G., and Kasamatsu, T. (1987). In preparation.

Nelson, S. B., Schwartz, M., and Daniels, J. D. (1985). *Dev. Brain Res.* **23,** 39–50.

Olson, C. R., and Freeman, R. D. (1975). *J. Neurophysiol.* **38,** 26–32.

Onténiente, B., König, N., Sievers, J., Jenner, S., Klemm, H. P., and Marty, R. (1980). *Anat. Embryol.* **159,** 245–255.

Paradiso, M. A., Bear, M. F., and Daniels, J. D. (1983). *Exp. Brain Res.* **51,** 413–422.

Perry, G. W., and Cronly-Dillon, J. R. (1978). *Brain Res.* **142,** 374–378.

Pettigrew, J. D., and Kasamatsu, T. (1978). *Nature (London)* **271,** 761–763.

Rose, S. P. R., Sinha, A. K., and Jones-Lecointe, A. (1976). *FEBS Lett.* **65,** 135–139.

Sachs, C., and Jonsson, G. (1973). *J. Neurochem.* **21,** 1517–1524.

Sachs, C., and Jonsson, G. (1975). *Brain Res.* **99,** 277–291.

Schmidt, R. H., and Bhatnagar, R. K. (1979). *Brain Res.* **166,** 293–308.

Schneider, G. E. (1973). *Brain Behav. Evol.* **8,** 73–109.

Schneider, G. E. (1981). *TINS* **4,** 187–192.

Shatz, C. J., and Stryker, M. P. (1978). *J. Physiol. (London)* **281,** 267–283.

Sherman, S. M., and Spear, P. D. (1982). *Physiol. Rev.* **62,** 738–855.

Shirokawa, T., and Kasamatsu, T. (1986). *Neuroscience* **18,** 1035–1046.

Singer, W. (1982). *Hum. Neurobiol.* **1,** 41–43.

Stent, G. S. (1973). *Proc. Natl. Acad. Sci. U.S.A.* **70,** 997–1001.

Trombley, P., Allen, E. E., Soyke, J., Blaha, C. D., Lane, R. F., and Gordon, B. (1986). *J. Neurosci.* **6,** 266–273.

Tsumoto, T., and Freeman, R. D. (1981). *Exp. Brain Res.* **44,** 347–351.

Tsumoto, T., and Suda, K. (1979). *Brain Res.* **168,** 190–194.

CHAPTER 15

DEVELOPMENT OF THE NORADRENERGIC, SEROTONERGIC, AND DOPAMINERGIC INNERVATION OF NEOCORTEX

Stephen L. Foote

DEPARTMENT OF PSYCHIATRY
UNIVERSITY OF CALIFORNIA, SAN DIEGO
SCHOOL OF MEDICINE
LA JOLLA, CALIFORNIA 92093

and

John H. Morrison

DIVISION OF PRECLINICAL NEUROSCIENCE AND ENDOCRINOLOGY
RESEARCH INSTITUTE OF SCRIPPS CLINIC
LA JOLLA, CALIFORNIA 92037

I. Introduction

A. SCOPE OF THIS REVIEW

This review attempts to place in perspective the numerous developmental questions raised by the presence of extensive innervation of neocortex by at least three distinct sets of extrathalamic afferents. The three afferent systems which have been most intensively studied, the dopaminergic, serotonergic, and noradrenergic, each arise from a distinct set of cell bodies in the brainstem, and each system exhibits a unique pattern of innervation of neocortex. Two other such systems are known to exist; a cholinergic system arising in the basal forebrain and a peptidergic/histaminergic system arising in the hypothalamus. However, very limited data are available concerning their development, and they will not be discussed in this review. The initial section of this chapter will briefly describe the known principles of neocortical organization which will serve as a framework for the descriptive and conceptual issues raised by the data presented subsequently. Current knowledge about the innervation patterns exhibited by the three systems in adult animals will then be briefly summarized in order to make clear the neuroanatomical phenomena which must be explained by developmental studies.

CURRENT TOPICS IN
DEVELOPMENTAL BIOLOGY, VOL. 21

Following these descriptions of neocortex and its monoaminergic innervation, the available developmental data will be reviewed. These data are much less comprehensive than the data dealing with adult innervation patterns in three basic ways. First, in most developmental studies, the details of the developing patterns are difficult to discern, as the time points are not closely spaced or comprehensive. Second, while numerous neocortical regions have been studied in detail in adult organisms, a much smaller number have been studied in detail during development. Third, nearly all studies of the development of monoaminergic innervation patterns in neocortex have been performed in rodents or carnivores. Thus, the data available for primates are limited to a few time points and very few neocortical areas. In the concluding sections of this chapter, the conceptual issues raised by the available data are discussed. There are two major issues: (1) How do monoamine axons, arising at a considerable distance from the neocortex, find their way into the cortical mantle, distribute themselves in the manner appropriate to each transmitter system, and find their appropriate classes of target neurons? and (2) What developmental influence(s) do these fibers and their transmitters exert on the development of intrinsic cortical circuitry or the distribution of other cortical afferents, particularly with reference to the issue of plasticity? Each issue is discussed in a separate section at the end of the chapter.

B. Organizational Principles of Neocortex

The cerebral cortex develops from the telencephalon, the most rostral of the five major subdivisions of the mammalian brain. Phylogenetically, the cortex reaches its peak development in primates, most dramatically in the human species. Based on ontogenetic and phylogenetic principles as well as lamination schemes, the cerebral cortex is subdivided into archicortex, paleocortex, and neocortex. In lower mammals, a large portion may be archi- or paleocortex; however, in monkeys and man, the vast majority of cortex is 6-layered and thus classified as neocortex.

In a Nissl-stained section, two striking characteristics of neocortex are evident: (1) it is a laminated structure, in that certain cell types and bands of myelinated fibers occupy distinct zones that are oriented parallel to the cortical surface, and (2) the pattern of lamination varies systematically from one region to another. The lamination of neocortex focuses one's attention on the horizontal plane and the fact that it is a horizontally continuous sheet of gray matter. Early cytoarchitectonic studies emphasized variations in lamination patterns across the cortical surface. With meticulous care, investigators managed to subdivide the cortex into from fifty to several hundred independent cortical regions. Although some viewed the subdivisions as excessive and somewhat imaginative, the major cytoarchitectonic boundaries of these maps have subsequently been found to reveal important functional boundaries.

The earliest localization of function to particular subdivisions of neocortex was based on systematic relationships between specific behavioral deficits and the relatively crudely localized neuropathology resulting from trauma, strokes, or tumors. The neuroanatomical substrate for localized cortical function became firmly established when Polyak (1957), Walker (1938), and Le Gros Clark (1932), as well as others, documented the precise relationship between specific thalamic nuclei and cortical subdivisions. These three sources of information (cytocarchitectonics, clinical, pathway tracing) established a basic tenet of cortical organization: a given cortical field could be defined by (1) its specific and distinctive cytoarchitectonic pattern, (2) its unique set of extrinsic afferents, and (3) its specific "function." Rose and Woolsey (1958) proposed this three-way definition of a cortical field in the late 1940s. Although many complex modifications [for example, the work of Van Essen et al. (1979, 1982) in the visual system] to their scheme of cortical parcellation have emerged, most of their original observations are still valid.

The lamination of neocortex not only reflects a specific laminar distribution of cell types but is also closely related to patterns of origin and termination of cortical efferents and afferents. All major afferents have preferred layers of termination. For example, corticocortical fibers terminate primarily in layers I–III, whereas specific thalamic projections are directed at layer IV. Cells of origin for cortical efferents exhibit laminar segregation based on their target site; layer VI pyramidal cells project to thalamus, layer V pyramidal cells project to brain stem, layer III neurons to other cortical regions. Even though cortical regions vary in the degree to which they are dominated by a specific set of afferents or efferents, these laminar principles are evident throughout the neocortex.

The intrinsic circuitry of neocortex is extremely complex, yet it possesses many common elements of connectivity which persist across cortical regions. Lorente de No (1938) was the first to appreciate this dichotomy: neocortex is simultaneously regionally heterogeneous and globally homogeneous. It is globally homogeneous in that all regions share several common elements of intrinsic organization, but these are elaborated and embellished in different ways in various regions. Based on Lorente de No's observations and the electrophysiologic studies of Mountcastle (1957) and Hubel and Wiesel (1970, 1977), it has been proposed that the basic information processing unit in neocortex is a radically oriented (i.e., orthogonal to the cortical surface) module of cells, possibly columnar in shape. This putative column is a complex of cell–cell interactions that occurs largely in the radial dimension, with minimal horizontal interaction, that is driven by extrinsic afferents which terminate in a point-to-point, radially restricted fashion. One should not view this cortical column as a restricted anatomic entity but rather a radial unit with floating boundaries which are depenent upon the exact spatial distribution of afferent activity. Also, more recent data have demonstrated that there is ample opportunity for horizontal interactions in neocortex, particularly through the tangentially organized monoamine projections which are the subject of this chapter.

C. Noradrenergic Innervation of Adult Neocortex

The noradrenergic (NA) innervation of neocortex arises solely from the nucleus locus ceruleus (LC) which is located in the pontine brainstem. This nucleus innervates every major region of the neuraxis even though it is composed of a relatively small number of neurons; approximately 1600 per hemisphere in rat, 5000 per hemisphere in monkey, and 13,000 per hemisphere in human. Individual LC neurons often innervate widely separated brain regions, indicating that these axons must be highly divergent. It has also been demonstrated that individual LC neurons innervate different cortical regions. Thus, after an LC axon has traveled through the brainstem to the frontal regions of cortex, it diverges as it sweeps in a rostrocaudal trajectory through the cortical hemisphere, sending off collaterals to many different cortical regions and to both superficial and deep cortical layers (see Fig. 1) (Morrison et al., 1978, 1979, 1981; Loughlin et al., 1982). These NA fibers are found to be primarily distributed in a tangential (i.e., parallel to the cortical surface) fashion, particularly in layer VI, where they are oriented predominantly in the anteroposterior plane, forming a continuous sheet of longitudinal fibers overlying the white matter. In rat, these very fine caliber fibers branch to innervate all six layers of neocortex. The pattern of NA axon distribution possesses a geometric orderliness and distinct laminar pattern that is consistent throughout lateral neocortex (see Fig. 2). Rostrocaudally oriented tangential fibers predominate in layer VI, whereas layer I possesses a lattice of rostrocaudal and mediolateral tangential fibers. The layer VI axons branch profusely into upper layer V and layer IV, possibly representing terminal axons. The fibers in layers II and III are predominantly radial in orientation. The layer I fibers do not necessarily all arise from local fibers in deeper layers; some fibers may run long distances within this lamina (see below).

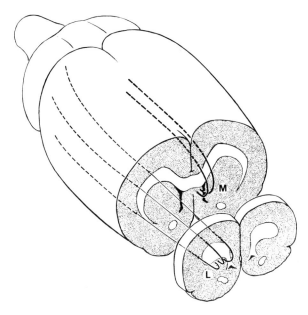

FIG. 1. Intracortical trajectory of NA fibers in rat. The majority of these ascending fibers follow one of two routes to the neocortex: a medial group (M) ascends through the septal area, and a lateral group (L) continues rostrally through the ventral telencephalon. Once they enter neocortex, these fibers form a continuous sheet of axons, largely within layer VI, that proceeds caudally throughout the longitudinal extent of the medial and dorsolateral cortex, supplying the cortical NA innervation throughout their trajectory. From Morrison *et al.* (1979), with permission.

The NA innervation of primate cortex exhibits a high degree of regional specialization in both density and laminar pattern of innervation (see Fig. 3) (Morrison *et al.*, 1982a,b, 1984; Levitt *et al.*, 1984a,b; Morrison and Foote, 1986). For example, primary somatosensory and motor regions are densely innervated in all six laminae while temporal cortical regions are very sparsely innervated. In primary visual cortex the density of innervation is intermediate, but there is a striking absence of fibers in lamina IV. As in rat, a strong tangential, intracortical trajectory is a dominant feature of the NA innervation of this much more convoluted cortex (Morrison *et al.*, 1982b). As in rat, the functional significance of these patterns is not yet understood. We have recently studied in detail the NA innervation of several cortical visual areas in monkey using immunohistochemical methods (Morrison and Foote, 1986). Occipital areas subserving visual functions are moderately innervated, and temporal areas involved in visual pattern analysis are sparsely innervated. Parietal areas primarily involved in spatial analysis of visual information display a much more dense net-

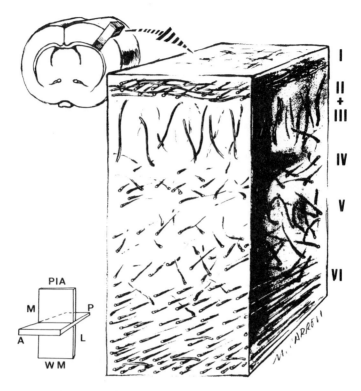

FIG. 2. This schematic three-dimensional illustration shows the orderly geometric pattern of noradrenergic innervation common to all areas of neocortex in the rat. The prevailing orientation of the NA axons within each cortical layer is emphasized. Axon diameter is exaggerated in order to enhance the three-dimensional relationship. Layer I contains long, tangential fibers oriented at right angles to each other. Layers II and III are typically traversed by numerous straight, radial fibers which rarely extend through layer I or IV. Layers IV and V contain many short, oblique axon segments, while layer V also has occasional radial and tangential fibers. The NA fibers in layer VI are, with few exceptions, tangential and have an anteroposterior orientation. Cortical layers are indicated with roman numerals. From Morrison *et al.* (1978), with permission.

work of NA fibers (see Fig. 4). Thus, the relative densities of regional innervation by NA fibers may have distinct functional implications, but their exact logic is not yet clear. Also, the functional implications of laminar differentiation are not yet understood. It is not known whether these laminar patterns reflect selective innervation of specific classes of neurons.

D. SEROTONERGIC INNERVATION OF ADULT NEOCORTEX

The serotonergic [5-hydroxytryptamine (5-HT)] innervation of rat neocortex arises primarily from the dorsal raphe nucleus. The 5-HT

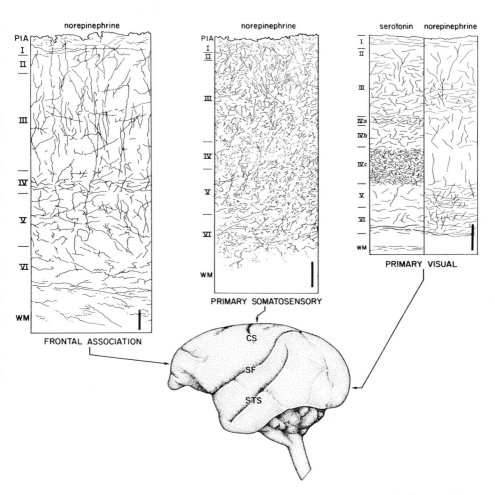

FIG. 3. NA innervation of three different regions of primate neocortex. In the case of visual cortex, both NA and 5-HT innervation patterns are shown. The three areas depicted are dorsolateral frontal association cortex, primary somatosensory cortex, and primary visual cortex. The frontal and primary somatosensory regions are similar in that fibers are present in all six layers. However, these regions differ with respect to laminar patterns of fiber orientation and relative density, and in that the primary somatosensory cortex is far more densely innervated than the dorsolateral frontal cortex. The laminar pattern of NA innervation in primary visual cortex differs fundamentally from these areas in that the NA innervation exhibits a high degree of laminar complementarity with the 5-HT innervation. WM, White matter; CS, central sulcus; SF, sylvian fissure; STS, superior temporal sulcus; PIA, pial surface. Bars, 200 μm. From Morrison and Magistretti (1983), with permission.

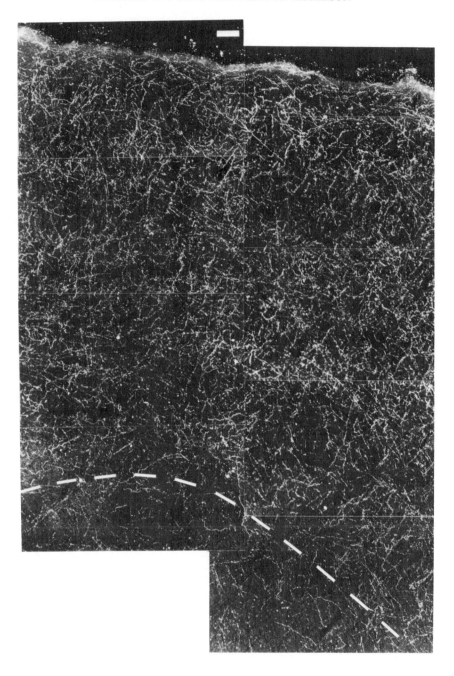

innervation of adult rat neocortex is more dense than is the NA innervation (Lidov *et al.*, 1980). The entire neocortical mantle is penetrated
by these fine, varicose, and highly convoluted fibers which appear in
relatively uniform density across all cortical layers. It has been noted
that the density and distribution of these fibers are such that every
neuron in neocortex might be innervated by a 5-HT fiber (Lidov *et al.*,
1980).

The 5-HT innervation of adult monkey neocortex is very dense, as
is the case in rat, but in monkey the innervation of different neocortical regions exhibits differences in density and laminar distribution
of axons (Morrison *et al.*, 1982a; Takeuchi and Sano, 1983; Kosofsky *et
al.*, 1984; Morrison and Foote, 1986). The 5-HT innervation of primary
visual cortex (area 17) has been characterized in greater detail than
the 5-HT innervation of any other region. The 5-HT fibers here are
very dense and are distributed in a strictly laminated fashion (see Fig.
5). The density and pattern of this innervation is strikingly different
from that exhibited by NA fibers in this area. The preference of 5-HT
fibers for layer IV in area 17 is also evident in other cortical areas.
This is especially interesting since this is the lamina that is the primary recipient of thalamocortical afferents. The morphology of 5-HT
fibers in cortex is quite heterogeneous with a mixture of thick non-
varicose fibers, large varicose fibers, and extremely fine varicose fibers.
Since the very thick fibers are most evident in white matter and the
deep cortical laminae, and are more evident in young animals, it is
highly likely that they are fibers of passage. These fibers are usually
tangential in orientation. As shown in Fig. 5, there are differences
between primate species in the distribution and density of 5-HT fibers
in area 17. There are also striking differences between the distributions and densities of 5-HT and NA fibers, the most striking being that
lamina IV contains the highest density of 5-HT fibers and the lowest
density of NA fibers (Morrison *et al.*, 1982a; Kosofsky *et al.*, 1984;
Foote and Morrison, 1984). We have examined the 5-HT innervation of
several cortical regions subserving visual functions (Morrison and
Foote, 1986). There is often a degree of complementarity between the

---

FIG. 4. Darkfield photomontage of NA fibers in area 7 (inferior parietal lobe) of
squirrel monkey neocortex. The dashed line at the bottom of the montage indicates the
boundary between layer VI and white matter. Note the uniform density of axons across
laminae and the relatively high degree of arborization. This visual area is more densely
innervated than primary visual cortex but less densely innervated than primary
somatosensory cortex. The 5-HT innervation of this area is less dense than that seen in
most other neocortical areas. Bar, 100 μm. From Morrison and Foote (1986).

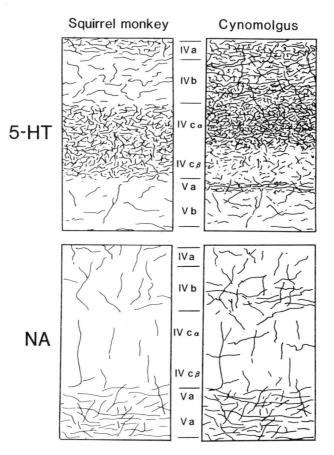

FIG. 5. Schematic illustrations of the laminar distributions of 5-HT and NA fibers in primary visual cortex of the squirrel monkey (New World) and the cynomolgus monkey (Old World). (Top) 5-HT innervation. Note that each sublamina of layers IVc and V in area 17 of cynomolgus has a distinct terminal pattern and density of innervation, whereas in the squirrel monkey the innervation of each of these laminae is homogeneous. (Bottom) NA innervation. Note the higher density in layer IVb of cynomolgus compared to layer IVb of the squirrel monkey. From Foote and Morrison (1984), with permission.

density of NA and 5-HT innervation in these areas. For example, as mentioned earlier, the NA innervation of temporal lobe visual areas is very sparse. 5-HT innervates these same areas very densely, as shown in Fig. 6.

E. DOPAMINERGIC INNERVATION OF ADULT NEOCORTEX

The dopaminergic (DA) innervation of neocortex has been studied most extensively in rat (see Lindvall and Björklund, 1984, for review).

This projection arises from cells in the ventral portion of the midbrain, close to the midline, which send ascending projections to distinct, circumscribed regions of cortex. In rat, the DA projection to neocortex is coextensive with the thalamocortical projection arising from the mediodorsal thalamic nucleus. Both afferents terminate in three distinct regions: the pregenual medial cortex, the anterior cingulate cortex, and the suprarhinal area. The pregenual area is largely rostral to the genu of the corpus callosum and constitutes the most anterior portion of the medial surface of the cerebral hemisphere, while the suprarhinal area can be considered a lateral and caudal extension of this dorsomedial termination zone. These two subareas of the prefrontal DA system arise from distinct subgroups of DA cell bodies in the ventral tegmental area. In both subareas, DA fibers are predominantly restricted to the deep layers of cortex, although in certain restricted zones fibers are also observed in more superficial layers. In contrast, the DA innervation of the anterior cingulate cortex is directed at layers I and III. This terminal plexus consists of extremely fine, highly arborized fibers and honors the cytoarchitectonic boundaries of the anterior cingulate cortex in that it terminates abruptly at the border between anterior and posterior cingulate cortex. In addition, the DA innervation of anterior cingulate cortex exhibits laminar complementarity with the NA innervation of this area; DA densely innervates the superficial layers and is largely absent from deep layers, while NA is primarily directed at the deep layers with a very sparse innervation of layers I–III (Lewis et al., 1979).

Biochemical data from monkey indicate that the DA innervation of primate neocortex may extend over a much larger area than does DA innervation in rodent (Brown and Goldman, 1977; Brown et al., 1979; Goldman-Rakic and Brown, 1982). For example, high DA levels are observed throughout the frontal lobe, although there is a clear decrease in DA concentrations in more posterior frontal areas. The temporal lobe also exhibits substantial levels of DA, while levels in parietal and occipital lobes are quite low. Fibers exhibiting catecholamine histofluorescence, some of which possess the same morphological characteristics as DA axons in rat, have been demonstrated in monkey frontal and temporal lobes (Levitt et al., 1984a,b). The fact that the technique used to visualize these axons creates the same fluorescent reaction product in both NA and DA fibers has prevented detailed analysis of the laminar and regional distribution of DA fibers. Our recent immunohistochemical observations (Lewis et al., 1986) indicate that DA axons are widespread in the frontal, temporal, and parietal lobes of monkey, and they exhibit a high degree of regional and lami-

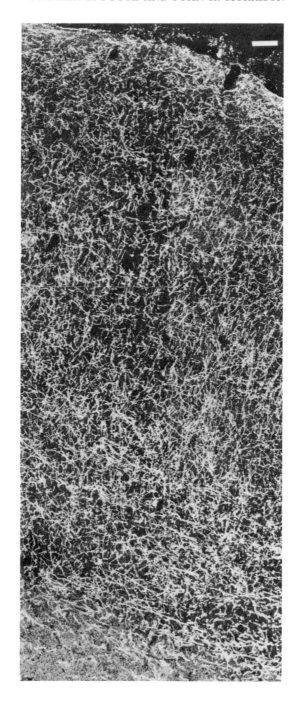

nar specificity (see Fig. 7). The highest densities are observed in the dorsomedial prefrontal areas, anterior cingulate and motor cortices, and inferior parietal lobule (Lewis *et al.*, 1986).

F. PRINCIPLES OF MONOAMINERGIC INNERVATION OF ADULT NEOCORTEX

The data just summarized reveal certain organizational principles which characterize the NA, 5-HT, and DA innervation of adult primate neocortex: (1) there is regional specialization of innervation density and pattern for all three systems, (2) each system exhibits a unique pattern of regional specialization, (3) 5-HT is, in general, much more dense than NA in neocortex, and (4) there are substantial species similarities, but distinct species differences, in laminar innervation patterns for each of these systems. These principles indicate that the NA and 5-HT systems exhibit a previously unsuspected degree of specialization in their innervation of neocortex and that the two systems are seeking out quite different targets upon which to terminate. Given the tangential nature of these systems, it is quite likely that a single axon may collateralize to innervate several cytoarchitectonic regions. This implies that an individual axon might have several collaterals, each of which exhibits a different pattern of arborization depending upon the exact cortical region it innervates. The characteristics of the monoamine innervation of primate neocortex suggest that, coincident with the extensive phylogenetic development and differentiation of neocortex in primates, there is a parallel elaboration and differentiation of the monoamine afferents, perhaps reflecting increased functional specialization of these systems in primate. In fact, this is also reflected in the pattern of DA innervation in rat and primate cortex. In rat, the DA projection is directed to three relatively small, circumscribed regions. The fact that the DA system is far more widespread in the primate should not be taken as a shift in the high degree of regional specificity present in rat. On the contrary, it may be that the DA system is directed at functionally homologous regions in both species;

FIG. 6. Darkfield photomontage of 5-HT-immunoreactive fibers in the caudal, inferior temporal lobe of adult cynomolgus monkey. The brain surface is evident at the top and subcortical white matter can be seen as a gray region in the lower left corner. Note the relatively uniform high density of fine-caliber, highly arborized fibers across all laminae, and the concentration of tangential, thick, poorly arborized fibers in the lower layers. This neocortical visual region is among those most densely innervated by 5-HT fibers and least densely innervated by NA fibers. Bar, 100 μm. From Morrison and Foote (1986).

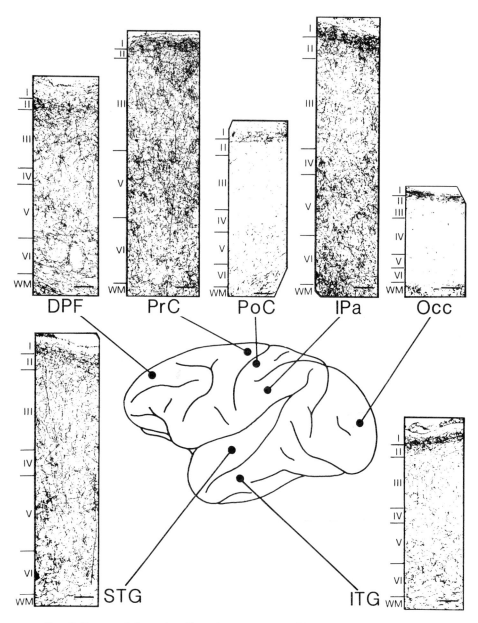

FIG. 7. Presumed dopamine fibers in seven cytoarchitectonic areas of cynomolgus monkey cortex. The fibers are visualized using an antiserum directed against tyrosine hydroxylase. In monkey cortex, this antiserum appears to specifically react with dopamine axons. Note the extensive regional heterogeneity in the density and laminar

however, in primate, the expansion and elaboration of the DA system reflects the tremendous increase in structural and functional specialization that has occurred in primate frontal and temporal lobes compared to rodent cortex.

II. Development of Noradrenergic Innervation

A. RODENT AND CARNIVORE

The neurons of the rat LC undergo their final division between embryonic day 11 (E11) and E13 (Lauder and Bloom, 1974), and norepinephrine synthesis appears as early as E13. Thus, these cells are biochemically competent very shortly after their appearance. The development of the NA innervation of neocortex has been studied with histochemical techniques in rat (Seiger and Olson, 1973; Levitt and Moore, 1979; Schlumpf et al., 1980; Lidov et al., 1978; Specht et al., 1981a,b; Verney et al., 1984; Berger and Verney, 1984) and mouse (Caviness and Korde, 1981). Schlumpf et al. (1980), using CA fluorescence, observed superficial cortical fibers as early as E16. These enter chiefly at the ventrorostral portion of neocortex then bifurcate into the deep and superficial layers of the cortex. Over the next few days, the innervation progresses in ventral-to-dorsal and rostral-to-caudal directions, covering the entire cortical hemisphere. The segregation between superficial and deep fibers is generally maintained until birth, with fibers only rarely crossing the cortical plate. Similar observations have been made by Levitt and Moore (1979), who found that the adult pattern and density of innervation is evident at the end of the first postnatal week. In the 6-day-old rat, Lidov et al. (1978) observed CA fibers in all layers and regions, although the superficial and deep plexi were still evident with relatively few fibers traversing the cortical plate. These CA fibers were eliminated by a midbrain lesion placed caudal to the DA cell bodies of the substantia nigra and ventral tegmental area, an observation which suggests that these fibers are NA and not DA in nature. These various observations have now been verified using immunohistochemical methods which specifically reveal NA fibers (Berger and Verney, 1984).

---

distribution of labeled fibers. Photographs are reversed images of darkfield photomontages. DPF, Dorsomedial prefrontal cortex (area 9); PrC, precentral (primary motor cortex); PoC, postcentral (primary somatosensory cortex); IPa, inferior parietal cortex (area 7); Occ, occipital (primary visual cortex); STG, rostral superior temporal gyrus (auditory association cortex); ITG, rostral inferior temporal gyrus (visual association cortex). Bars, 200 μm. From Lewis et al., J. Neurosci. in press, with permission.

In summary, the data presently available indicate that NA innervation is immature but present throughout all regions of the cortex at birth and that it becomes anatomically mature within the following week. This postnatal maturation, like the initial penetration, occurs along a rostral-to-caudal gradient. This is compatible with the notion that in the adult the major "trunk lines" run in the rostrocaudal dimension (Morrison *et al.*, 1981) and with biochemical data in rat (Loizou, 1972) which also indicate that there are continued increases in cortical norepinephrine levels after anatomically mature innervation patterns are evident. This suggests that the biochemical machinery of these axons continues to mature and become more robust after morphological development is complete. There is ultrastructural evidence for a dense monoaminergic innervation of neocortex in the newborn rat (Molliver and Kristt, 1975; Coyle and Molliver, 1977; Kristt, 1979), although this is difficult to reconcile with light microcsopic observations which indicate a limited innervation. The ultrastructurally visualized innervation is directed primarily at the primordium of layer IV. HRP injections into infant rat somatosensory cortex have been found to label only LC neurons and not raphe cells (Kristt and Silverman, 1980), suggesting that in rat, unlike monkey, this early innervation of layer IV is noradrenergic.

Caviness and Korde (1981) have examined the development of the monoamine innervation of mouse neocortex in normal and mutant animals. In the mutant animals particular cell types occupy different laminar positions than in normal animals. Thus, it is possible to determine whether ingrowing fibers innervate a particular lamina or seek out a particular cell type. These studies (Caviness and Korde, 1981) indicate that presumed NA axons seek out particular cell classes independent of the lamina in which they reside and independent of what terrain the axons must cross to arrive at their targets.

Itakura *et al.* (1981) studied the monoaminergic innervation of area 17 (primary visual cortex) in 6- to 8-week-old kittens with glyoxylic acid histofluorescence and performed ultrastructural studies on glyoxylic acid-treated tissue postfixed with permanganate. These authors suggest that NA fibers (as opposed to 5-HT or DA fibers) were preferentially visualized by these methods. Fluorescent fibers were observed in all layers and appeared to arborize most densely in layers II and III. Boutons containing dense-core vesicles were also observed in all layers. Jonsson and Kasamatsu (1983) have also studied the postnatal development of monoamine levels and receptor binding in cat occipital cortex. Receptor binding for 5-HT and NA developed at a slightly earlier age than did transmitter levels, and the 5-HT system seemed to mature earlier than did the NA system.

B. PRIMATE

Studies of the neurogenesis of LC neurons in primates (Levitt and Rakic, 1982) indicate that, as in rat, LC neurons are generated early in development (E30–E33). Very little is known about the development of the monoaminergic innervation of neocortex in primates. Although there have been studies of neocortex in human fetal material using the "smear" technique, no published studies have used histochemical methods to study the prenatal development of the monoamine innervation of neocortex in intact tissue sections. We have recently obtained preliminary data from one prenatal age and one neocortical area (Morrison and Foote, unpublished). These data suggest that the NA innervation of area 17 is essentially nonexistent at day 120 of gestation in rhesus monkey. Such "negative" evidence must be viewed with caution. In this study, NA fibers were visualized using an antiserum directed against dopamine β-dydroxylase (DBH), and it is possible that such a method will not detect the initial arrival of axonal processes originating from the LC. In DA projection fields, however, immunohistochemical methods reveal tyrosine hydroxylase (TH) immunoreactivity in the axonal growth cone (Berger *et al.*, 1982; Specht *et al.*, 1981a,b) so it is entirely possible that DBH immunoreactivity is a very sensitive indicator of ingrowing NA axons. At any rate, present evidence (Morrison and Foote, unpublished observations; Foote and Morrison, 1984) indicates that NA fibers invade area 17 during the last half of gestation. If there exists a rostrocaudal gradient in the development of NA innervation in primates that is similar to that demonstrated in rats, NA innervation of more rostral areas would occur at earlier stages. No data are available to address this issue.

The postnatal development of the NA innervation of area 17 has been studied in detail in monkey (*Macaca fascicularis;* Foote and Morrison, 1984). At birth, there is a very low density of poorly arborized NA fibers evident in all cortical layers (see Figs. 8 and 9). Over the next 60 days, these fibers increase in density and arborize extensively (see Fig. 10). By 60 days of age, the adult patterns of laminar distribution of fibers are present, although the density of innervation is somewhat lower. There are no anatomical studies describing the development of NA innervation in other brain regions. Biochemical data (Goldman-Rakic and Brown, 1982) indicate that, as in rat, there are continuing increases in norepinephrine storage capacity and rate of

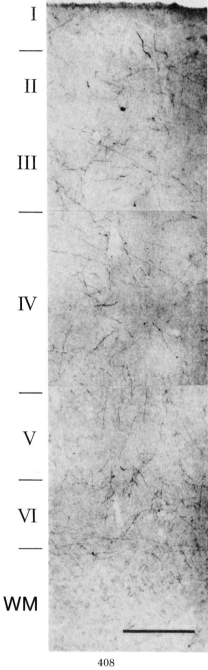

I

II

III

IV

V

VI

WM

408

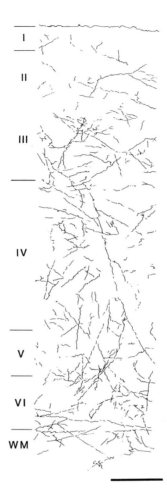

FIG. 9. Distribution of NA fibers in primary visual cortex of a 1-day-old cynomolgus monkey. This illustration is an accurate tracing of the DBH-immunoreactive fibers in the photomontage shown in Fig. 4 except that the caliber of fibers has been slightly enlarged for clarity. The method used to produce this drawing was also used to produce those shown in Figs. 6 and 7. Cortical laminae are indicated on the left. WM, Subcortical white matter. Bar, 200 μm. From Foote and Morrison (1984), with permission.

FIG. 8. NA fibers in primary visual cortex of a 1-day-old cynomolgus monkey. DBH-immunoreactive fibers are shown in a brightfield image of a 40-μm thick section. Cortical laminae are indicated at the left. WM, White matter. Bar, 200 μm. From Foote and Morrison (1984), with permission.

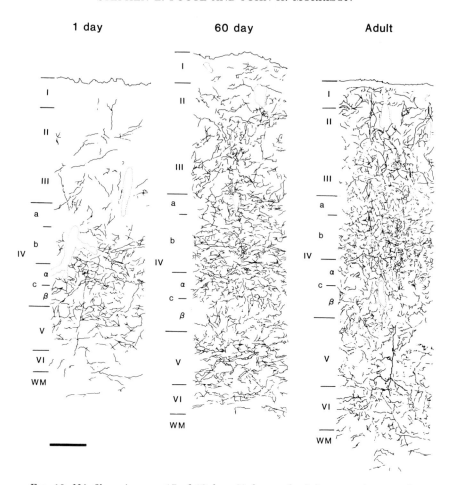

1 day   60 day   Adult

FIG. 10. NA fibers in area 17 of 18-day, 60-day, and adult cynomolgus monkeys. These illustrations are accurate tracings of all DBH-immunoreactive fibers visible in photomontages. Note the increased density and arborization with increasing age and the correlated appearance of a decreased density of fibers in layer IVc. WM, White matter. Bar, 200 μm. From Foote and Morrison (1984), with permission.

synthesis long after morphological maturity is achieved. In frontal areas, these biochemical data indicate that there may be a temporary increase in transmitter levels at 5–8 months of age, a subsequent reduction, and then a maturation to adult levels over the next several months.

III. Development of Serotonergic Innervation

A. RODENT

The appearance and differentiation of raphe 5-HT cells have been shown to occur early in embryonic life in rat (e.g., Lauder and Bloom, 1974; Lidov and Molliver, 1982c; Wallace and Lauder, 1983). By E19 the neurons have migrated into subgroups that are similar to those observed in the adult. The development of 5-HT innervation in rat neocortex has been studied with immunohistochemical methods (Lidov and Molliver, 1982a). There are three phases of ontogeny: a period of initial axon elongation (E13–E16), the differentiation of selective pathways (E15–E19), and a period of terminal field innervation and elaboration (E19–P21). Like NA afferents, 5-HT fibers initially enter the cortical anlage as two tangential and parallel sheets, one above and one below the cortical plate. Also like NA afferents, 5-HT fibers gradually extend in a caudal direction so that the occipital region is among the last to be innervated. Finally, like NA afferents, 5-HT fibers gradually penetrate the middle layers of cortex from above and below to generate terminal arborizations (see Fig. 11). Although this sequence of events is similar for the two transmitters, these events occur earlier for the NA system. NA innervation achieves adult density and pattern by the 7th postnatal day (Lidov *et al.*, 1978; Levitt and Moore, 1979), whereas the 5-HT innervation approaches the adult density by 3 weeks of age (Lidov and Molliver, 1982a; Molliver, 1982). Thus, in the rat, 5-HT terminals may be one of the last extrinsic afferents to innervate the cortex.

B. PRIMATE

At birth, 5-HT fibers in area 17 are dense, highly arborized, and concentrated in layer IV (see Fig. 12). Similar observations have been made in Old World (Foote and Morrison, 1984) and New World (Morrison and Foote, unpublished) monkeys. Over the following 60 days, these fibers increase dramatically in density and arborize extensively (see Fig. 13). By 60 days of age, the adult patterns of laminar distribution of fibers are present, although the density of innervation is somewhat lower. There are no anatomical studies describing the development of 5-HT innervation in any other brain region in primates.

The immunohistochemical data (Foote and Morrison, 1984; Morrison and Foote, unpublished observations) are compatible with the biochemical data of Goldman-Rakic and Brown (1982) from rhesus monkeys which indicates that, at birth, there is a much higher level of

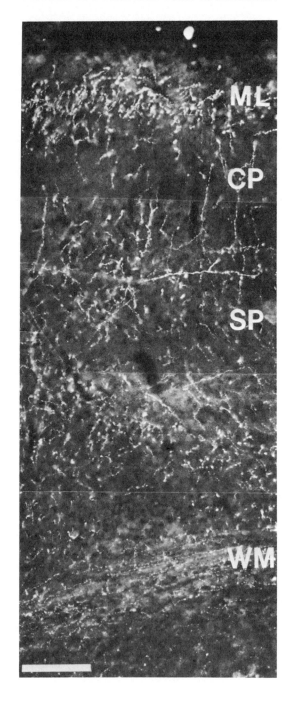

5-HT than of norepinephrine in occipital cortex. The concentrations of both transmitters increase rapidly over the next 60 days, by which time they are approaching adult levels. Throughout life, 5-HT levels are approximately three times the norepinephrine levels in occipital cortex (see also Brown and Goldman, 1977). The immunohistochemical data also reflect the biochemical observation of an inverse relationship between NA and 5-HT synthesis rates in a region-by-region analysis (Goldman-Rakic and Brown, 1982). Thus, the biochemical and immunohistochemical data both reflect the immaturity of these systems at birth, their rapid development over the next 60 days, their relative maturity at this age, and the earlier and greater density of the 5-HT system.

As has been demonstrated in several studies (Morrison *et al.*, 1982a; Takeuchi and Sano, 1983; Kosofsky *et al.*, 1984), layer IV is the primary recipient of the 5-HT projection to area 17 in the adult primate. Immunohistochemical studies of area 17 from 120-day postgestational (60-day prenatal) rhesus monkeys have shown that the 5-HT innervation of layer IV is quite dense (Morrison and Foote, unpublished). While layer IV of area 17 already possesses a rather dense terminal plexus at this age, few fibers were evident in other layers or in any layer of the adjacent area 18 (Morrison and Foote, unpublished). Thus, it appears that the 5-HT innervation of layer IV in area 17 arrives at an early stage of development and is, in fact, present when the thalamocortical fibers arriving from the lateral geniculate nucleus begin to form connections.

IV. Development of Dopaminergic Innervation

The development of the DA innervation of neocortical regions has been studied with anatomical methods only in rat (Verney *et al.*, 1982; Berger and Verney, 1984). Immunohistochemical methods reveal the first sign of DA fibers at E16 in the recently formed intermediate zone of the anterior frontal cortex. By E18, two distinct bundles of fibers are evident, one in a lateral and one in a medial position. Thus, DA axons reach the neocortical anlage one day prior to the arrival of NA fibers. However, unlike NA fibers which penetrate the cortical plate from E18 onward, the DA fibers, like thalamocortical fibers which also begin to

---

Fig. 11. 5-HT-immunoreactive fibers in lateral (suprarhinal) neocortex; coronal section from a 1-day-old rat. Note that the majority of immunoreactive axons is above and below the cortical plate (CP). A dense band of axons follows the white matter (WM) of the external capsule at the bottom. SP, Subplate; ML, molecular layer. Bar, 100 μm. From Lidov and Molliver (1982a), with permission.

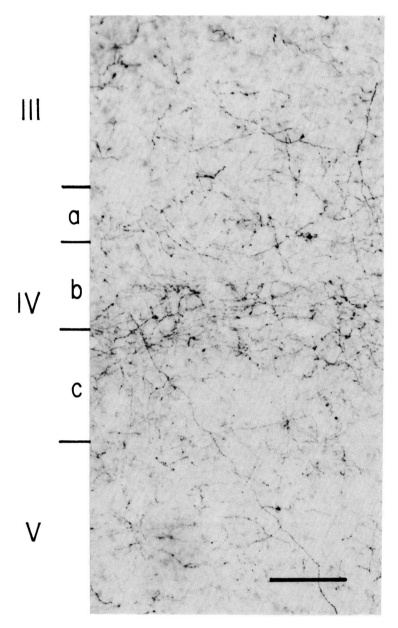

III

IV a

b

c

V

Fig. 12. 5-HT fibers in layers III, IV, and V of area 17 in a 1-day-old cynomolgus monkey. The greatest density of fibers is centered on the boundary between layers IVb and IVc. Bar, 100 μm. From Foote and Morrison (1984), with permission.

1 day   60 days   Adult

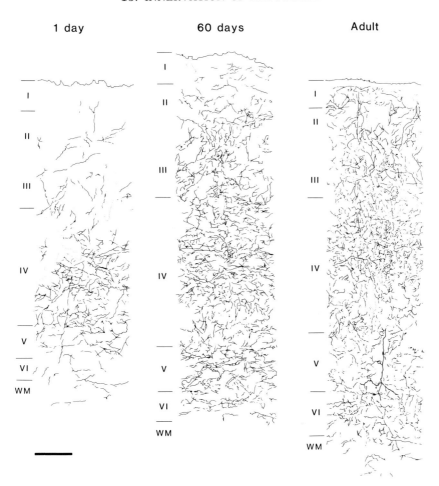

FIG. 13. Distribution of 5-HT fibers in area 17 of 1-day-old, 60-day-old, and adult cynomolgus monkeys. These illustrations are accurate tracings of all 5-HT-immunoreactive fibers visible in photomontages. Cortical laminae are labeled for each illustration. WM, Subcortical white matter. Bar, 200 μm. From Foote and Morrison (1984), with permission.

arrive on E16, remain confined to the intermediate zone for several days before arborizing into the cortical plate. The DA innervation of frontal and suprarhinal zones is well developed by P3, with entorhinal innervation lagging behind. In cingulate cortex, the innervation of deep layers proceeds apace with the development of other cortical systems, but the innervation of more superficial layers is delayed by many days.

Some studies have suggested that there may be a much earlier maturation of the prefrontal DA projection, or perhaps a development to above-adult levels with a subsequent reduction (Schmidt *et al.*, 1982; see Levitt and Moore, 1979, for an analogous phenomenon in the NA system). Although there has been no anatomical characterization of the development of DA innervation of primate neocortex, biochemical studies indicate that DA levels in different subregions of the frontal lobe show complex developmental sequences (Goldman-Rakic and Brown, 1982). Prefrontal and premotor regions, for example, show evidence of a peak at an early stage of development, a subsequent decrease, and then a gradual maturation to adult levels (Goldman-Rakic and Brown, 1982). By contrast, DA concentrations in precentral motor cortex show a peak at 5–8 months of age and then a decline to adult levels, producing the anterior-to-posterior gradient in concentrations that is seen in adult animals (Goldman-Rakic and Brown, 1982).

V. Issues Raised by the Existing Data about the Development of Monoamine Innervation Patterns in Neocortex

The most important issues concerning the development of monoaminergic innervation patterns in neocortex are evident from examination of adult innervation patterns. First, what are the factors determining which particular monoamine neurons within the brainstem nuclei will project to neocortex? Retrograde transport studies (Loughlin *et al.*, 1986a,b; Waterhouse *et al.*, 1983) indicate that only a specific subset of LC neurons, a circumscribed group of ventral tegmental area neurons, and certain raphe neurons project into neocortex. It is not clear whether it is the physical location of these subsets or whether it is a differentiation process that is independent of position that results in their projecting to neocortex. Second, how do monoaminergic axons, arising at great distances in the brainstem, guide themselves to the neocortical mantle? Third, what factors determine differences in regional densities of innervation? This problem is most clearly evident in primate studies, where there are striking differences in the density with which NA fibers penetrate different cortical areas. In one sense, this is a more critical problem for tangenitally organized systems than for radially organized afferents. Most thalamic nuclei project largely to one cortical region and avoid all others. The NA system projects to all cortical areas and arborizes to an extent characteristic for each region. Thus, an individual axon may express very different degrees of arborization in different cortical areas, implying that there are important cues in the target region which determine the terminal pattern. Fourth, how is laminar specialization of arborization guided. This is an especially pressing problem since different laminae are preferentially innervated in different functional areas of neocortex. This question may be the same or distinct from a possible fifth major question; that is, how does the afferent axon find and terminate on its appropriate class of target cell? Finally, it should be clear that each monoamine system must have unique mechanisms for most of these problems, since each exhibits a unique distribution of axons between and within the various functional regions of the neocortex.

The data from rodents differ substantially from observations in primates in two regards: (1) for 5-HT, and possibly for NA, the primate data do not reveal the "outside-in" pattern so readily evident in rodents, and (2) in primates, 5-HT innervation developes earlier than NA innervation and is more dense at all ages examined. In primates, 5-HT and NA differ quite strikingly from each other in developmental pattern in area 17. Initially, 5-HT fibers densely innervate layer IVb and gradually expand their domain. NA fibers, by contrast, initially penetrate all layers sparsely and gradually arborize in those laminae destined to receive substantial innervation. These interpretations must be tempered by possible differences between the two primary antisera in their ability to reveal developing fibers. The available data probably reveal fibers beginning at the onset of transmitter synthesis in terminal regions, an event which may lag behind the structural development of these axons.

The results presented above suggest the following major conclusions. First, there are substantial differences between NA and 5-HT innervation density and pattern throughout the period studied. Second, these patterns develop from an obviously immature state to an essentially mature level within the 60 days after birth. Third, there is a smooth transition between birth and adulthood in this development. Fourth, the distinctive sublaminar specialization of area 17 in *M. fascicularis* is evident in monoamine innervation patterns in both adult and young animals.

These observations on the ingrowth of 5-HT and NA fibers must obviously be viewed in the context of other developmental events in primary visual cortex. Although neurogenesis is probably complete at birth (Rakic, 1985), the first 60 days of postnatal life are a period of rapid maturation for other aspects of cortical anatomy. It is possible, for example, that the development of monoaminergic innervation patterns accompanies the progressive maturation of termination sites for other inputs on target neurons. Golgi studies of developing neurons in monkey area 17 indicate that the number of dendritic spines on various types of neurons increases from birth to postnatal week 8 (Lund *et al.*, 1977; Boothe *et al.*, 1979). This increase is followed in some types of neurons by a less rapid decrease in spine number and in others by a prolonged plateau and then a decrease (Boothe *et al.*, 1979). Ultrastructural studies in cat (Cragg, 1975) describe an increase in the number of synapses per neuron in area 17 between postnatal day 8 and 7 weeks of age. There is then a reduction in the number of synapses. This suggests that spine counts reflect changes in the number of synapses. Other studies in monkey indicate that lateral geniculate nucleus terminals in layer IV initially show overlap between the two eyes and that there is a gradual retraction to well-defined ocular dominance columns which begins just before birth and is completed by 3 to 6 weeks postnatally (Rakic, 1976; Hubel *et al.*, 1977). Thus, the first several weeks after birth are a period during which spine and synapse numbers peak, thalamic afferents become segregated by eye-of-origin, ocular dominance and other features exhibit plasticity, and monoaminergic fibers show maturation into adult densities and patterns. Insufficient data are available to characterize the interrelationships among these events. For example, the hypothesis that the elaboration of monoaminergic innervation patterns is dependent upon maturation of cortical target neurons must be tempered by studies in rat of the effects of antimitotic agents. These studies indicate that monoaminergic fibers produce the same quantity of nerve terminal arborizations even when cortical development is stunted by such agents (Johnston and Coyle, 1979; Johnston *et al.*, 1979; Jonsson and Hallman, 1982).

VI. Do Monoamines Influence the Development of Other Cortical Circuitry?

It has been suggested that the monoamines, especially the LC/NE system, influence the development of other neural structures. This is partly because of their precocious development in rat (Olson and Seiger, 1972; Lauder and Bloom, 1974), in monkey (Levitt and Rakic, 1982), and in human (Nobin and Bjorklund, 1973; Olson *et al.*, 1973). They have been shown to develop the ability to synthesize their putative transmitters at an early stage of development (Olson and Seiger, 1972; Golden, 1973; Lauder and Bloom, 1974; Choi *et al.*, 1975), and these transmitters have been shown to have complex, long-lasting effects on postsynaptic biochemical processes which might mediate developmental influences (McMahon, 1974). Levitt and Rakic (1982) have demonstrated that in rhesus monkeys the peak genesis of locus coeruleus and many raphe neurons precedes the peak genesis of visual cortex neurons by approximately 45 days (visual cortex data from Rakic, 1974).

There have been reports of specific anatomical changes in neocortical target areas of the NA system in rats after perinatal lesions of the LC or its efferents (Maeda et al., 1974; Blue and Parnavelas, 1982; Felten et al., 1982; Parnavelas and Blue, 1982; see also Amaral et al., 1980), although others have reported no such changes (Wendlandt et al., 1977; Ebersole et al., 1981; Lidov and Molliver, 1982b). The most intensively investigated claim of dependence of a developmental process on NA innervation is the demonstration that postnatal plasticity in the primary visual cortex of the cat is dependent upon the integrity of the NA innervation of this cortical region (reviewed by Kasamatsu, this volume; Kasamatsu and Pettigrew, 1976; Pettigrew and Kasamatsu, 1978; Kasamatsu et al., 1981; Daw et al., 1983), although this finding has been controversial (Bear and Daniels, 1983; Daw et al., 1985). Daw et al. (1983) have demonstrated that 6-OHDA administration into the visual cortex prevents both ocular dominance plasticity, as previously shown by Kasamatsu and Pettigrew (1976), and plasticity of directional selectivity. This is of interest since the critical periods for these two types of plasticity differ (Daw and Wyatt, 1976; Berman and Daw, 1977), and the synaptic basis of ocular dominance is different from that of directional sensitivity (Sillito, 1977; Sillito et al., 1981). It should be noted that there has been no demonstration of NA effects on visual cortex plasticity in primates.

Normal development of ocular dominance slabs in area 17 of the monkey depends partially on binocular competition before birth (Rakic, 1983). Our observations on primary visual cortex of prenatal rhesus monkeys indicate that a dense 5-HT innervation of layer IV is evident at E120. The early presence of this dense 5-HT innervation at an age when the NA innervation is still extremely sparse, and the fact that 5-HT innervation is, initially and continually, most dense in layer IV, suggests that 5-HT may be much more likely to participate in early plasticity, especially of ocular dominance which seems to depend crucially upon synaptogenic events in this lamina (Rakic, 1976; Hubel et al., 1977; Le Vay et al., 1978, 1980; Shatz and Stryker, 1978). Le Vay et al. (1980) found that in macaque monkeys ocular dominance plasticity is maximal from birth to 6 weeks of age, coincident with the normal segregation of geniculocortical fibers into ocular dominance zones. Reduced plasticity is evident for 3 additional weeks. Thus, plasticity is also inversely proportional to the maturity of monoaminergic innervation patterns, with plasticity no longer being evident once the monoamines have become mature. This is difficult to reconcile with the proposed role for NE in mediating plasticity.

Although it is possible that the NA and 5-HT systems mediate some aspect of morphological development, another, not mutually exclusive possibility, is that the development of these inputs to area 17 mediates the gradual imposition of brainstem control on cortical electrophysiological activity. Substantial effects of sleep and arousal on cortical visual receptive fields have been described (Livingstone and Hubel, 1981). These effects are very similar to those produced by iontophoresis of norepinephrine onto visual, auditory, or somatosensory cortex neurons (Foote *et al.*, 1975; Waterhouse and Woodward, 1980; Kasamatsu and Heggelund, 1982; reviewed in Foote *et al.*, 1983). No data are available to determine the age at which this type of brainstem control is imposed on neocortical neuronal activity.

REFERENCES

Amaral, D., Avendano, G. C., and Cowan, W. M. (1980). *J. Comp. Neurol.* **194**, 171–191.

Bear, M. F., and Daniels, J. D. (1983). *J. Neurosci*, **3**, 407–416.

Berger, B., and Verney, C. (1984). *Neurol. Neurobiol.* **10**, 95–121.

Berger, B., Di Porzio, U., Daguet, M. C., Gay, M., Vigny, A., Glowinski, J., and Prochian, A. (1982). *Neuroscience* **7**, 193–205.

Berman, N., and Daw, N. W. (1977). *J. Physiol. (London)* **265**, 249–259.

Blakemore, C., Garey, L. J., and Vital-Durand, F. (1978). *J. Physiol. (London)* **283**, 223–262.

Blue, M. E., and Parnavelas, J. G. (1982). *J. Comp. Neurol.* **205**, 199–205.

Boothe, R. G., Greenough, W. T., Lund, J. S., and Wrege, K. (1979). *J. Comp. Neurol.* **186**, 437–490.

Brown, R. M., and Goldman, P. S. (1977). *Brain Res.* **124**, 576–580.

Brown, R. M., Crane, A. M., and Goldman, P. S. (1979). *Brain Res.* **168**, 133–150.

Caviness, V. S., Jr., and Korde, M. G. (1981). *Brain Res.* **209**, 1–9.

Choi, B. H., Antanitus, D. S., and Lapham, L. W. (1975). *J. Neuropathol. Exp. Neurol.* **34**, 507–516.

Coyle, J. T., and Molliver, M. E. (1977). *Science* **196**, 444–447.

Cragg, B. G. (1975). *J. Comp. Neurol.* **160**, 147–166.

Daw, N. W., and Wyatt, H. J. (1976). *J. Physiol. (London)* **257**, 155–170.

Daw, N. W., Rader, R. K., Robertson, T. W., and Ariel, M. (1983). *J. Neurosci.* **3**, 907–914.

Daw, N. W., Videen, T. O., Parkinson, D., and Rader, R. K. (1985). *J. Neurosci.* **5**, 1925–1933.

Ebersole, P., Parnavelas, J. G., and Blue, M. E. (1981). *Anat. Embryol.* **162**, 489–492.

Felten, D. L., Hallman, H., and Johnson, G. (1982). *J. Neurocytol.* **11**, 119–135.

Foote, S. L., and Morrison, J. H. (1984). *J. Neurosci.* **4**, 2667–2680.

Foote, S. L., Freedman, R., and Oliver, A. P. (1975). *Brain Res.* **86**, 229–242.

Foote, S. L., Bloom, F. E., and Aston-Jones, G. (1983). *Physiol. Rev.* **63**, 844–914.

Frigon, R. P., O'Connor, D. T., and Levine, G. L. (1981). *Mol. Pharmacol.* **19**, 444–450.

Golden, G. S. (1973). *Dev. Biol.* **33**, 300–311.

Goldman-Rakic, P. S., and Brown, R. M. (1982). *Dev. Brain Res.* **4**, 339–349.

Hubel, D. H., and Wiesel, T. N. (1970). *J. Physiol. (London)* **206**, 419–436.

Hubel, D. H., and Wiesel, T. N. (1977). *Proc. R. Soc. London Ser. B* **198**, 1–59.

Hubel, D. H., Wiesel, T. N., and LeVay, S. (1977). *Philos. Trans. R. Soc. London Ser. B* **278**, 377–409.

Hubel, D. H., Wiesel, T. N., and Stryker, M. P. (1978). *J. Comp. Neurol.* **177**, 361–379.

Humphrey, A. L., and Hendrickson, A. E. (1983). *J. Neurosci.* **3**, 345–358.

Itakura, T., Kasamatsu, T., and Pettigrew, J. D. (1981). *Neuroscience* **6**, 159–175.

Johnston, M. V., and Coyle, J. T. (1979). *Brain Res.* **170**, 135–155.

Johnston, M. V., Grzanna, R., and Coyle, J. T. (1979). *Science* **203**, 369–371.

Jonsson, G., and Hallman, H. (1982). *Dev. Brain Res.* **2**, 513–530.

Jonsson, G., and Kasamatsu, T. (1983). *Exp. Brain Res.* **50**, 449–458.

Kasamatsu, T. (1983). *Prog. Psychobiol.* **10**, 1–112.

Kasamatsu, T., and Heggelund, P. (1982). *Exp. Brain Res.* **45**, 317–327.

Kasamatsu, T., and Pettigrew, J. D. (1976). *Science* **194**, 206–209.

Kasamatsu, T., Pettigrew, J. D., and Ary, M. (1981). *J. Neurophysiol.* **45**, 254–266.

Kosofsky, B. E., Molliver, M. E., Morrison, J. H., and Foote, S. L. (1984). *J. Comp. Neurol.* **230**, 168–178.

Kristt, D. A. (1979). *Brain Res.* **178**, 69–88.

Kristt, D. A., and Silverman, J. D. (1980). *Neurosci. Lett.* **16**, 181–186.

Lauder, J. M., and Bloom, F. E. (1974). *J. Comp. Neurol.* **155**, 469–482.

Le Gros Clark, W. E. (1932). *Brain* **55**, 406–470.

Le Vay, S., Stryker, M. P., and Shatz, C. J. (1978). *J. Comp. Neurol.* **179**, 223–244.

Le Vay, S., Wiesel, T. N., and Hubel, D. H. (1980). *J. Comp. Neurol.* **191**, 1–52.

Levitt, P., and Moore, R. Y. (1979). *Brain Res.* **162**, 243–259.

Levitt, P., and Rakic, P. (1982). *Dev. Brain Res.* **4**, 35–57.

Levitt, P., Rakic, P., and Goldman-Rakic, P. (1984a). *J. Comp. Neurol.* **227**, 23–36.

Levitt, P., Rakic, P., and Goldman-Rakic, P. (1984b). *Neurol. Neurobiol.* **10**, 41–59.

Lewis, M. S., Molliver, M. E., Morrison, J. H., and Lidov, H. G. W. (1979). *Brain Res.* **164**, 328–333.

Lewis, D. A., Campbell, M. J., Foote, S. L., Goldstein, M., and Morrison, J. H. (1985). *Soc. Neurosci. Abstr.* **11**, 203.

Lewis, D. A., Campbell, M. J., Foote, S. L., and Morrison, J. H. (1986). *Human Neurobiol.* **5**, 181–188.

Lidov, H. G. W., and Molliver, M. E. (1982a). *Brain Res. Bull.* **8**, 389–430.

Lidov, H. G. W., and Molliver, M. E. (1982b). *Dev. Brain Res.* **3**, 81–108.

Lidov, H. G. W., and Molliver, M. E. (1982c). *Brain Res. Bull.* **9**, 559–604.

Lidov, H. G. W., Molliver, M. E., and Zecevic, N. R. (1978). *J. Comp. Neurol.* **181**, 663–680.

Lidov, H. G. W., Grzanna, R., and Molliver, M. E. (1980). *Neuroscience* **5**, 207–227.

Lindvall, O., and Björklund, A. (1984). *Neurol. Neurobiol.* **10**, 9–40.

Livingstone, M. S., and Hubel, D. H. (1981). *Nature (London)* **291**, 554–561.

Loizou, L. (1972). *Brain Res.* **40**, 395–418.

Lorento de No, R. (1938). *In* "Physiology of the Nervous System" (J. F. Fulton, ed.), pp. 291–339. Oxford Univ. Press, New York.

Loughlin, S. E., Foote, S. L., and Fallon, J. H. (1982). *Brain Res. Bull.* **9**, 287–294.

Loughlin, S. E., Foote, S. L., and Bloom, F. E. (1986a). *Neuroscience* **18**, 291–306.

Loughlin, S. E., Foote, S. L., and Grzanna, R. (1986b). *Neuroscience* **18**, 307–319.

Lund, J. A. (1973). *J. Comp. Neurol.* **147**, 455–496.

Lund, J. S., Boothe, R. G., and Lund, R. D. (1977). *J. Comp. Neurol.* **176**, 149–188.
McMahon, D. (1974). *Science* **185**, 1012–1021.
Maeda, T., Tohyama, M., and Shimizu, N. (1974). *Brain Res.* **70**, 515–520.
Molliver, M. E. (1982). *Neurosci. Res. Program Bull.* **20**, 492–507.
Molliver, M. E., and Kristt, D. A. (1975). *Neurosci. Lett.* **1**, 305–310.
Morrison, J. H., and Foote, S. L. (1986). *J. Comp. Neurol.* **243**, 117–138.
Morrison, J. H., and Magistretti, P. J. (1983). *Trends Neuro Sci. (Pers. Ed.)* **6**, 146–151.
Morrison, J. H., Grzanna, R., Molliver, M. E., and Coyle, J. T. (1978). *J. Comp. Neurol.* **181**, 17–40.
Morrison, J. H., Molliver, M. E., and Grzanna, R. (1979). *Science* **205**, 313–316.
Morrison, J. H., Molliver, M. E., Grzanna, R., and Coyle, J. T. (1981). *Neuroscience* **6**, 139–158.
Morrison, J. H., Foote, S. L., Molliver, M. E., Bloom, F. E., and Lidov, H. G. W. (1982a). *Proc. Natl. Acad. Sci. U.S.A.* **79**, 2401–2405.
Morrison, J. H., Foote, S. L., O'Connor, D., and Bloom, F. E. (1982b). *Brain Res. Bull.* **9**, 309–319.
Morrison, J. H., Foote, S. L., and Bloom, F. E. (1984). *Neurol. Neurobiol.* **10**, 61–75.
Mountcastle, V. B. (1957). *J. Neurophysiol.* **20**, 408–434.
Nobin, A., and Bjorklund, A. (1973). *Acta Physol. Scand. Suppl.* p. 388.
O'Connor, D. T., Frigon, R. P., and Stone, R. A. (1979). *Mol. Pharmacol.* **16**, 529–538.
Olson, L., and Seiger, A. (1972). *Z. Anat. Entwicklungsgesch.* **137**, 301–316.
Olson, L., Boreus, L. O., and Seiger, A. (1973). *Z. Anat. Entwicklungsgesch.* **139**, 259–282.
Parnavelas, J. G., and Blue, M. E. (1982). *Dev. Brain Res.* **3**, 140–144.
Pettigrew, J. D., and Kasamatsu, T. (1978). *Nature (London)* **271**, 761–763.
Polyak, S. (1932). "The Main Afferent Systems of the Cerebral Cortex in Primates." Univ. of California Press, Berkeley, California.
Rakic, P. (1974). *Science* **183**, 425–427.
Rakic, P. (1976). *Nature (London)* **261**, 467–471.
Rakic, P. (1983). *Science* **214**, 928–931.
Rakic, P. (1985). *Science* **227**, 1054–1056.
Rakic, P., and Goldman-Rakic, P. S. (1982). *Neurosci. Res. Prog. Bull.* **20**.
Rose, J. E., and Woolsey, C. N. (1958). *In* "Biological and Biochemical Bases of Behavior" (H. F. Harlow and C. N. Woolsey, eds.). Univ. of Wisconsin Press, Madison.
Schlumpf, M., Shoemaker, W. J., and Bloom, F. E. (1980). *J. Comp. Neurol.* **192**, 361–376.
Schmidt, R. H., Bjorklund, A., Lindvall, O., and Loren, I. (1982). *Dev. Brain Res.* **5**, 222–228.
Seiger, A., and Olson, L. (1973). *z. Anat. Entwicklungsgesch.* **140**, 281–318.
Shatz, C. J., and Stryker, M. P. (1978). *J. Physiol. (London)* **281**, 267–283.
Sherman, S. M., and Spear, P. D. (1982). *Physiol. Rev.* **62**, 738–855.
Sillito, A. M. (1977). *J. Physiol. (London)* **271**, 699–720.
Sillito, A. M., Kemp, J. A., and Blakemore, C. (1981). *Nature (London)* **291**, 318–320.
Specht, L. A., Pickel, V. M., Joh, T. H., and Reis, D. J. (1981a). *J. Comp. Neurol.* **199**, 233–253.
Specht, L. A., Pickel, V. M., Joh, T. H., and Reis, D. J. (1981b). *J. Comp. Neurol.* **199**, 255–276.
Steinbusch, H. W. M., Verhofstad, A. A. J., and Joosten, H. W. J. (1978). *Neuroscience* **3**, 811–819.
Takeuchi, Y., and Sano, Y. (1983). *Anat. Embryol.* **166**, 155–168.

Takeuchi, Y., and Sano, Y. (1984). *Anat. Embryol.* **169,** 1–8.

Van Essen, D. C. (1979). *Annu. Rev. Neurosci.* **2,** 227–263.

Van Essen, D. C., Newsome, W. T., and Bixby, J. L. (1982). *J. Neurosci.* **2,** 265–283.

Verney, C., Berger, B., Adrien, J., Vigny, A., and Gay, M. (1982). *Dev. Brain Res.* **5,** 41–52.

Verney, C., Berger, B., Baulac, M., Helle, K. B., and Alvarez, C. (1984). *Int. J. Dev. Neurosci.* **2,** 491–503.

Walker, A. E. (1938). "The Primate Thalamus." Univ. of Chicago Press, Chicago.

Wallace, J. A., and Lauder, J. M. (1983). *Brain Res. Bull.* **10,** 459–479.

Waterhouse, B. D., and Woodward, D. J. (1980). *Exp. Neurol.* **67,** 11–34.

Waterhouse, B. D., Lin, C.-S., Burne, R. A., and Woodward, D. J. (1983). *J. Comp. Neurol.* **217,** 418–431.

Wendlandt, S., Crow, T. J., and Stirling, R. V. (1977). *Brain Res.* **125,** 1–9.

Wiesel, T. N., and Hubel, D. H. (1963). *J. Neurophysiol.* **26,** 1003–1017.

INDEX